Biohazards in Biological Research

Biohazards in Biological Research

Proceedings of a Conference
held at the Asilomar Conference Center
Pacific Grove, California
January 22-24, 1973

Edited by

A. HELLMAN
National Cancer Institute

M. N. OXMAN
Harvard Medical School

R. POLLACK
Cold Spring Harbor Laboratory

Cold Spring Harbor Laboratory   1973

Biohazards in Biological Research

QR
64
.7
.C66
1973

Orders should be addressed to:
Cold Spring Harbor Laboratory, P. O. Box 100
Cold Spring Harbor, New York 11724

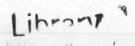

Preface

    With the widespread interest and growing participation of many laboratories in the problems of animal cell biology and tumor viruses, there is a growing need for consideration of potential health hazards. Much of the experience and knowledge concerning such hazards, imaginary as well as real, is known to only a few people and is not widely publicized. Consequently there is strong interest on the part of many investigators in holding a series of conferences at which these matters may be reviewed and discussed.

    Through the combined sponsorship of the National Science Foundation, the National Cancer Institute and the American Cancer Society, the first such meeting, the Conference on Biohazards in Cancer Research, was held at the Asilomar Conference Center, Pacific Grove, California, on January 22-24, 1973. This book contains the proceedings of that conference.

Contents,

viii   Contents

Session I
**Laboratory Infections Introduced
by Experimental Animals
or Animal Cell Cultures**

BIOHAZARDS ASSOCIATED WITH SIMIAN VIRUSES

Robert N. Hull
Biological Sciences Research Administration
Lilly Research Laboratories
Indianapolis, Indiana

Nearly all subhuman primates displayed in zoos, acquired as pets, or purchased for laboratory experimentation are captured in the wild. In their natural habitats these animals are exposed to infectious diseases, many of which are exotic to the United States. In addition monkeys carry numerous indigenous latent viruses. The incidence of infection with some of these viruses (with or without recognizable clinical disease) increases as a result of capture, crowding and stress induced during collection and transportation to the laboratory. A similar situation also exists with respect to certain nonviral infectious agents. These animals, therefore, present a potential hazard to laboratory personnel and they must be housed and handled accordingly. Cell cultures prepared from monkey tissues must, likewise, be considered infectious. It is the purpose of this report to review those simian viruses which have been transmitted from monkeys or monkey cell cultures to man, to examine the factors involved in such transmission, and to assess the consequences of the resulting infections. In conclusion some measures for prevention of human infection will be discussed.

In any discussion of zoonoses involving monkeys, two viruses immediately come to mind; namely B virus (herpes B, monkey B, Herpesvirus simiae) and Marburg virus. These, as well as other simian viruses which have been implicated in human infections, are listed in Table 1 and are categorized with respect to the severity

3

Table 1.   Simian Viruses Transmitted to Man

Produced Fatal Disease

B virus (Herpesvirus simiae)
Marburg virus

Produced Nonfatal Disease

Monkeypox virus
Yaba virus
YLD virus--Tanapox
SV23

Produced Infection but No Clinical Disease

SV40

Recovered From Man or Implicated by Serologic Data

SV5
SV20
Herpesvirus tamarinus

---

of the infections produced. These viruses will be
discussed in the order presented in the table.

B VIRUS

B virus was the first recognized simian virus, as
well as the first known to be infectious for man (1).
Until the Marburg virus incident of several years ago,
B virus was the only virus of major concern to physicians
and scientists charged with responsibility for the health
and safety of persons working with monkeys, monkey
tissues or cultures of monkey cells. B virus continues
to be of major concern since it is indigenous in several
species of Asiatic macaques and may be encountered at
any time in these animals. Marburg virus disease, on
the other hand, has occurred but one time to date.

NATURAL HOST RANGE. B virus was recognized by Sabin (2)
to be a member of the herpesvirus family, and it is now
known that many species of monkeys are latently infected
with their own serologically unique herpesviruses (3-6).
When possibilities for cross-infecting species did not
exist, B virus was recovered only from monkeys of the
Macaca genus. Not all species of this genus were

incriminated, but sufficient numbers from various geo-
graphical areas have been found to be naturally infected
with B virus so that all Asiatic macaques should be
regarded with suspicion. Other Asiatic, African, and New
World species were not found to carry B virus, although
as noted many did carry other herpesviruses, the pathogenicity of which for man remains unknown. B virus has
been transmitted to African green and patas monkeys (7),
in which it produced subclinical infection. Little is
known about the susceptibility of other monkey species to
B virus, but in one test the author was unable to produce
either disease or infection in a squirrel monkey by the
subcutaneous inoculation of 1800 TCID$_{50}$ (tissue culture
infective doses) of B virus. Thus although all monkeys
should be handled as though infected, available information incriminates only the Macaca monkeys as the natural
hosts for B virus.

PATHOGENICITY IN MACACA MONKEY. In the Macaca monkey B
virus produced little evidence of disease. Vesicles and
ulcerations on the tongue, or in and about the oral
cavity, are seen in an occasional animal (8) and a
substantial number of cases of viral conjunctivitis in
these animals are of B virus etiology (7). However more
extensive disease has been observed in monkeys following
immunosuppression with steroids or X-irradiation (8,9).
This appears to have resulted from the activation of
latent infections. In general, however, it is more
likely that a latently infected, or even a virus-shedding,
monkey will show no clinical evidence of disease. The
virus is widespread in such animals and has been recovered
from most organs, as well as from blood, exudates, mouth,
and throat swabs. On at least 20 occasions in one
laboratory B virus has been recovered as a latent infection in rhesus and cynomolgus monkey kidney cell cultures
(10).

EPIDEMIOLOGY IN MONKEYS. The incidence of B virus
infection in rhesus monkeys under various circumstances
has been assessed by several authors primarily through
serologic studies (11-13). The results can be summarized
as follows: Monkeys in the wild have a lower incidence
of B virus antibody than monkeys in laboratory colonies,
and in the wild the incidence increases with age. Under
adverse conditions of capture and transportation, the
number of antibody positive animals may rise from
approximately 10% at the time of capture to 70% or more
by time of arrival in the United States. These increases
in the incidence of antibody suggest that, under the
conditions of capture and transportation, many animals
experience their primary infection with B virus.

Presumably with the development of antibody these
infections subside, although most animals probably remain
latently infected and from time to time shed virus.
These data imply that a higher percentage of newly
infected virus-shedding animals will be present in a
group of recently captured new arrivals in the laboratory
than in established colonies of monkeys. For this reason
strict isolation and quarantine of new animals is
imperative.

Although not clearly demonstrated in these studies,
it is probable that some animals experience recurrent
infections, much like those which occur with herpes
simplex virus in man. This would account for the
occasional animals in established colonies which have
been found to be infected or shedding virus. In summary,
the greatest potential risk appears to be from recently
colonized monkeys; however any Macaca monkey should be
regarded as infectious at all times. Other species
which may have been exposed to macaques at any time must
also be so regarded.

PROPERTIES OF THE VIRUS. B virus is a herpesvirus,
possessing all of the physical and chemical properties
generally associated with this family of viruses. It is
not particularly stable and is readily inactivated by
sterilizing temperatures; high-titered virus in tissue
culture media can be inactivated in less than 30 minutes
by heating to 60°. Like most viruses it is readily
inactivated by uncontrolled drying. B virus grows in
tissue cultures of monkey, rabbit, human, rat, and mouse
cells, and probably in cells of other species. It is
highly pathogenic for the rabbit, varies in its
infectivity for mice, and produces infection and local
skin lesions in guinea pigs. It may be cultivated in
embryonated eggs, in which it produces pock-like lesions
on the chorioallantoic membrane. The rabbit, the animal
of choice for most experimental procedures, can be infected by any route except through intact skin.

HUMAN INFECTIONS. Twenty suspected cases of B virus
infection in man have been reported over the past 40
years, with only 3 possible survivors (14-16,6).
Unfortunately all of these cases were not well documented
in the literature, nor confirmed by laboratory diagnosis.
In 10 cases the virus was isolated and identified, and in
3 additional ones serological data supported a diagnosis
of B virus infection. In the remaining cases B virus
infection was suggested by history and clinical symptoms.
Rhesus, cynomolgus, and African green monkeys were
involved in these 20 human infections. The one African

Table 2. Mode of Transmission and Results of
Confirmed Human Infections with B Virus

Case	Mode of Transmission	Results
Sabin No. 1	Bite	Death
Sabin No. 2	Accident	Death
Ottawa	Unknown	Death
Lilly	Unknown	Death
Pfizer, England	Unknown	Death
University California	Accident	Death
Merck, Sharpe & Dohme	Accident	Death
Lederle	Unknown	Death
Wyeth	Unknown	Death
Scripps Howard Inst.	Unknown	Survived

monkey, however, had been exposed to rhesus monkeys.
Detailed descriptions of these cases are beyond the scope
of this paper, but they may be found in the references
cited.

Of particular concern to the present discussion is
the mode of transmission of B virus to man. The
original case reported by Sabin and Wright (1) followed
the bite of a rhesus monkey, and this led to the belief
that monkey bites are the chief source of infection.
While this is a very potent and likely method of transmission, only 1 of the 10 cases, confirmed by virus
isolation, could actually be attributed to a specific
incident of monkey bite (Table 2). In addition to these
10 cases, at least 3 others were reasonably well
documented by serologic and symptomatic evidence.
These 3 patients received bites or scratches and two of
them died. Information pertaining to the remaining 7
possible cases is too fragmentary to permit analysis.
Of the 13 fairly certain cases, 4 appeared to have been
associated with monkey bites, 3 were associated with
laboratory accidents, and in 6, the source of infection
was undetermined. In all but one case, however, the
victims either had current or recent exposure to monkeys
or monkey cell cultures.

The 3 cases associated with laboratory accidents
warrant description here. One case (17) resulted from
the contamination of a minor cut on the hand with saliva
from a rhesus monkey. No bite was involved. In another
case (18) a technician cut his hand on a broken culture
vessel containing primary monkey kidney cells, which
apparently were latently infected with B virus. Soper

(14) describes a third case which occurred in a post-
doctoral student who was preparing a monkey skull for
study.   The exact mode of infection was not determined.

With the extensive use of monkeys during recent
years, it seems surprising that only 20 possible human
infections with B virus have been observed.   Various
reasons have been postulated for this low rate of morbidity, but only one reasonably sound fact will be presented
at this time.   The findings of serologic surveys of human
populations for B virus antibody were reported by several
investigators (19-22).   The percentage of people with
neutralizing antibody for B virus varied from 27.5 to
69%, with an average of about 50%.   This antibody
probably represents a heterologous response to infection
with herpes simplex virus.   Since antibody to B virus
produced in the rabbit by hyperimmunization with herpes
simplex virus can protect the animal from direct
challenge with B virus, it has been suggested that
similar antibody in man may also be protective (23).
However herpes simplex antibody alone, regardless of its
titer, affords no protection to the rabbit and probably
none to man.   One human B virus infection occurred in an
individual subject to recurrent herpes simplex infection.
This person undoubtedly had herpes simplex neutralizing
antibody (24), but preinfection antibody titers to
either herpes simplex or B virus were not reported.   In
the few cases reported in which preinfection sera were
available and were assayed for B virus neutralizing antibody, none were found to be positive.

MARBURG VIRUS

The marburg incident, in which a number of
laboratory workers and associated personnel fell ill
following contact with certain African green monkeys,
is now well known.   Numerous reports describe the
circumstances of the outbreak, the disease syndrome
in man and experimental animals, and the properties of
the virus.   The findings of many of these reports were
reviewed by Siegert (25) and appear in the text of a
symposium edited by Martini and Siegert (26).   In brief,
a febrile hemorrhagic disease was observed during the
late summer of 1967 in 31 persons in restricted areas of
Germany and Yugoslavia.   Seven of these terminated in
death.   This disease syndrome was previously unrecognized,
and the etiological agent was not immediately identified,
although monkeys were incriminated as the source of
infection.

NATURAL HOST RANGE. In spite of extensive investigations the natural host for Marburg virus remains undetermined. Experimental infections are uniformly fatal in rhesus, African green, and squirrel monkeys, regardless of the route of injection or the dose administered (27). In view of these findings it is unlikely that any of these three species of monkeys are the natural host for Marburg virus. The results of sero-epidemiologic studies performed by various investigators on simian populations in diverse geographic areas have been confusing, first, because high percentages of positive sera were found in some studies, and second, because of the lack of agreement among laboratories performing assays on the same specimens. This confusion now appears to have been cleared by Slenczka (28) who developed a more specific CF antigen from infected Vero cell cultures. Using this antigen, which showed specificity in tests with acute and convalescent human and animal sera, these authors found no evidence of Marburg virus infection in wild-living monkeys. The information gathered to date, therefore, suggests that Marburg virus probably is not a simian virus and implies that the particular monkeys that transmitted the infection to man were accidental hosts for the virus.

The source of these monkeys and the itinerary of their trip from Uganda, Africa to West Germany was studied in an attempt to determine possible sources of the infection. Taking many factors into consideration, Siegert (25) concluded that the original infection(s) in these monkeys was acquired prior to arrival in Germany. The monkeys traveled to Germany and Yugoslavia via London. At the London holding facility, where they spent from 9 to 36 hours, they were potentially exposed to 48 different species of animals from many parts of the world. However a definite source of the virus was not found in this place. Thus the source of the Marburg virus and the identity of its natural host(s) remain unknown.

PATHOGENICITY FOR MONKEYS. As previously stated, Marburg virus produced 100% fatality in experimentally infected monkeys. Simpson (27) produced fatal infection in African green monkeys with as little as 0.02 guinea pig LD_{50} given subcutaneously. He was also able to produce infection with a 10^{-10} dilution of a 10% suspension of infected monkey liver (28a). Haas and Maass (29) produced infection in monkeys by the inoculation of 1.0ml of a 10^{-10} dilution of infected monkey blood. The infectivity of the same sample when assayed in primary monkey kidney cell cultures was 100-fold lower. Thus the monkey is not only a highly sensitive host, but when

infected produces huge quantities of virus. Haas and
Maass (29) observed little clinical symptomology in their
animals other than apathy, refusal to eat, and failure
to react to stimuli for about 2 days prior to death.
Simpson's observations were similar except that a
petechial rash was seen in some animals one or two days
prior to death. Clearly then, the clinical expression
of Marburg virus disease in monkeys might well be overlooked or not recognized by laboratory personnel or
others charged with the health care of monkey colonies.

EPIDEMIOLOGY IN MONKEYS. Since Marburg virus does not
appear to be a simian virus and has, to date, been
encountered only this one time, nothing is known of its
natural transmission. Laboratory studies suggested that
transmission of experimental infection probably occurred
by direct contact among monkeys or via virus-contaminated
fomites. Virus is excreted by infected monkeys in the
saliva, urine, blood, and feces (28) and virus titers in
urine and saliva are quite high. In the study of Haas
and Maass (29), control monkeys held in cages adjacent to
infected monkeys became infected, but others housed a
few meters away did not. Simpson also reported spread
of infection to normal monkeys housed in the same room
and suggested that this might have been by aerosol. In
view of the large quantities of virus excreted by these
experimentally infected animals, it does not seem
necessary to postulate some unique mode of spread.

Transmission of the virus occurred in the laboratory
under conditions that eliminated the possibility of an
arthropod vector. However Marburg virus has some properties in common with the arboviruses, and in view of
the high concentration of virus in the blood, it seems
possible that mosquitoes, ticks, or some other arthropod
may play a role in its natural transmission. Kunz and
Hoffman (30) succeeded in infecting Aedes egypti
mosquitoes with Marburg virus by intrathoracic inoculation, but failed to do so with Anopheles mosquitoes or
ticks of the genus Ixodes. Virus transmission studies
with infected mosquitoes were not reported, but these
findings do allow the possibility that Aedes mosquitoes
might be a potential vector of the virus in nature.

PROPERTIES OF THE VIRUS. The causative agent of Marburg
disease was originally isolated by Smith (31) and others
by inoculation of guinea pigs with acute-stage human
blood or post-mortem tissue suspensions. Early microscopic observation of infected tissues and tissue culture
cells revealed bodies that suggested a relationship to
rickettsia or to the psittacosis-lymphogranuloma group of

viruses.   Filtration through Gradocol membranes suggested
a size of between 340 and 790 mμ.   However subsequent
investigations have left little doubt that the causative
agent was a virus, the morphology of which is unlike
that of any known family of viruses.   The electron
microscopic studies of Almeida (32) suggest that 3
morphologic forms of the virus exist.   These include a
naked helix of varying length but with a central core
diameter of 28 mμ, a sinuous membrane-covered highly
pleomorphic form, and the mature virion which was
frequently seen in ring form.   The outside diameter of
these ring forms was 300-400 mμ.   The morphology of
Marburg virus in some respects is similar to that of
vesicular stomatitis and rabies viruses.   Kissling (33)
further demonstrated that Marburg virus is ether
sensitive, heat labile, and not inhibited by 5-bromodeoxy-
uridine (BrdU).   The genetic material in the virus,
therefore, may be assumed to be RNA.

The sensitivity of Marburg virus to various physical
and chemical factors was described by Bowen (34).   These
authors found that it required one hour at 56° to
completely inactivate a sample containing 5.5 logs of
virus.   Exposure to ultraviolet light under the conditions
of their test destroyed the infectivity in 2 minutes.
The virus was again shown to be sensitive to ether as
well as to sodium desoxycholate, thus indicating the
presence of essential lipid.   Disinfectants such as
formalin, acetone, chloros, methyl alcohol, and Tego
MGH completely inactivated the virus after one hour of
exposure.   Overnight fixation of infected tissues in
osmium tetraoxide also rendered the tissue noninfectious.
However phenol, Cetramide and trypsin did not completely
destroy infectivity.   Virus-inoculated guinea pigs were
treated with a number of recognized antibiotics, but
none of these altered the course of the disease.

Smith (31) originally isolated the virus by
inoculation of guinea pigs with infectious human blood.
In the first or early passages guinea pigs became
febrile 4-10 days after inoculation, showed some signs
of illness, but survived.   Virulence increased on passage
as evidenced by shorter incubation periods and increased
mortality.   After 8 passages the virus was uniformly
fatal to inoculated guinea pigs.   Detailed descriptions
of the pathology of the disease in guinea pigs are
available in the literature (31,35,36), and since these
animals are susceptible by various routes of inoculation,
guinea pigs appear to be the animal of choice for
laboratory investigation.

Suckling hamsters inoculated either intracerebrally
(IC) or intraperitoneally (IP) showed a degree of

susceptibility to virus which had been passaged in
either guinea pigs or monkeys. In one study (37)
virulence increased on continued hamster passage, and by
the 8th transfer the mortality was 80% in IC-inoculated
hamsters and 40% in those inoculated IP. Brains and
livers from these animals contained 5.0-5.5 x 10^5 guinea
pig LD_{50} of Marburg virus. Virus passaged 8-10 times
in suckling hamsters developed virulence for both
weanling and 5- to 6-week-old hamsters. A unique feature
of the infection in hamsters was the high concentration
of virus in the brains of infected animals. This was
not seen in guinea pigs or monkeys, even after IC
inoculation. Hamsters and guinea pigs that survived
infection were immune to subsequent challenge.

In preliminary studies Smith (31) was unable to
produce infection in mice or in embryonated eggs with
either human blood or guinea pig-passaged virus. More
recently Simpson (27) reported some success in passing
virus that had been serially passaged in guinea pigs or
monkeys to suckling mice by IC inoculation. The virus
was further passaged to adult mice in which some indication of paralysis was observed.

The growth of Marburg virus in cell cultures has
been reviewed by Hoffman and Kunz (38) and by Siegert
(25). In brief, many cells derived from monkeys,
guinea pigs, hamsters, and humans were found to support
the growth of Marburg virus, but the recognition of such
growth frequently required back passage in animals,
special histologic preparations, or immunofluorescent
technics. Cytopathology was not a consistent observation. Furthermore it was demonstrated that primary
monkey kidney cells infected with Marburg virus could
still be subcultured and that chronic infections could
be established in some cell lines. Thus the presence
of Marburg virus in monkey kidney cells would probably
not be detected by ordinary light microscopic observation.

HUMAN INFECTIONS. Thirty-one human cases of Marburg
virus disease occurred over a 2-month period in two
cities in West Germany and in Belgrade, Yugoslavia.
Twenty-five of these occurred in laboratory personnel
who were in direct contact with specific shipments of
African green monkeys, their blood or their tissues.
Six additional cases resulted from human dissemination
of the virus. These cases are tabulated according to
the tasks performed by the patients which led to their
exposure, or to their contact with infected patients
(Table 3). In two instances infection was attributable
to an accident: In one a physician pricked herself with

Table 3. Possible Sources of Infection in
31 Cases of Marburg Virus Disease

Occupation or Exposure	No. Cases
Monkey sacrifice, autopsy	11
Cleaning contaminated equipment	5
Exposure to patients (med. personnel)	4
Trephination of monkey skulls	3
Animal handling	3
Kidney removal, culture prep.	2
Pathology assistant	1
Morgue attendant	1
Sexual intercourse	1

Prepared from the published reports of Martini (39,40),
Stille and Rohle (41), and Todorovitch et al. (42).

a needle used to obtain blood from an infected patient,
and in the other, a technician broke a test tube containing infected material. The incubation periods in
these two cases were 5 and 7 days, respectively, which
fell within the estimated range of 3-9 days for most
patients. Another case occurred in the wife of a
convalescent patient who apparently became infected
through sexual intercourse. Virus was demonstrated in
the semen by immunofluorescence. At least one case
occurred in an individual (a veterinarian) who took all
the usual precautions in performing an autopsy on an
infected animal. The mode of infection in most of the
patients was unknown, but in view of the huge quantities
of virus probably present in the infected monkeys and
their tissues, transmission to laboratory workers and
other personnel is not surprising. It was fortunate
that no more cases occurred, and this was probably due to
the early recognition of monkeys as the source of the
infection, and the prompt action taken to control or to
eliminate the source.

MONKEYPOX VIRUS

Monkeypox virus (MPV) is one of several poxviruses
recovered from monkeys. The disease produced by MPV in
the monkey was first observed by von Magnus in 1958 (43)
in an outbreak involving 6 of 32 cynomolgus monkeys
imported from Singapore. The disease was not observed,
however, until 62 days after the monkeys arrived in the
laboratory. A second outbreak occurred later that same
year and involved 30% of another shipment of cynomolgus

monkeys. Again clinical symptoms were not seen until 51
days after arrival. The disease consisted of a petechial
rash, followed by a maculopapular eruption. The animals
appeared otherwise to be healthy and at autopsy showed
no evidence of generalized disease. During the period
of virus activity in this colony, MPV appeared in tissue
cultures prepared from the kidneys of apparently healthy
animals.

Prier and Sauer (44) described another outbreak of
MPV in which both rhesus and cynomolgus monkeys were
involved. Two clinical forms of disease were recognized;
the first was similar to that reported by von Magnus;
but in the second form, in addition to the rash, facial
edema extending to the neck was noted. This was followed
by difficulty with respiration and death by asphyxiation.
These latter symptoms were seen only in cynomolgus
monkeys. Clinical disease was observed in all of 6
nursing infant monkeys. Sauer (45) estimated that this
outbreak involved about 10% of the animals in a colony of
2000.

A third episode of monkeypox was observed by Stewart
(46) in cynomolgus monkeys 45 days after whole body
irradiation (350r). Two of 9 treated animals developed
the disease and one died. The other 7, as well as 19
non-irradiated monkeys housed in the same room, remained
asymptomatic. As previously noted in respect to B
virus, this probably represented a situation in which
latent virus infection was activated in association with
immunosuppression.

In a later review article Arita and Henderson (47)
described a total of 9 outbreaks of monkeypox including
those described above. In 5 of these MPV was determined
to be the causative agent on the basis of virus isolation
and identification. The remaining cases were assumed to
be epidemics of monkeypox. A 10th outbreak, this time
in chimpanzees, was mentioned in a recent report (48).

von Magnus readily isolated the virus by inoculation
of embryonated eggs with scrapings from the papular
lesions and also by inoculation of monkey kidney, HeLa
or human amnion cell cultures. Lesions were produced on
the chorioallantoic membrane of the eggs, even with 10^{-7}
dilutions of the inoculum. Cytopathic effects (CPE)
occurred in inoculated tissue cultures within 2-3 days.
Prier and Sauer employed cultures of rabbit kidney cells
and were able to observe CPE within 48 hours. Thus the
virus is readily isolated from the skin lesions of
infected monkeys.

VIRULENCE FOR MONKEYS. The virulence of MPV for rhesus
and cynomolgus monkeys is described above; of the two
species, the cynomolgus appears to be most susceptible
to clinical disease. In reviewing an outbreak in Holland,
Espana (49) noted that the primary cases occurred in 2
giant anteaters housed in a zoo. The disease spread to
orangutans, marmosets, gibbons, squirrel and African
green monkeys. Fatalities occurred in a number of these
species. This epidemic was originally reported by
Peters (50). Espana also described an epidemic in his
laboratory which started in one species of Indian
langurs and spread to a second species, as well as to
rhesus and stumptail macaques. Deaths occurred in all
species. Thus MPV appears to be virulent for a wide
variety of subhuman primates from diverse geographic
areas.

EPIDEMIOLOGY. Monkeypox has been seen only in colonies
of captive animals, but serologic surveys of captive
monkeys and apes of numerous species have failed to
indicate that monkeypox is a natural disease of any
species of subhuman primate (51,48). As was the case
for Marburg virus, the natural host for MPV remains in
doubt. The virus, as noted, did spread through groups
of laboratory or zoo animals and was probably transmitted
by direct contact or through exposure to virus-infected
equipment. Serologic data indicated that the degree of
spread in some colonies was greater than suggested by the
number of clinical cases.

Although MPV is closely related to variola and
vaccinia viruses, there is no evidence that MPV is a
strain of variola virus modified by monkey passage.
Noble and Rich (51,52), for example, found that monkeys
from South America, Africa, and Asia were relatively
resistant to experimental infection with either variola
or alastrim. Cynomolgus monkeys were most susceptible
to variola but the induced infections were mild, and
serial passage of the virus was not possible in this
species. Thus it seems unlikely that monkeys are a
significant reservoir for smallpox or that MPV represents
a modified form of smallpox virus. MPV is not identical
to any previously known poxvirus.

PROPERTIES OF THE VIRUS. Wilner (53) placed MPV in the
vaccinia subgroup of the poxvirus family, along with
vaccinia, variola, alastrim, and 4 other animal poxviruses.
This implied that the virus has a double stranded DNA
core and certain morphological features, replicates only
in the cytoplasm where it produces inclusion bodies, is
acid-labile, and is relatively heat resistant.

Sensitivity to ether varies within this group, but
von Magnus (43) reported MPV to be ether resistant.   Both
von Magnus and Prier and Sauer (44) described the
morphology of their isolates of MPV as brick-shaped,
which is typical of poxviruses.   The size was found to be
200-250mµ.

The production of pock-type lesions on the chorioallantoic membrane of embryonated eggs by MPV was
mentioned.   The pocks produced by MPV more closely
resembled those of variola than of vaccinia.   Bedson and
Dunbell (54) determined the maximum temperatures at
which various poxviruses could grow in embryonated eggs
and employed these data to distinguish MPV from variola
and vaccinia viruses.   MPV had a ceiling growth temperature of 39° and thus was intermediate between variola
(at 38.5°) and vaccinia (at 40.5°).   Rabbits were
susceptible to MPV by inoculation of scarified skin, and
the virus could be serially propagated in rabbits by
this method.   According to von Magnus (43) and other
investigators, serial passage in rabbits provides another
characteristic of MPV which distinguishes it from variola
virus.

Adult mice were susceptible to IC inoculation with
MPV, but not to IP or intradermal inoculation.   MPV
again differed from variola in that serial passage in
mice by IC injection was possible.   Suckling mice were
infected by and succumbed to intranasal inoculation.
von Magnus found that guinea pigs and chickens were
resistant to intravenous, IP, and subcutaneous injection,
but Prier and Sauer reported that guinea pigs developed
swelling and granulomatous lesions following footpad
inoculation.   Both groups reported that MPV agglutinated
chicken, but not mouse, red blood cells.   The virus could
also be propagated in a wide variety of simian, human,
and other animal cells in tissue culture (55,56).

The antigenic similarity of MPV to vaccinia was
recognized early by both von Magnus (43) and by Prier
and Sauer (44).   This was demonstrated by neutralization,
hemagglutination inhibition (HI), complement fixation
(CF), and animal protection tests.   In the latter,
rabbits immunized with either virus were resistant to
both viruses.   von Magnus found by immunodiffusion that
both viruses produced 4 identical bands, but that
vaccinia produced one additional band.   More recently
Rondle and Sayeed (57) reported that although they
could not distinguish MPV from variola virus by HI and
neutralization tests, they could detect some antigenic
differences between MPV and vaccinia virus.   They also
pointed out other characteristics which served to
distinguish MPV from variola.   The findings of these

and other investigations indicate that MPV, vaccinia, and variola constitute a closely related antigenic group of viruses.

HUMAN INFECTIONS.   Infection with monkeypox virus has not occurred in laboratory personnel.   Foster (58) however described 6 human cases of monkeypox in residents of West and Central Africa.   Five of these occurred in children, age 4 to 9, and the other in a 24-year-old male.   The virus was not communicated to any of 24 susceptible household contacts.   None of these patients had received smallpox immunization, and smallpox vaccination during convalescence gave immune reactions. Five of the 6 patients lived in an area where monkeys were hunted for food.   No direct evidence was obtained, however, for patient contact with sick or freshly killed animals within 3 weeks of onset of disease.   The 6th patient lived in a village where monkeys were occasionally seen, but where they were not eaten. Another case (59) occurred in a non-vaccinated 9-month-old child.   This child also lived in a village where monkeys were used as food; but again no direct contact with monkeys was established.   The parents and other residents of the village had been immunized against smallpox and no further cases occurred.   The patient survived the MPV infection but died of measles 6 days after discharge from the hospital.   Information gained from these human cases, as well as from laboratory observations, suggests that immunization with vaccinia confers immunity to infection with MPV.

YABA VIRUS

Yaba is another poxvirus which produces tumor-like growths (histiocytomas) primarily in Asiatic monkeys. It was first recognized (60) in rhesus monkeys housed in outdoor pens at Yaba, near Lagos in Nigeria. Following the first appearance in this colony, the disease spread to 20 of its 35 animals during a 6-month period.   The tumors were benign and histologically pleomorphic.   Following the emergence of the first tumor, multiple secondary tumors occurred along routes of lymphatic drainage, but metastases to internal organs did not occur and the infected animals remained otherwise healthy.   The growths eventually sloughed and healed. Acidophilic bodies 1-5μ in diameter were seen in the cytoplasm of infected cells.   The virus was isolated and studied in more detail by Niven (61).

NATURAL HOST RANGE. Little is known about the virus in free-ranging monkeys, but limited serological surveys have revealed a high incidence of antibody in Asian monkeys (62). Only 5 of 57 African baboon sera were antibody positive.

VIRULENCE FOR MONKEYS. The general picture of the disease in the rhesus monkey is described above. In the original outbreak at Yaba the infection spread to one dog-faced baboon, and Niven (61) was able to produce tumors in cynomolgus and African green monkeys, but not in patas, Cerocebrus, or capucin monkeys. Ambrus (63) produced infection in Asiatic Macaca monkeys, but not in several species of African and South American monkeys.

EPIDEMIOLOGY IN MONKEYS. The virus appeared to spread in the Yaba colony, which was housed out of doors. In contrast Ambrus and associates (64) stated that very little, if any, spread occurred in their laboratory animal room, even when many experimentally infected animals were present. This was true even when an infected animal was caged with a normal control. Ambrus reported further that when the colony was moved to new quarters, in which housekeeping chores were not done as thoroughly and routinely as in the original quarters, 11 "spontaneous" cases of Yaba tumors occurred. Both flies and mosquitoes were seen in these quarters. When these insects were eliminated by spraying and use of insecticides, no further infections were seen in uninoculated animals. These observations suggested the possibility of an insect vector.

Wolfe (65) produced infection in 6 of 15 monkeys exposed to an aerosol in a closed system. Only one animal developed clinical disease, but others were found to have lung tumors at necropsy. The exposure in this experimental setting was probably far greater than that which might occur under natural conditions, but the results point to possible dangers from virus aerosols.

PROPERTIES OF THE VIRUS. Virus particles seen in the cytoplasm of infected cells were said (61) to be morphologically and cytochemically similar to those of vaccinia virus. The dense particles were 280 mμ in size. The elementary bodies were shown to contain DNA by staining with acridine orange. However Yaba virus, unlike vaccinia and MPV, was readily inactivated by ether. The virus was not infectious for embryonated eggs, rabbits, dogs, guinea pigs, hamsters, rats, or mice (63), further distinguishing it from vaccinia or MPV. Growth of Yaba virus in tissue culture was limited

to a few simian and human cell types. Niven (61)
observed some virus replication in monkey kidney cells,
but none in HeLa or chick embryo cells. Yohn and coworkers (66-69) obtained growth in two monkey kidney
cell strains, LLC-MK$_2$ and BSC-1, and in a human embryo
kidney strain, MA-1D. BSC-1 cells proved most sensitive.
The CPE of Yaba virus appeared as foci of rounded and
piled up cells which were described as "microtumors."
Virus titer was determined by counting the number of
foci produced. These in vitro growth properties of Yaba
virus varied considerably from those of vaccinia and MPV.

HUMAN INFECTIONS. Naturally occurring human infection
with Yaba virus has not been reported, but infection has
occurred following accidental and experimental inoculation (70). The accidental infection happened when a
laboratory worker stuck his hand with a virus-contaminated
needle. Four months later a nodule appeared at the site
and grew to a size of 2cm before it was removed. The
nodule was histopathologically similar to Yaba monkey
tumors and CF antibody was detected in the patient's
serum. Six volunteers inoculated intradermally with Yaba
virus developed nodules within 5-7 days. These grew to
about 2cm, then sloughed. Virus was recovered from the
nodules and CF antibody was produced. The virus was
further passaged in volunteers. Although man was demonstrated to be susceptible to the virus, there appeared
to be no dire consequences of the infection.

YLD

     YLD virus (Yaba-like-disease) is another monkey pox-
like virus and has been referred to in the literature by
several other names, including agent No. 1211, Y-R,
OrTeCa, and BEMP. In addition it was found to be
identical to Tanapox virus. The disease in monkeys was
first observed by Espana (71), who isolated the virus
from skin lesions. Espana (49) recently reviewed
outbreaks of YLD that have occurred in monkey colonies
in this country. The disease, which is characterized
by epidermal growths on the face and extremities, was
first noted in 3 rhesus monkeys at the Davis, California,
primate center. The infection spread to other Macaca
monkeys and to one species of langur during two waves of
infection which occurred over a year's time. African,
Cercopithecus, and Cercocebus monkeys did not become
infected. A total of 337 monkeys in a colony of about
4000 animals contracted the disease. A similar outbreak
of YLD occurred at the Oregon Primate Center (72).
Morbidity was high in Macaca monkeys, but African and
South American species were unaffected. A third episode

occurred in rhesus monkeys at Brook Air Force Base in
Texas (73). In the latter case the YLD lesions appeared
on 15 of 150 animals 7 days after they were tattooed for
identification purposes.

NATURAL HOST RANGE. Downie and Espana (74) suggested
that African monkeys were the natural host for YLD virus
and demonstrated that YLD and Tanapox viruses were
identical. Tanapox is a mild pox disease of humans
observed only in the Tana River valley in Kenya. It was
not clearly demonstrated, however, that natives of the
area contracted the disease from monkeys. The rhesus
monkeys in which YLD first occurred in the three U. S.
laboratories were supplied by one importer who housed
Asiatic and African monkeys in the same crowded quarters
prior to shipment. Thus Downie and Espana proposed that
the rhesus monkeys acquired the disease from infected
African monkeys while held in the importer's compound.
This was not proven but appears to be the best explanation now available for the source of the virus.

Pathogenicity in monkeys. The disease, as seen in rhesus
and cynomolgus monkeys, consists of tumor-like skin
lesions with very little other symptomology. YLD lesions
involve the epidermis, whereas Yaba tumors involve the
dermis. McNulty (75) described the lesions as flat,
elevated papules, 1-3mm in height and 10mm in diameter.
The centers were crater-like in appearance. Eosinophilic
cytoplasmic inclusions were seen in tissue sections.
The lesions reached a peak size within a week and then
regressed, leaving no scar. Convalescent rhesus monkeys
were immune to reinfection. Although disease was not
seen in African monkeys during laboratory outbreaks,
antibody-free African green monkeys were susceptible to
experimental infection (74).

Epidemiology in monkeys. The possibility that YLD is a
natural disease in some African monkeys was mentioned
above, and it was postulated that when fully susceptible
Asiatic monkeys were brought into contact with them, an
epidemic ensued. The infectious nature of the virus was
well demonstrated by the degree of spread that occurred
in laboratory colonies. This happened in some cases in
spite of attempts to control the spread by isolation of
infected animals. The actual mechanism of virus
dissemination in monkey colonies has not been elucidated.

PROPERTIES OF THE VIRUS. YLD virus was described (76) as
a bricked-shaped or elongated oval structure, 220-310 mμ

in length and 125-175 mμ in width. The particle contained a central core surrounded by electron-dense
material of uniform thickness. The mature virion had a
double membrane external coat. Like Yaba virus, YLD
was ether sensitive, in contrast to MPV and vaccinia
viruses. No hemagglutination was observed when a variety
of animal cells were tested. The virus was originally
isolated in BSC-1 cell cultures and has also been
cultivated in primary green monkey kidney, human embryo
skin-muscle and kidney, and WI38 cell cultures. Growth
was poor in HeLa and rabbit embryo kidney cells, and
no growth occurred in the LLC-MK$_2$ or RK$_{13}$ cell lines (3).
Monkey infectivity was retained following 12 cell culture
passages. The virus grew in foci, as did Yaba virus, but
cell lysis occurred on continued incubation. Crandell
(76) described multinucleated cells or syncytia with
vacuolated cytoplasm as a characteristic of the virus/CPE
after several tissue culture passages. Both intranuclear
and cytoplasmic inclusions were observed.

The virus had little or no virulence for embryonated
eggs or laboratory animals other than primates. However
Crandell (76) did obtain skin lesions in the German
checker rabbit, but not in New Zealand white rabbits.

Some serological relationship between YLD and Yaba
virus was noted in early studies, and this has been more
clearly delineated by Nicholas (55). In cross neutralization tests, YLD antiserum neutralized Yaba virus at a
4-fold lower dilution than the homologous virus, whereas
Yaba antiserum neutralized Yaba virus at a 32-fold higher
dilution than it did YLD. Cross-CF tests with 2 units of
complement revealed a strong, one-way cross reaction in
which YLD antiserum fixed complement to the same level
with both antigens, whereas Yaba antiserum reacted only
weakly with YLD antigen. It appeared, therefore, that
the two viruses were related and that YLD was antigenically dominant. In the studies of Downie and Espana
(74) several of the YLD isolates were compared to each
other and to Tanapox virus by a variety of immunologic
tests, including monkey cross protection, CF, neutralization and precipitation tests. The viruses were found to
be indistinguishable in all of these test systems. These
authors suggested that the virus should be referred to as
Tanapox, since Tanapox virus was recognized prior to the
outbreaks of this disease in the United States.

HUMAN INFECTIONS. The demonstrated identity of YLD with
Tanapox virus indicates its potential pathogenicity for
man. This was confirmed by the human cases that occurred
during each of three outbreaks of disease in monkey
colonies in the United States. Espana (71) reported 16

cases at the Davis primate center, mainly in animal handlers. McNulty (75) described 5 human infections acquired during the Oregon Primate Center outbreak, and Crandell (76) mentions 2 cases.   Thus a total of 23 human infections resulted from these epidemics in laboratory monkeys. Skin lesions identical to those seen in monkeys appeared in most instances after monkey bites, scratches, or some other local trauma. In some patients these lesions regressed in a few weeks, leaving a flat scar, and there were no other manifestations of the disease. Other patients, however, experienced more severe disease characterized by regional lymphadenopathy and fever. All patients survived and those tested developed CF antibody. The virus was readily isolated from the human skin lesions.

SV23

SV23 is one of a number of simian adenoviruses originally isolated from rhesus and cynomolgus monkey kidney cell cultures (10). It is included in the present discussion because of one accidental human infection in which it was involved (7). In this case the needle on a syringe containing SV23 virus became plugged, and in the course of dealing with the problem, a laboratory worker received a drop of virus in his eye. Evidence of irritation was seen within 24 hours. This increased in intensity and was accompanied by lacrimation. A severe conjunctivitis developed by 48 hours. The infection spread to the other eye and persisted for 5 weeks. SV23 virus was recovered from the eye, but no serologic conversion occurred. The infection was limited to the conjunctiva and recovery was complete. Several of the simian adenoviruses were found to be oncogenic for newborn hamsters by Hull and associates (77) and thus were used in many laboratories for studies of oncogenesis. The above-described incident demonstrates the need for extreme care in handling any of these simian viruses.

SV40

SV40 is a well-known oncogenic simian papovavirus carried latently by rhesus and cynomolgus monkeys. The rhesus monkey appears to be the principal host for the virus (78,79,3). The virus produces no apparent disease in the natural host but is generally detected as a latent infection in kidney cell cultures prepared from rhesus or cynomolgus monkeys. The infection is so prevalent in rhesus monkeys that all primary cultures of rhesus kidney cells must be considered infectious. No CPE occurs in these cultures; thus the virus remains

undetected unless special efforts are made to detect
its presence. SV40 is an extremely stable, ether-
resistant DNA virus, which survives for weeks at 37° even
in the presence of 0.25% formaldehyde.

The oncogenicity of SV40 for newborn hamsters was
discovered by Eddy (80) not long after its original
isolation. This property of the virus has made it a
very popular laboratory model for studies of viral
oncogenesis, and it has been widely distributed for this
and other purposes.

The first evidence that SV40 was infectious for man
came from serological studies of persons immunized with
poliomyelitis vaccine that contained SV40 virus. Melnick
and Stinebaugh (81) further reported that SV40 had been
recovered from the stools of children for as long as 5
weeks after ingestion of an experimental live poliomyelitis vaccine contaminated with SV40. Virus multiplication seemed necessary to support this prolonged
secretion of virus. Morris and coworkers (82) observed
infection via the respiratory tract when volunteers were
inadvertently inoculated with SV40-contaminated
respiratory syncytial virus. SV40 was recovered from
throat swabs of some of these volunteers, and seroconversion occurred in more than half. Shah (83)
suggested the natural transmission of SV40 to man by the
demonstration of SV40 antibody in persons in India who
were in close association with rhesus monkeys, but who
had not received vaccines prepared in monkey kidney
cells. None of these incidences of human infection with
SV40 was associated with any immediate overt clinical
illness. Furthermore, follow-up studies of children
immunized with SV40-contaminated vaccines have, to date,
revealed no increase in their incidence of malignant
disease (84).

Some evidence of human infection with SV40 or a
related virus has been obtained in serologic studies.
Shah has assayed over two thousand sera from people whom
he listed in 8 different categories for antibody to
SV40. The highest percent of positive sera (27%) was
detected in a group of adult monkey handlers in India
who were not exposed to SV40 through immunizations.
Only 8.7% of a group of the average residents of Northern
India were found to possess SV40 antibody. The second
highest incidence of SV40 antibody (19.8%) was found in
children who had received inactivated poliomyelitis
vaccine prior to the discovery and elimination of SV40
from such products. In contrast, another group of sera
from children born after SV40 contamination of vaccines
was eliminated, contained only 3.2% positive specimens.
The incidence of antibody in sera from other groups,

including cancerous or non-cancerous patients from both
India and the United States, ranged from 2 to 5% and
there were no significant group differences. However all
groups contained some antibody-positive individuals, and
some of these had no history of exposure to monkeys and
had not been immunized with SV40-contaminated vaccines.
Thus it appears not only that man may become infected
with SV40 through close association with rhesus monkeys,
or as a result of inoculation with monkey-derived SV40
virus, but also that another source of SV40 or of an
antigenically similar virus must exist both in India and
the United States. This subject has been pursued further
by Shah (85).

In addition to the above evidence, several reports
have recently appeared in the literature describing
papovavirus-like particles in sections of human brain
tissue from patients with demyelinating diseases or other
central nervous system disorders. In some instances
SV40-like viruses were isolated from these tissues. The
electron micrographic evidence for such viruses was
first reported by ZuRhein (86) and subsequent reports by
other investigators confirmed these observations. Two,
or possibly three, types of viruses were isolated (see
Takemoto and Mullarkey, this volume). Padgett (87)
isolated a papovavirus from the brain of a patient with
progressive multifocal leucoencephalopathy (PML) by
inoculation of tissue extracts into cultures of human
fetal glial cells. This virus failed to grow in green
monkey kidney cells and did not react in fluorescent
antibody (FA) tests with SV40 antiserum. However Weiner
(88,89) isolated viruses from two patients with PML
which, on the basis of cross-neutralization tests,
appeared to be identical to each other and to SV40.
These viruses grew in both primary green monkey kidney
and BSC-1 cells, with the production of typical SV40
CPE. A third isolation was made by Gardner (90) from
the urine of a patient following kidney transplantation.
The virus was excreted in the urine for over a month,
and this was accompanied by a rise in antibody. This
virus was related only slightly to SV40 serologically,
and differed further by its production of a hemagglutinin
for human type O and guinea pig red blood cells. It
grew slowly in the Vero green monkey kidney cell line.

Both patients from whom Weiner isolated SV40-like
virus had received commercial oral poliomyelitis vaccine
several years earlier, but SV40 contamination of vaccine
was eliminated prior to release of commercial oral poliomyelitis vaccine. Neither patient had a history of
parenterally administered monkey kidney-prepared vaccines.
The immunization histories of the patients from whom the
other papovaviruses were isolated were not stated. If

any of these viruses were, in fact, of monkey origin,
the manner of transmission to man was not apparent. In
view of all the data presented above, it is obvious that
man is susceptible to infection by SV40 of monkey origin,
but the consequences of such infection are difficult to
assess at this time. In addition it appears that man
experiences infection with the same or antigenically
similar viruses, the origin of which is difficult to
trace to monkeys. In view of these findings, as well
as the other properties of SV40, this virus should be
handled with great care.

SV5

SV5 is a paramyxovirus frequently present as a
latent infection in kidney cultures of both Asiatic and
African monkeys. It was originally isolated and
described by Hull (91). Antibody to SV5 was demonstrated
in human sera (92,93), but this was not surprising in
view of the serologic relationship of SV5 to mumps and
to the human parainfluenza viruses (3). SV5 or a similar
virus has been isolated from man on several occasions,
and Binn (94) isolated it from dogs with respiratory
disease. Schultz and Habel (95) isolated a virus which
they called "SA" from nasal washing of a patient with a
respiratory disease. Chanock (96) later found this virus
to be identical to SV5. A virus referred to as "DA" was
recovered from post-mortem blood of a patient who had
infectious hepatitis (97). This agent was also found to
be serologically identical to SV5. A further isolation
was made by von Euler (98) from a throat swab taken from
a 7-year-old patient with high fever and pneumonia.
Pneumococci were also isolated from this patient, but an
antibody rise from 1:40 to >1:320 to SV5 (referred to as
DA) was seen between the acute and convalescent stages
of the infection. Viruses less closely related to SV5
were isolated from serum and urine of 4 patients during
the incubation period of infectious hepatitis (99). All
isolates appeared to be identical. Although 8-fold
differences were observed in cross-neutralization and
Hi tests with SV5, these isolates were more closely
related to SV5 than they were to parainfluenza type 2
virus. While no evidence has been obtained that SV5 was
the etiological agent of disease in any of the humans
from whom it was isolated, its potential virulence for
man must not be overlooked. Whether or not monkeys are
the natural host for SV5 is debatable, but it remains
well established that monkeys are frequently infected
with this virus.

SV20

SV20 should also be mentioned, since it is one of the more highly oncogenic simian adenoviruses (77) and because serological data suggested that SV20, or an antigenically related virus, has caused infection in man. Aulisio (93) found a high percentage of SV20 antibody-positive sera in samples obtained from natives of New Guinea. Seventy percent of the sera from children 5 years or less of age contained SV20 antibody, and the incidence further increased with age. SV20 antibody was not detected in sera from young children in the United States, but 20-30% of teenage and adult sera were positive. Inoculation of SV20 can produce lymphoma-like tumors in newborn hamsters, and similar tumors occur naturally in the natives of New Guinea. However there is no monkey population in New Guinea and there is no direct evidence linking SV20 or a similar virus to malignant disease in this population.

HERPESVIRUS TAMARINUS

H. tamarinus (marmoset herpesvirus, Herpes T, H. platyrrhinae, H. saimiri) is a herpesvirus occurring naturally in South American squirrel monkeys. The virus is highly virulent for marmosets and was first isolated from marmosets during a laboratory epidemic in this species (100). One report of a possible human infection with H. tamarinus has appeared (101). In this case an investigator who employed squirrel monkeys in his laboratory developed a nonfatal encephalitis. A rising titer to H. tamarinus virus was observed in serial bleedings. No attempt was made to isolate the causative agent, and no other evidence was reported to support the possible role of H. tamarinus virus in the etiology of the patient's illness.

OTHER VIRUSES OF POSSIBLE CONCERN

Although all viruses should be handled with care and should be considered potentially dangerous to man, there are some specific agents, in addition to those discussed above, that laboratory workers should consider when handling sub-human primates or their tissues. Rabies virus infection has been diagnosed in monkeys on 8 occasions in the United States (102), and 5 of these occurred between 1955 and 1966. No human cases resulted from these simian infections. Sixteen cases of laboratory confirmed rabies in nonhuman primates had been reported in the United States during the past 43 years (103). In a case of rabies in an imported chimpanzee, who bit or scratched two technicians shortly before it died, the brain was positive for rabies by the FA test and the virus was recovered by mouse inoculation. Both patients received post-exposure anti-rabies immunization.

Numerous herpesviruses, other than those specifically discussed, are known to be carried primarily by New World monkeys, but also by African species. These are of concern not only as possible infectious agents for man, but also because of their oncogenicity. Melendez and his colleagues (104) have isolated a number of simian herpesviruses from South American monkeys which produced malignant disease in primates. In one instance they observed the horizontal transmission of one of these viruses (H. ateles) to a marmoset, and this resulted in a fatal malignant disease (Med. World News, No. 32, 13, 18-19, 1972).

Simian sarcoma virus was first observed in woolly monkeys by Theilen (105) and was found to be capable of producing fibrosarcoma when inoculated into marmosets. Wolfe (106) further reported that human cells (WI38) were susceptible to infection with this virus and that infected cells showed evidence of "transformation."

An RNA "C"-type tumor virus, known as Mason-Pfizer virus, has been isolated from a rhesus monkey mammary carcinoma (107) The susceptibility of human cells to this virus was reported by Fine (108). A virus of the same type and with the capacity to replicate in human cells was recovered from a gibbon with lymphosarcoma (109).

There is as yet no evidence that any of these viruses produce malignancies in man, but available knowledge of their oncogenicity in sub-human primates and of their ability to infect and transform human cells in culture provides sufficient evidence of their potential danger. Investigators working with these viruses, or with their natural hosts, should be aware of the possible dangers involved.

ASSESSMENT OF RISKS AND GENERAL PROTECTIVE MEASURES

In the previous sections, at least 83 cases of human disease caused by simian viruses, or by viruses transmitted from monkeys to man, were noted. Twenty-four of these infections terminated in death, while in some others extensive sequelae, or long periods of convalescence followed the acute infections. Other instances of infection by simian viruses without overt clinical disease were also mentioned, and serological evidence of possible infection with still other members of this

group of viruses was presented. It is difficult to
assess the over-all risks involved in numerical terms,
since the extent of possible exposure cannot be determined. Concern over zoonoses involving monkeys developed
following the first human case of B virus infection in
1934, and the tabulation of human infections began at
that time. Thus the cases cited occurred over a period
of nearly 40 years. However all but two occurred since
1950, a period marked by the accelerated use of monkeys
for laboratory purposes. It seems, therefore, that the
risk of acquiring infection at any one time is probably
small, but that the consequences may be great, as
evidenced by the 24 deaths which occurred.

Strict adherence to standard aseptic or sterile
operative techniques, together with some added precautions in the housing and handling of monkeys, should
further reduce the risk of human disease caused by
viruses associated with subhuman primates. Several
authors have suggested precautions against infection with
B virus (110,111,7). These precautions deal essentially
with the handling of monkeys, and if followed, would
reduce the chances of human infection not only by B
virus, but by most viruses known to be carried by monkeys.
Although it is beyond the scope of this review to provide
detailed instructions for laboratory safety, some general
precautions are itemized and discussed below:

1. Reduce the incidence of active infections in
monkeys.
2. Prevent cross-infection of species after capture.
3. Minimize human exposure to captive monkeys.
4. Maintain barriers between monkeys and operators.
5. Regard as infectious all monkeys, their tissues
and excreta, monkey cell cultures, and all materials and
equipment brought into contact with any of the afore-
mentioned.
6. Be familiar with the viruses and diseases known
to occur in the species under study.
7. Provide medical surveillance of all persons at
risk of exposure to monkey diseases.

Items 1 through 3 pertain to the capture, transportation and housing of monkeys. Animal procurement
was at its worst in the 1950's, when large numbers of
monkeys were captured, housed in gang cages in crowded
compounds, and rapidly shipped by air to the laboratories
where they were scheduled for research or vaccine
production. Some laboratories held these animals in
gang cages, and in most laboratories the monkeys were
used shortly after arrival. As noted in the discussion
of B virus, such practices appeared to foster the
dissemination of virus, as well as activate latent

infections due to the stress involved. Thus they
increased the incidence of active infection in the
colony. When new arrivals were housed in the same
quarters with other monkeys already on hand, the posibil-
ity for further spread of the virus(es) within the
laboratory existed. Monkey procurement and husbandry
practices have improved since that time in some
laboratories, but they vary considerably among institutions importing and using these animals.

Laboratory purchasers of monkeys can influence or
control procurement of animals in the wild only through
their requests for certain qualities in the animals
which they receive. Some dealers will provide special
handling and shipping upon request. Direct control over
the animals begins at the time of arrival in the
laboratory. New arrivals should be placed in individual
cages and isolated in quarantine quarters for at least
6 weeks, as recommended in the USPHS Regulations (Pub.
Hlth. Ser. Reg. Manufact. Biol. Proc. Title 42, Part 73,
73.501(F)(2), 1971). The animals should be handled as
little as possible during this period and used for
experimental purposes only after the satisfactory
completion of the quarantine period. In retrospect it
appears that the extent of the Marburg virus disease
episode might have been greatly reduced had these
animals been held in such a quarantine prior to use. At
the completion of the quarantine period, the monkeys may
be moved to other quarters, preferably apart from other
animals, and especially from other simian species.
Access to monkey quarters should be restricted to those
persons immediately concerned with the care of the
animals and to trained personnel involved in experimental
procedures and observations.

The maintenance of barriers between monkeys and
operators pertains primarily to the wearing of protective
clothing. This should not be overlooked, even for
minor procedures such as drawing blood or palpating an
animal during a physical examination. The use of a
plastic face shield, along with other protective clothing
during autopsy and surgical procedures, is recommended.
This is especially important for such operations as bone
cutting or sawing or during the use of a power-operated
trephine. Many types of restraining equipment are
available for use with subhuman primates, and these can
provide added ease and safety in handling the animals.
Whenever possible, monkeys should be tranquilized or
anaesthetized before being examined or manipulated, as
this will reduce the possibility of bites, scratches,
and other accidents brought on by a struggling animal.

Item 5 is self-explanatory, but a few comments,

particularly on the use of monkey cell cultures, seem
warranted. Some laboratories prepare their own cell
cultures from monkey tissues. Personnel in these
laboratories are exposed to the monkeys and to the
hazards of tissue procurement and processing, as well as
to the cultures themselves. Those who purchase or
obtain cultures from other sources have only the latter
concern. In these laboratories people may be a little
less diligent in their observations of safety practices
than are those in the laboratories where cultures are
produced. Commercial suppliers of monkey cell cultures
presumably use healthy animals as tissue donors and they
do perform screening tests for viral contaminants on
each lot of cell cultures produced. By necessity, however, these cultures are shipped to the purchasers and
are used before fully adequate tests can be completed.
Thus, these cultures, the supernatant medium, and all
equipment used in their manipulation must be decontaminated before being discarded. Nearly all cell culture
media contain sodium bicarbonate as a buffer, which
results in some pressure build-up in sealed tubes or
culture vessels. This causes an outward rush of gas
when the seal is broken, producing virus aerosols from
infected cultures. For this reason it is wise to place
the mouth of the vessel towards, and close to, a flame
when removing a stopper or cap. For most purposes one
of the several virus-free continuous monkey kidney cell
lines now available can be substituted for primary monkey
kidney cells. Such cell lines should be used whenever
feasible in order to avoid many of the possible dangers
associated with primary monkey cell cultures.

A great deal is now known about the latent viruses
indigenous in monkeys of various species. Much of this
information can be found in the references cited in this
review, as well as in other papers in the literature.
Investigators planning programs in which monkeys are to
be used should consult the literature, or other authori-
tative sources, for information pertaining to the
particular species to be studied. Furthermore, such
information should play an important role in the selection of the particular species to be used when such a
choice is possible. The threat of B virus infection,
for example, can be greatly reduced if a species other
than certain of the Asiatic Macaca monkeys is selected.

Persons working with monkeys should be under some
form of medical surveillance involving periodic examinations, and the responsible physician should be aware of
the potential infections which might be acquired as
occupational diseases. All injuries should be reported
immediately and treated if necessary. Unusual illnesses
should be carefully monitored. It is frequently helpful

in making a diagnosis when an infection does occur to have pre-exposure serum. For this reason, a bank of frozen serum samples from persons at risk of exposure to monkey viruses should be maintained.

There is little to offer at present in the way of specific immunization against virus diseases of monkeys. Both clinical and laboratory observations have suggested that immunization with vaccinia affords protection against infection with monkeypox virus. This probably accounts for the absence of laboratory-acquired infections during outbreaks of MPV in monkey colonies. An experimental B virus vaccine has been prepared and tested in man (23) but is not presently available for general distribution. Rabies has been observed in laboratory nonhuman primates, and in one instance direct human exposure resulted from a bite of an infected animal. Pre-exposure rabies immunization should thus be considered. Monkeys coming into the United States from areas of endemic yellow fever must be held in specified quarantine by the importer before distribution. Yellow fever, thus, has not been encountered in laboratory monkeys and immunization of personnel against yellow fever does not appear to be necessary.

References

1. Sabin, A. B. and A. M. Wright. 1934. Acute ascending myelitis following a monkey bite, with the isolation of a virus capable of reproducing the disease. J. Exp. Med. 59:115.
2. Sabin, A. B. 1934. Studies on the B virus. I. The immunological identity of a virus isolated from a human case of ascending myelitis associated with visceral necrosis. Brit. J. Exp. Pathol. 15:248.
3. Hull, R. N. 1968. The simian viruses. Virol. Monogr. 2:2. Springer-Verlag, New York.
4. Hunt, R. D. and L. V. Melendez. 1969. Herpesvirus infections of nonhuman primates: A review. Lab. Animal Care 19:221.
5. Melendez, L. V., M. D. Daniel, H. H. Barahona, C. E. O. Fraser, R. D. Hunt and F. G. Garcia. 1971. New herpesviruses from South American monkeys. Preliminary report. Lab. Animal Sci. 21:part 2, 1050.
6. Hull, R. N. 1973. The simian herpesviruses (in press). Academic Press, New York.
7. Hull, R. N. 1969. The significance of simian viruses to the monkey colony and the laboratory investigator. Ann. N. Y. Acad. Sci. 162:472.
8. Keeble, S. A. 1960. B virus infection in monkeys. Ann. N. Y. Acad. Sci. 85:960.
9. Kirschstein, R. L. and G. L. Van Hoosier. 1961.

Virus-B infection of the central nervous system of
monkeys used for the poliomyelitis vaccine safety
test. Amer. J. Pathol. 38:119.

10. Hull, R. N., J. R. Minner and C. C. Mascoli. 1958.
New viral agents recovered from tissue cultures of
monkey kidney cells. III. Recovery of additional
agents both from cultures of monkey tissues and
directly from tissues and excreta. Amer. J. Hyg.
68:31.

11. Hull, R. N. and J. C. Nash. 1960. Immunization
against B virus infection. I. Preparation of an
experimental vaccine. Amer. J. Hyg. 71:15.

12. Shah, K. V. and C. H. Southwick. 1965. Prevalence
of antibodies to certain viruses in sera of free-
living rhesus and of captive monkeys. Ind. J. Med.
Res. 53:488.

13. Shah, K. V. and J. A. Morrison. 1969. Comparison
of three rhesus groups for antibody patterns to
some viruses: Absence of active SV40 transmission
in the free-ranging rhesus of Cayo Antiago. Amer. J.
Epidemiol. 89:308.

14. Soper, W. T. 1959. Monkey B virus. A review of
the literature. Technical study 17, Biological
Warfare Laboratories, Fort Detrick, Maryland.

15. Love, F. M. and E. Jungherr. 1962. Occupational
infection with virus B of monkeys. J. Amer. Med.
Ass. 179:804.

16. Sumner-Smith, G. 1966. B virus in association with
a monkey colony at a department of psychology.
Lab. Primate Newsletter 5:No. 1, 1.

17. Sabin, A. B. 1949. Fatal B virus encephalomyelitis
in a physician working with monkeys. J. Clin.
Invest. 28:808.

18. Hummeler, K., W. L. Davidson, W. Henle, A. C.
LaBoccetta and H. G. Ruch. 1959. Encephalomyelitis
due to infection with Herpesvirus simiae (Herpes B
virus). A report of two fatal laboratory-acquired
cases. New Eng. J. Med. 261:64.

19. Melnick, J. L. and D. D. Banker. 1954. Isolation of
B virus from the central nervous system of a rhesus
monkey. J. Exp. Med. 100:181.

20. Hull, R. N., F. B. Peck, Jr., T. G. Ward and J. C.
Nash. 1962. Immunization against B virus infection.
II. Further laboratory and clinical studies with an
experimental vaccine. Amer. J. Hyg. 76:239.

21. Van Hoosier, G. L. and J. L. Melnick. 1961.
Neutralizing antibodies in human sera to Herpesvirus
simiae (B virus). Tex. Rep. Biol. Med. 19:376.

22. Cabasso, V. J., W. A. Chappell, J. E. Avampato and
J. L. Bittle. 1967. Correlation of B virus and
herpes simplex virus antibodies in human sera. J.
Lab. Clin. Med. 70:170.

23. Hull, R. N. 1971. B virus vaccine. Lab. Animal

Sci. 21:part 2, 1068.
24. Nagler, F. P. and M. Klotz. 1958. A fatal B virus infection in a person subject to recurrent Herpes labialis. Can. Med. Ass. J. 79:743.
25. Siegert, R. 1972. Marburg virus. Virol. Monogr. 11:98. Springer-Verlag, New York.
26. Martini, G. A. and R. Seigert. 1971. Marburg virus disease. Springer-Verlag, New York.
27. Simpson, D. I. H. 1970. Marburg virus disease: Experimental infection of monkeys. Lab. Animal Handb. 4:149.
28. Slenczka, W., G. Wolff and R. Siegert. 1971. A critical study of monkey sera for the presence of antibody against the Marburg virus. Amer. J. Epidemiol. 93:496.
28a. Simpson, D. I. H. 1969. Marburg agent disease: In monkeys. Trans. Roy. Soc. Trop. Med. Hyg. 63:303.
29. Haas, R. and G. Maass. 1971. Experimental infection of monkeys with the Marburg virus. Marburg virus disease, ed. G. Martini and R. Siegert, p. 136. Springer-Verlag, New York.
30. Kunz, C. and H. Hoffman. 1971. Some characteristics of the Marburg virus. Marburg virus disease, ed. G. A. Martini and R. Siegert. Springer-Verlag, New York.
31. Smith, C. E. G., D. I. H. Simpson, E. T. W. Bowen and I. Zlotnik. 1967. Fatal human disease from vervet monkeys. Lancet 2:1119.
32. Almeida, J. D., A. P. Waterson and D. I. H. Simpson. 1971. Morphology and morphogenesis of the Marburg agent. Marburg virus disease, ed. G. Martini and R. Siegert, p. 84. Springer-Verlag, New York.
33. Kissling, R. E., R. Q. Robinson, F. A. Murphy and S. G. Whitfield. 1968. Agent of disease contracted from green monkeys. Science 160:888.
34. Bowen, E. T. W., D. I. H. Simpson, W. F. Bright, I. Zlotnik and D. M. R. Howard. 1969. Vervet monkey disease: Studies on some physical and chemical properties of the causative agent. Brit. J. Exp. Pathol. 50:400.
35. Simpson, D. I. H., I. Zlotnik and D. A. Rutter. 1968. Vervet monkey disease. Experimental infection of guinea pigs and monkeys with the causative agent. Brit. J. Exp. Pathol. 49:458.
36. Zlotnik, I. 1969. Marburg agent disease: Pathology. Trans. Roy. Soc. Trop. Med. Hyg. 63:310.
37. Simpson, D. I. H. 1969. Vervet monkey disease. Transmission to the hamster. Brit. J. Exp. Pathol. 50:389.
38. Hoffman, H. and C. Kunz. 1971. Cultivation of the Marburg virus (Rhabdovirus simiae). Marburg virus disease, ed. G. Martini and R. Siegert, p. 112.

Springer-Verlag, New York.

39. Martini, G. A. 1969. Vervet monkey disease: Human cases. Trans. Roy. Soc. Trop. Med. Hyg. 63:295.

40. Martini, G. A. 1971. Marburg virus disease, ed. G. Martini and R. Siegert, p. 1. Springer-Verlag, New York.

41. Stille, W. and E. Bohle. 1971. Clinical course and prognosis of Marburg virus ("green-monkey") disease. Marburg virus disease, p. 10. Springer-Verlag, New York.

42. Todorovitch, K., M. Mocitch and R. Klansnja. 1971. Two cases of Cercopithecus-monkeys-associated haemorrhagic fever. Marburg virus disease, ed. G. Martini and R. Siegert. Springer-Verlag, New York.

43. von Magnus, P., E. G. Andersen, K. B. Petersen and A. Birch-Andersen. 1960. A pox-like disease in cynomolgus monkeys. Acta Pathol. Microbiol. Scand. 46:156.

44. Prier, J. E. and R. M. Sauer. 1960. A pox disease of monkeys. Ann. N. Y. Acad. Sci. 85:951.

45. Sauer, R. M., J. E. Prier, R. S. Buchanan, A. A. Creamer and H. C. Fegley. 1960. Studies on a pox disease of monkeys. I. Pathology. Amer. J. Vet. Res. 21:377.

46. Stewart, J. M., Y. F. Herman, D. E. Mattson and L. Erickson. 1962. Monkey pox disease in irradiated cynomolgus monkeys. Nature 195:1128.

47. Arita, I. and D. A. Henderson. 1968. Smallpox and monkeypox in non-human primates. Bull. World Health Org. 39:277.

48. Arita, I., R. Gispen, S. S. Kalter, L. T. Wah, S. S. Marennikova, R. Netter and I. Tagaya. 1972. Outbreaks of monkeypox and serological surveys in nonhuman primates. Bull. World Health Org. 46:625.

49. Espana, C. 1971. Review of some outbreaks of viral disease in captive nonhuman primates. Lab. Animal Sci. 21:1023.

50. Peters, J. C. 1966. An epizootic of monkeypox at Rotterdam zoo. The Intl. Zoo Yearbook 6:274.

51. Noble, J., Jr. 1970. A study of New and Old World monkeys to determine the likelihood of a simian reservoir of smallpox. Bull. World Health Org. 42:509.

52. Noble, J., Jr. and J. A. Rich. 1969. Transmission of smallpox by contact and by aerosol routes in Macaca irus. Bull. World Health Org. 40:279.

53. Wilner, B. I. 1969. A classification of the major groups of human and other animal viruses, 4th ed., p. 97. Burgess, Minneapolis, Minn.

54. Bedson, H. S. and K. R. Dunbell. 1961. The effect of temperature on the growth of pox viruses in chick embryo. J. Hyg. (London) 59:475.

55. Nicholas, A. H. 1970. Poxvirus of primates.

II. Immunology. J. Nat. Cancer Inst. 45:907.

56. Cho, C. T. and H. A. Wenner. 1972. In vitro growth characteristics of monkeypox virus. Proc. Soc. Exp. Biol. Med. 139:916.

57. Rondle, C. J. M. and K. A. R. Sayeed. 1972. Studies on monkeypox virus. Bull. World Health Org. 46:577.

58. Foster, S. O., E. W. Brink, D. L. Hutchins et al. 1972. Human monkeypox. Bull. World Health Org. 46:569.

59. Ladnyj, I. D., P. Ziegler and E. Kima. 1972. A human infection caused by monkeypox virus in Basankusu Territory, Democratic Republic of the Congo. Bull. World Health Org. 46:593.

60. Bearcroft, W. G. C. and M. F. Jamieson. 1958. An outbreak of subcutaneous tumors in rhesus monkeys. Nature 182:195.

61. Niven, J. S., J. A. Armstrong, C. H. Andrewes, H. G. Pereira and R. C. Valentine. 1961. Subcutaneous "growths" in monkeys produced by a poxvirus. J. Pathol. Bact. 81:1.

62. Kalter, S. S. and R. L. Heberling. 1971. Comparative virology of primates. Bact. Rev. 35:310.

63. Ambrus, J. L., E. T. Feltz, J. T. Grace, Jr. and G. Owens. 1963. Virus induced tumor of primates. Nat. Cancer Inst. Monogr. 10:447.

64. Ambrus, J. L., H. V. Strandstrom and W. Kawinski. 1969. "Spontaneous" occurrence of Yaba tumor in a monkey colony. Experientia 25:64.

65. Wolfe, L. G., R. A. Griesemer and R. L. Farrell. 1968. Experimental aerosol transmission of Yaba virus in monkeys. J. Nat. Cancer Inst. 41:1176.

66. Yohn, D. S., J. T. Grace, Jr. and V. A. Haendiges. 1964. A quantitative cell culture assay for Yaba tumor virus. Nature 202:881.

67. Yohn, D. S., V. A. Haendiges and J. T. Grace, Jr. 1966. Yaba tumor poxvirus synthesis in vitro. I. Cytopathological, histochemical and immunofluorescent studies. J. Bact. 91:1977.

68. Yohn, D. S., V. A. Haendiges and J. T. Grace, Jr. 1966. Yaba tumor poxvirus synthesis in vitro. II. Adsorption, inactivation and assay studies. J. Bact. 91:1953.

69. Yohn, D. S., V. A. Haendiges and E. de Harven. 1966. Yaba tumor poxvirus synthesis in vitro. III. Growth kinetics. J. Bact. 91:1986.

70. Grace, J. T., Jr., E. A. Miraud, S. J. Millian and R. S. Metzgar. 1962. Experimental studies of human tumors. Fed. Proc. 21:(suppl.)32.

71. Espana, C. 1966. A Yaba-like disease in primates. Lab. Primate Newsletter 5:ii.

72. Hall, A. S. and W. P. McNulty, Jr. 1967. A contagious pox disease in monkeys. J. Amer. Vet. Med. Ass. 151:833.

73. Casey, H. W., J. M. Woodruff and W. I. Butcher. 1967. Electron microscopy of a benign epidermal pox disease of rhesus monkeys. Amer. J. Pathol. 51:431.
74. Downie, A. W. and C. Espana. 1972. Comparison of tanapox virus and yaba-like viruses causing epidemic disease in monkeys. J. Hyg. 70:23.
75. McNulty, W. P., W. C. Lobitz, Jr., F. Hu, C. A. Maruff and A. S. Hall. 1968. A pox disease in monkeys transmitted to man. Arch. Dermatol. 97:286.
76. Crandell, R. A., H. W. Casey and W. B. Brumlow. 1969. Studies of a newly recognized poxvirus of monkeys. J. Inf. Dis. 119:80.
77. Hull, R. N., I. S. Johnson, C. G. Culbertson, C. B. Reimer and H. F. Wright. 1965. Oncogenicity of the simian adenoviruses. Science 150:1044.
78. Sweet, B. H. and M. R. Hilleman. 1960. The vacuolating virus, SV40. Proc. Soc. Exp. Biol. Med. 105:420.
79. Meyers, H. M., Jr., H. E. Hopps, N. G. Rogers, B. E. Brook, B. C. Bernheim, W. P. Jones, A. Nisalak and R. G. Douglas. 1966. Studies on SV40. J. Immunol. 88:796.
80. Eddy, B. E., G. S. Borman, G. Grubbs and R. D. Young. 1962. Identification of the oncogenic substance in rhesus monkey kidney cell cultures as SV40. Virology 17:65.
81. Melnick, J. L. and S. Stinebaugh. 1962. Excretion of vacuolating SV40 virus (Papovavirus group) after ingestion as contaminant of oral poliovaccine. Proc. Soc. Exp. Biol. Med. 109:965.
82. Morris, J. A., K. M. Johnson, C. G. Aulisio, R. M. Chanock and V. Knight. 1961. Clinical and serologic responses in volunteers given vacuolating virus (SV40) by respiratory route. Proc. Soc. Exp. Biol. Med. 108:56.
83. Shah, K. V. 1966. Neutralizing antibodies to SV40 in human sera from India. Proc. Soc. Exp. Biol. Med. 121:303.
84. Fraumeni, J. F., Jr., F. Ederer and R. W. Miller. 1963. An evaluation of the carcinogenicity of SV40 in man. J. Amer. Med. Ass. 185:713.
85. Shah, K. V. 1972. Evidence for an SV40-related papovavirus infection of man. Amer. J. Epidemiol. 95:199.
86. ZuRhein, G. M. 1969. Association of papova-virions with a human demyelinating disease (progressive multifocal leucoencephalopathy). Prog. Med. Virol. 11:185.
87. Padgett, B. L., D. L. Walker, G. M. ZuRhein and R. J. Eckroade. 1971. Cultivation of papova-like virus from human brain with PML. Lancet 1:1257.
88. Weiner, L. P., R. M. Herndon, O. Narayan, R. T.

Johnson, K. Shah, L. J. Rubinstein, T. J. Preziosi
and F. K. Conley. 1972. Virus related to SV40 in
patients with PML. New Eng. J. Med. 286:385.
89. Weiner, L. P., R. M. Herndon, O. Narayan and R. T.
Johnson. 1972. Further studies of a SV40-like
virus isolated from human brain. J. Virol. 10:147.
90. Gardner, S. D., A. M. Field, D. V. Coleman and B.
Hulme. 1971. New human papovavirus (BK) isolated
from urine after renal transplantation. Lancet 1:
1253.
91. Hull, R. N., J. R. Minner and J. W. Smith. 1956.
New viral agents recovered from tissue cultures of
monkey kidney cells. I. Origin and properties of
cytopathogenic agents SV1, SV2, SV4, SV5, SV6,
SV11, SV12, and SV15. Amer. J. Hyg. 63:204.
92. Hsiung, G. D., P. Isacson and G. Tucker. 1963.
Studies of parainfluenza viruses. II. Serologic
interrelationships in humans. Yale J. Biol. Med.
35:534.
93. Aulisio, C. G., D. C. Wong and J. A. Morris. 1964.
Neutralizing antibodies against simian viruses, SV5
and SV20 in human sera. Proc. Soc. Exp. Biol. Med.
117:6.
94. Binn, L. N., E. C. Lazar, M. Rogul, V. S. Shepler,
L. J. Swango, T. Claypoole, D. W. Hubbard, S. G.
Asbill and A. D. Alexander. 1968. Upper respiratory
disease in military dogs: Bacterial, mycoplasma
and viral studies. Amer. J. Vet. Res. 29:1809.
95. Schultz, E. W. and K. J. Habel. 1959. SA virus--a
new member of the myxovirus group. J. Immunol. 82:
274.
96. Chanock, R. M., K. M. Johnson, M. K. Cook, D. C.
Wong and A. Vargosko. 1961. The hemadsorption
technique with special reference to the problem of
naturally occurring simian parainfluenza virus.
Amer. Rev. Resp. Dis. 83:No. 2, pt. 2, 125.
97. Hsiung, G. D., P. Isacson and R. W. McCollum. 1962.
Studies of a myxovirus isolated from human blood.
I. Isolation and properties. J. Immunol. 88:284.
98. von Euler, L., F. S. Kantor and G. D. Hsiung.
1963. Studies on parainfluenza viruses. I.
Clinical, pathological and virological observations.
Yale J. Biol. Med. 35:523.
99. Liebhaber, H., S. Krugman, D. McGregor and J. P.
Giles. 1965. Studies of a myxovirus recovered
from patients with infectious hepatitis. I.
Isolation and characterization. J. Exp. Med. 122:
1135.
100. Holmes, A. W., R. E. Dedmon and F. Deinhardt. 1963.
Isolation of a new herpes-like virus from South
American marmosets. Fed. Proc. 22:334.
101. Schrier, A. M. 1966. Editor's notes. Primate Newsletter 5:No. 4, ii.

102. Richardson, J. H. and G. L. Humphrey. 1971. Rabies in imported nonhuman primates. Lab. Animal Sci. 21:part 2, 1083.
103. Moit, M. R. and R. K. Sikes. 1973. Rabies in a chimpanzee. J. Amer. Vet. Med. Ass. 162:54.
104. Melendez, L. V., H. Castellanos, H. H. Barahona, M. D. Daniel, R. D. Hunt, C. E. O. Fraser, F. G. Garcia and N. W. King. 1972. Two new herpesviruses from spider monkeys (Ateles geoffroyi). J. Nat. Cancer Inst. 49:233.
105. Theilen, G. H., D. Gould, M. Fowler and D. L. Dungworth. 1971. C-type virus in tumor tissue of a woolly monkey (Lagothrix spp.) with fibrosarcoma. J. Nat. Cancer Inst. 47:881.
106. Wolfe, L. G., R. Smith and F. Deinhardt. 1972. Simian sarcoma virus: Focus assays and susceptibility of human cells. Fed. Proc. 31:619.
107. Jensen, E. M., I. Zelljadt, H. C. Chopra and M. M. Mason. 1970. Isolation and propagation of a virus from a spontaneous mammary carcinoma of a rhesus monkey. Cancer Res. 30:2388.
108. Fine, D. L., J. C. Landon and M. T. Kubicek. 1971. Simian tumor virus isolate: Demonstration of cytopathic effects in vitro. Science 174:420.
109. Kawakami, T. G., S. D. Huff, P. M. Buckley, D. L. Dungworth, S. P. Snyder and R. V. Gilden. 1972. C-type virus associated with gibbon lymphosarcoma. Nature New Biol. 235:170.
110. Perkins, F. T. and E. G. Hartley. 1966. Precautions against B virus infection. Brit. Med. J. 1:899.
111. Cole, W. C., R. E. Bostrom and R. A. Whitney. 1968. Diagnosis and handling of B virus in a rhesus monkey (Macaca mulatta). J. Amer. Vet. Med. Ass. 153:894.

DISCUSSION

CASALS: The question about antibodies for the Marburg
agent has been discussed enough as it is, but I think it
is extremely important because workers in South Africa
and in Germany claimed that there were no antibodies in
most of the monkeys from Africa. Then it is very curious
that among the published results are reported antibodies
against Marburg in a number of Macaca fuscata from Japan;
this is certainly disconcerting. The matter of antibody
distribution is a very important question that should be
settled once and for all, not only by CF test, which with
monkeys is very bad business at best, but by a neutralization test.

SHAH: We found that rhesus monkeys born in the laboratory
and raised individually in cages were essentially free of
B virus antibodies. This was in spite of the fact that
these animals shared rooms with over a hundred other
imported monkeys, most of which had B virus antibodies.
It seems as if individual caging alone may prevent or
markedly reduce B virus infection. In primate centers
individual caging may be a routine practice. I wonder if
such individually raised animals have been checked for B
virus antibodies.

OXMAN: Especially in the case of herpesviruses, it may
not always be safe to assume that a negative serologic
test guarantees freedom from latent infection. For
example, in cytomegalovirus infection, antibody levels
may eventually fall below the threshold of the commonly
used CF test. Nevertheless, virus may persist in such
"antibody-negative" individuals and can be recovered from
them at later times. Thus it is important to make sure
that the particular test employed will detect latent
infection. In this regard, I wonder what testing
procedures are employed by commercial laboratories to
insure that the primary monkey cells we buy are free of
B virus.

HULL: It is quite possible that laboratory born rhesus
monkeys would be B virus antibody-negative. The incidence
of antibody in immature animals captured in the wild is
about 10%, but this may rise to 70% by the time the same
group of animals reaches the laboratory. I have had
rhesus monkeys captured, put into individual crates and
shipped to my laboratory so that arrival was only 72
hours after capture. In one group of 6 animals, one
monkey had B virus antibody in serum taken upon arrival
which rose in titer during the next two weeks. The other

five, held in individual cages in separate rooms, failed
to develop antibody during a year of observation.

BERG: Is there an agency which regulates or monitors
commercial suppliers of monkeys and their tissues for
research use?

KALTER: There are a number of regulations controlling
the importation of animals. However, these are relatively superficial as they are based upon cursory physical
examinations by veterinarians without any laboratory
support. WHO has made recommendations (Technical Report
#470) on the health aspects of the supply and use of
nonhuman primates for biomedical purposes. If these
recommendations are adhered to, it would provide the
scientific community with a far better quality of animal
than is currently available. It is, however, the feeling
of many knowledgeable individuals that the only way to
upgrade the nonhuman primate is by developing breeding
colonies and providing a standardized animal.

HULL: No period of quarantine will completely eliminate
the hazards associated with recently captured monkeys,
but a 6-weeks quarantine, such as required by the D.B.S.
for vaccine manufacturers, will reduce the risks. Most
likely, primary B virus infections would subside within
6 weeks, or agents of the Marburg type would kill most
of the animals during the quarantine period.

     There is no single (or multiple) test, that can
be performed on recently captured monkeys to identify
those that may be latently or actively infected, which
does not entail a lot of time and effort and does not
enhance exposure of personnel to possible infectious
agents carried by the monkeys.

HILLEMAN: Perhaps I should make a philosophic comment
at this point. The tissues of wild-caught animals, and
certainly monkeys, are commonly infected with wild
viruses. The simplest way to solve the monkey problem
is to eliminate the monkeys and this is being done using
diploid cells. Monkey kidney came into use by historic
decision to use monkey kidney to make poliovaccine.
There is no need to continue using monkeys when acceptable
alternatives are available.

LYMPHOCYTIC CHORIOMENINGITIS VIRUS INFECTION

Wallace P. Rowe
Laboratory of Viral Diseases
National Institute of Allergy and Infectious Diseases
Bethesda, Maryland

There are two main reasons why it is important to discuss lymphocytic choriomeningitis (LCM) virus in a meeting on biohazards. First, despite a large background of information and experience, LCM remains, and will long remain, a major problem to the laboratory animal user. Second, it provides a classical model illustrating how knowledge of the natural history of a virus may provide a basis for understanding the problems it presents and for instituting reasonable control measures.

The biology of LCM virus has been reviewed recently in an excellent monograph by Lehmann-Grube (1), and I have drawn heavily on his review for the material cited here.

The natural reservoir of LCM virus is exclusively the wild house mouse; a few sporadic isolations have been made from other wild rodents and complement-fixing (CF) antibody has been found on occasion, but there is no evidence that LCM is maintained in these species. In the wild, the virus is maintained by the same mechanisms that Traub (2,3) elucidated in laboratory mice. The cycle can be viewed as starting when a pregnant mouse becomes infected; if the infection occurs at a crucial stage of pregnancy, the fetuses become infected, and though they may develop normally, the resulting offspring are lifelong carriers and excreters of virus. The female offspring thenceforth regularly pass the virus to their offspring by vertical spread, and the congenital infection cycle continues generation after generation. The congenitally infected mice have high titers of virus in all tissues, in urine, feces, and nasal washings, and are negative for antibody.

Uninfected mice can acquire infection by horizontal spread; if the infection occurs during infancy, a chronic, high-titer, antibody-negative carrier state may occur (4). This is another way in which the vertical cycle can be initiated. In older animals the infection runs only a brief, asymptomatic course, most or all virus is eliminated, and complement-fixing antibodies develop.

Geographically LCM infection is widely distributed among wild mice, but it occurs as focal pockets of infected animals (5). In urban areas about 20% of house mice have been found to be carriers (6).

In laboratory mouse breeding colonies, the basic biology is as described above. The infection is acquired from incursions of wild mice and thereafter is permanently and stably maintained by the congenital infections. In infected colonies the prevalence of infection is variable, often involving only 5 or 10% of animals (Parker, pers. comm.), but with time it can extend throughout the entire colony (3). A very disturbing prospect is that if infection should get into a single female in the main pedigreed line of an inbred strain, where all future generations are descended from a single breeding pair, every future mouse of the strain would be a carrier.

Contrary to what seems to be the general impression, LCM-infected laboratory breeding colonies are relatively rare. However, within the past five years, LCM infection has been encountered in at least three breeding colonies, two in this country (Parker, pers. comm.) and one in England (7). In these the infection appears to have been eliminated, primarily by destruction of the infected strains and replacement with clean stock.

For the laboratory doing experimental work with LCM virus in mice, the biology and problem can be viewed somewhat differently. The mice can be categorized according to degree of risk. Experimental stocks of congenitally infected mice, such as the Haas strain (8), and neonatally infected mice are high risks. However, their contagiousness seems to be rather low, and prolonged contact is generally needed for infection of personnel to occur. The infections tend to be severe,

and if at all possible, mice of this type should be
kept in an isolator. In a second category are mice
inoculated as adults with field strains or viscerotropic
strains of LCM, or immunosuppressed mice infected with
any strain of the virus. This is an unknown risk
category, but since immunosuppressed mice may become
lifelong carriers, and since field and viscerotropic
strains are known to infect the kidneys, such animals
must be considered potentially dangerous and kept in
filter-top disposable cages. The third category, adult
mice inoculated with the usual laboratory-passaged
meningotropic strains, is a no-risk system since the
virus does not replicate in the kidney or other abdominal
viscera (9) and thus is not excreted into the environment; also, the infection is self-limited. These
animals should be kept in conventional, open-top cages.
For all work with LCM-infected animals it is desirable
to screen personnel for antibody and have only immune
persons handle the animals. Immunofluorescence is a
highly satisfactory screening test for antibody (10).

LCM IN OTHER SPECIES

     While most laboratory animal users have some
awareness that mice may carry LCM, a large part of the
LCM biohazard comes from its less well-known ability to
infect other species. Infections have been encountered
in laboratory guinea pigs, monkeys, dogs, and hamsters
(1).

     In guinea pigs, dogs, and monkeys, virus can spread
from one member of the species to another, but there is
no evidence that infection can become established as an
enzootic. Dogs are rarely infected but may present a
special hazard since two fatalities have occurred in
laboratory personnel working with a contaminated
distemper vaccine (11).

     Hamsters present the major problem (12,13) because
of their exceptionally high level of virus excretion.
This excretion is so efficient that the virus can
readily become established in a hamster colony, making
the hamster a very high risk to humans. Infected
sucklings excrete very large amounts of virus in the
urine throughout life; urine contains 10^7 ID_{50}/ml
(Parker and Rowe, in prep.). In contrast to mice, these
animals are uniformly positive for CF antibody. They
can transmit virus to their progeny in utero, and the
congenitally infected hamsters resemble neonatally
infected ones in having lifelong infection in the
presence of CF antibody. The vertical transmission has
been maintained for two generations, but it is not

known whether the breeding performance of such a line would allow indefinite propagation of vertical infection.

Even hamsters inoculated with virus as young adults show chronic excretion of virus in the urine for at least 6 months, and although the amount of virus is several logs lower than in the sucklings, it is sufficient to cause many infections in animal handlers. Therefore all LCM-inoculated hamsters should be kept in germ-free isolators.

Acquisition of infection by personnel from LCM-infected mice or hamsters seems to be primarily through inhalation of dust from infected bedding or by skin contact with infected urine. In some instances the mode of transmission is far from clear, since several cases have occurred where there was only very remote contact with infected animals. No infections have occurred from mouth-pipetting accidents or from working with the virus in tissue culture.

The consequences of LCM contamination of laboratory animals are many. There have been a small number of deaths, but many dozens of severe infections, mostly in persons who were not aware of their exposure to LCM. At least 12 cases have been encountered in persons who have bought hamsters for pets (14). Also there have been innumerable instances of interference with experiments by contamination of virus stocks, trans-planted tumors, and tissue cultures of mouse, guinea pig, dog, hamster, and monkey origin. Contamination of experimental materials is also a major source of intro-duction of virus into laboratory animal stocks. The introduction of virus into hamsters has been traced to contact with mice carrying contaminated tumor transplants, to the inoculation of contaminated hamster tumors, and to the inoculation of contaminated tissue culture cells.

Control of LCM depends in large part on mere aware-ness of the problem. Any laboratory using mice or hamsters, transplanted tumors or rodent-grown viruses must be aware that LCM may be present in its animal room. Prevention consists of rigid control against entrance of wild mice, procurement of animals only from sources known to be free of virus, and screening biologicals for virus before they are introduced into the animal rooms. Detection of LCM is simple and reliable, but has rarely been applied until illness or the disruption of experiments has signaled that the problem is there. Detection consists in periodic monitoring of a sample of the colony for CF or fluorescent antibody, or testing of pooled tissues for virus if there is any suspicion that the congenital type of infection is widespread. Another simple procedure for mouse colonies is to challenge a number of animals intracerebrally with a lethal dose of virus; both the congenitally and the horizontally infected mice will be totally refractory to the disease.

The National Cancer Institute provides LCM testing to the general scientific community, without charge, through a mouse virus diagnostic testing contract, under Dr. John C. Parker. If properly used, this service could do much to reduce the LCM problem.

References

1. Lehmann-Grube, F. 1971. Lymphocytic choriomeningitis virus. Virol. Monogr. Vol. 10. Springer-Verlag, New York.
2. Traub, E. 1936. The epidemiology of lymphocytic choriomeningitis in white mice. J. Exp. Med. 64:183-200.
3. Traub, E. 1939. Epidemiology of lymphocytic choriomeningitis in a mouse stock observed for four years. J. Exp. Med. 69:801-817.
4. Haas, V. H. 1941. Studies on the natural history of the virus of lymphocytic choriomeningitis in mice. Publ. Health Rep. (Wash.) 56:285-292.
5. Ackermann, R., H. Bloedhorn, B. Kupper, I. Winkens und W. Scheid. 1964. Uber die Verbreitung des Virus der Lymphocytaren Choriomeningitis unter den Mausen in Westdeutschland. I. Untersuchungen uberwiegend an Hausmausen (Mus musculus). Zbl. Bakt., I. Abt. Orig. 194:407-430.
6. Armstrong, C., J. J. Wallace and L. Ross. 1940. Lymphocytic choriomeningitis. Gray mice, Mus musculus, a reservoir for the infection. Publ. Health Rep. (Wash.) 55:1222-1229.
7. Skinner, H. H. and E. H. Knight. 1969. Studies on murine lymphocytic choriomeningitis within a partially infected colony. Lab. Anim. 3:175-184.
8. Haas, V. H. 1954. Some relationships between lymphocytic choriomeningitis (LCM) virus and mice. J. Infect. Dis. 94:187-198.
9. Rowe, W. Pl. 1954. Studies on pathogenesis and immunity in lymphocytic choriomeningitis infection of the mouse. Naval Med. Res. Inst., Res. Rep. NM 005 048.14.01.
10. Cohen, S. M., I. A. Triandaphilli, J. L. Barlow and J. Hotchin. 1966. Immunofluorescent detection of antibody to lymphocytic choriomeningitis virus in man. J. Immunol. 96:777-784.
11. Armstrong, C. 1942. Some recent research in the field of neurotropic viruses with especial reference

to lymphocytic choriomeningitis and herpes simplex.
Milit. Surg. 91:129-146.

12. Lewis, A. M., Jr., W. P. Rowe, H. C. Turner and
R. J. Huebner. 1965. Lymphocytic-choriomeningitis
virus in hamster tumor: spread to hamsters and
humans. Science 150:363-364.

13. Baum, S. G., A. M. Lewis, Jr., W. P. Rowe and R. J.
Huebner. 1966. Epidemic nonmeningitic lymphocytic-
choriomeningitis-virus infection. An outbreak in a
population of laboratory personnel. New Eng. J.
Med. 274:934-936.

14. CDC Veterinary Public Health Notes, June, 1971.

* * * * *

DISCUSSION

CASALS: Dr. Rowe, how important do you think is
infection of hamsters in this country? The reason I ask
this is that I had occasion to attend a meeting in
Hamburg just a few months ago and much to my astonish-
ment it seems that most of their breeder colonies are
infected; hundreds of thousand of hamsters seem to be
sold as pets in West Germany. Six of 12 breeder colonies
were infected with LCM and they maintained the infection.
Is the problem in this country similar to that in
magnitude?

ROWE: I don't think anyone has found LCM infection in
hamster colonies in this country, but there is every
reason to believe that it could, and will, happen.

OXMAN: It is worth emphasizing that apparently healthy
cell lines derived from hamster tumors may be latently
infected with LCM. Dr. Rowe, how would you screen such
cells for LCM infection?

ROWE: The fluorescent antibody test is extremely good.
The direct test with guinea pig antiserum is highly
sensitive and specific and the conjugated antiserum is
available from the NCI. Also infectivity test will give
a clearcut answer. For tissue cultures the best procedure is probably to inoculate weanling mice with an ml
of culture fluid intraperitoneally, and 2 to 3 weeks
later challenge intracerebrally with a large dose of
virulent virus. This would detect immunogenic variants
of low virulence.

POTENTIAL HAZARDS POSED BY NONVIRAL AGENTS

Edwin H. Lennette
Viral and Rickettsial Disease Laboratory
California State Department of Public Health
Berkeley, California

The potential threat of infection in the laboratory
has long been recognized by the medical microbiologist
as an ever-present occupational hazard. Periodic
published reports on instances of laboratory-acquired
infection have served as occasional reminders that this
potential could become an actuality and have pointed up
the need for unremitting adherence to precautionary
measures. Only in recent years have we come to appreciate
the magnitude of the problem of laboratory-acquired
infections and the many factors, both human and environmental, that are involved.

The latest tally (Sulkin, pers. com.) shows 2520
cases of laboratory infection attributed to nonviral
agents; 1522 were due to bacteria, 112 to chlamydia,
536 to rickettsiae, 93 to "parasites," and 257 to fungi.
In addition there were 41 cases in which the infectious
agent was not specified. Infection can occur through
a variety of exposures, e.g., working with the agent,
handling human or animal autopsy tissues, working with
clinical specimens, contact with infected animals and/or
their ectoparasites, handling discarded laboratory
instruments, glassware, etc. In many cases the exact
time and type of exposure is unknown, and in some cases,
as with tuberculosis, it may prove impossible to
determine whether the infection was acquired in the
laboratory or as a result of exposure elsewhere. It is
gratifying and encouraging that biomedical scientists

are aware of and concerned about the potential threat of
infection posed by the animals, tissues and cells with
which they are working as well as the agents, known or
unknown, which may be present in the human materials with
which they work.

TUBERCULOSIS

Tuberculosis is a dangerous disease not only in
primate colonies but also in laboratory monkey quarters.
The literature dealing with tuberculosis in primates is
concerned mainly with rhesus monkeys, since by far the
highest incidence of infection occurs in this species;
tuberculosis in other species apparently is infrequent
or rare. Stones (1), reporting the results of post-
mortem examinations at Sandwich, found 37 rhesus positive
for tuberculosis out of 2301 examined as compared to
only 5 vervets positive out of 21,233 examined. Whether
such differences reflect a true difference in species
susceptibility or whether environmental factors (e.g.,
ecology, exposure to infected humans or bovines,
shipping conditions, etc.) are responsible is still
unclear. Thus while baboon species are generally
considered to be highly resistant to tuberculosis,
Heywood and Hague (2) encountered 27 cases of the disease
over an 18-month period. In any event, animals suspected
clinically of having tuberculosis should be destroyed
because of the danger to both simian and human contacts.

Some idea of the hazard involved is provided by
data collected by the Center for Disease Control (3).
On the basis of conversions to reactivity to tuberculin
and the detection of active cases, Millar (cited in that
report) estimated the annual rate for tuberculous
infections in the general population (1970 and 1971) to
be 3 per 10,000. In 1970, of some 350 individuals in
direct contact with nonhuman primates, 6 developed a
positive tuberculin reaction. In 1971, of some 500
individuals similarly exposed, 16 developed a positive
reaction. In terms of infection rates, this represents
171 infections per 10,000 individuals in 1970 and 320
per 10,000 in 1971, rates from 60 to 100 times those for
the population at large.

Personnel assigned to duties which bring them into
contact with nonhuman primates should be given a
tuberculin test and have a chest roentgenogram taken.
Further recommendations in the CDC report suggest that
tuberculin-positive indiciduals who have not received
chemotherapy should not be employed to work with monkeys,
or if they are, isoniazid therapy for at least one year
should be given serious consideration.

Tuberculin-negative individuals should be retested every six months. Tuberculin convertors should be referred to appropriate medical specialists for treatment and follow-up; until they are shown to be non-infectious, such individuals should not work with monkeys. If there is no evidence of active disease concomitant with tuberculin conversion, consideration should be given to isoniazid therapy, since "...6-8% of all newly tuberculin-positive individuals will develop active tuberculosis during their lifetimes, with the greatest risk of developing active disease in the first few years following tuberculin conversion. Isoniazid therapy will reduce the risk of developing active disease by 50-85%..." (3). Individuals who become tuberculin positive but decline isoniazid therapy should be apprised of the possibilities of the infection developing into overt disease. The chemoprophylactic approach, however, has been complicated by the occurrence of cases of hepatitis associated with isoniazid prophylaxis (4); in the Capitol Hill outbreak, 19 cases of liver disease, 2 of them fatal, occurred among 2321 tuberculin reactors during the first 9 months following initiation of isoniazid chemoprophylaxis. In Baltimore, 30 cases of liver disease, 7 of them fatal, have occurred since February 1972 among 3170 persons receiving isoniazid under special surveillance.

Beattie (5, in discussion) describes the difficulty of definitively determining the source of infection. An animal technician working with baboons developed pulmonary tuberculosis; the baboon suspected of being tuberculous was thoroughly examined at autopsy but no lesions were found. Also although protective clothing such as gowns, masks and gloves is frequently resorted to, there is no evidence that it is effective in preventing infection via the respiratory route. That infection can occur by other routes is illustrated by the fact that an instrument used in post-mortem examination of a tuberculous monkey caused a puncture wound in the finger of one of the staff. Some months later the individual developed a tuberculous synovitis.

Tuberculosis in simians may be due to human, bovine or avian species of Mycobacteria. Infections due to Mycobacterium bovis in monkeys newly imported into the United States are only infrequently reported. However, a small outbreak due to this organism occurred at the State University of New York at Buffalo in the spring of 1971. M. bovis was isolated from 3 of the tuberculin reactors and M. intracellulare from a fourth. Although all the emphasis has been on the laboratory simian as a possible source of tuberculosis for his human contacts, other laboratory animals, specifically the dog and cat,

may constitute a similar potential hazard. Dogs and
cats in contact with cattle with bovine tuberculosis
have been known to develop histopathologic and
bacteriologic evidence of infection with M. bovis or
with atypical mycobacteria, suggesting that dogs and
cats coming from rural areas should be regarded as
possible reservoirs for human infection. Those who
work with canine and feline populations, or with tissues
from these animals (e.g., in search of etiologic agents
of cancer) may conceivably be exposed to a potential
hazard that has yet to be measured and evaluated.

PSEUDOTUBERCULOSIS (YERSINIOSIS)

Pseudotuberculosis is caused by a gram-negative
non-hemolytic coccobacillus now designated as Yersinia
(formerly Pasteurella) but recorded in the literature
under a variety of names such as Streptobaccilus
tuberculosis rodentium, Corynebacterium rodentium,
Corynebacterium pseudotuberculosis, Shigella pseudotuberculosis rodentium. Pasteurella pseudotuberculosis
and Pasteurella pestis have been placed into a separate
genus, Yersinia, because in some of their properties and
characteristics they are quite unlike typical Pasteurella.
With the exception of plague (caused by Y. pestis),
infections with other species within the genus Yersinia
are referred to as yersiniosis.

Pseudotuberculosis is primarily a disease of
guinea pigs, wild rodents and wild birds, but Feldman
and Karlson (6) place Y. pseudotuberculosis near the top
of the list with respect to the number of animal species
it can infect. The relative paucity of reports of the
disease would thus seem at variance with the broad
pathogenicity of the microorganism and perhaps reflects
failure to recognize the disease. This appears to be
primarily true in the United States, but a number of
reports have recently attracted attention to yersinial
disease. A recent summary of yersiniosis in mammals and
birds in the United States (7) reveals that foci of
infection are widespread in this country; reports of
infection have come from 19 states and include 19
mammalian and 5 avian species. The domestic animals
involved include cow, goat, sheep, swine, cat and
rabbit; the wild mammals included the black-tailed
jackrabbit (Lepus californicus), while-tailed antelope
squirrel (Citellus lecures), Eastern cottontail rabbits
(Sylvilagus floridanus) and three birds (canary, dove
and pigeon). Thirteen human cases were also tabulated.

In the naturally occurring disease in animals, the
lesions occur as grey or white nodules in the liver,

spleen, lungs and other organs. These contain purulent
or caseous material, and similar material may be found
in lymphoid tissue in the intestine and mesentery.

In man the disease is present in either of two forms:
as an acute mesenteric lymphadenitis, which is the most
frequently encountered form and occurs primarily in
children and adolescents (in whom it is frequently mis-
taken for appendicitis), or as a septicemia. The latter
form is infrequently seen and has generally been fatal.
Pseudotuberculosis is a familiar disease to personnel
of zoological gardens and one to which monkeys appear to
be highly susceptible (8). Guinea pigs may also pose a
hazard since naturally acquired infection may be present
in a colony, manifested occasionally either as sporadic
cases or as an epizootic.

The mode of transmission appears most commonly to
be via the fecal-oral route, although other possibilities
have been suggested depending upon the animal species.

PASTEURELLOSIS

Pasteurella multocida (formerly Pasteurella septica)
has long been recognized as a pathogen for domestic and
wild mammals and for fowl, among which it can give rise
to explosive outbreaks of highly fatal hemorrhagic
septicemia. It has only infrequently been reported as
an incitant of human disease, but it is gaining recognition as a pathogen which can produce a spectrum of
illnesses. In a survey of the medical literature up to
1965, Hubbert and Rosen (9,10) found 77 cases of P.
multocida infection following animal bite. To this
number they added data on 180 additional cases which
occurred in the United States over a 30-year period (early
1965 to early 1968) and followed trauma from animals.
In a very recent report, Tindall and Harrison (11) added
an additional 11 cases of infection following animal
bites or scratches, and the reader is referred to their
paper for a comprehensive listing of the literature.
Four of the 11 patients described by these authors had
received bites or scratches while working with experimental animals (cats and dogs). It is of interest, and of
some import, that in the case of the cat, infection may
follow injury inflicted either by a scratch or a bite,
whereas with the dog, infection may result from a bite
but apparently not from scratches (11). Infection
consequent upon trauma inflicted by animals is heralded
within a few hours of the injury by redness, swelling,
tenderness and pain accompanied by a serous or purulent
discharge from the injured area; regional lymphadenopathy
may appear, as may an axillary abcess if a limb is the

site of the wound.   The infection is generally low
grade and is often slow to resolve even with adequate
therapy with antibiotics, to which P. multocida is
sensitive.

Cats and dogs have been used in appreciable numbers
in recent years in studying the counterparts to human
cancer which are seen in these species, viz., feline
leukemia and canine lymphoma.   Working with these animals
constitutes a hazard from possible P. multocida infection.
Other animals, e.g., rabbit, rat and oppossum, have
also been shown to be carriers and hence may constitute
a source of infection for their handlers.

It should be pointed out that about one-half of the
cases reported in the literature give no history of
animal bite or scratch.   Thus data from the reports
mentioned above (9,11) give a total of 268 human
infections arising from injuries inflicted by animals,
whereas Hubbert and Rosen (10) lists a total of 253
cases up to 1968 as being unrelated to animal trauma.
In these non-bite related cases, the disease may manifest
itself in a variety of ways; it may involve (most
frequently) the respiratory tract, the gastrointestinal
tract, the central nervous system, or various organs.
Pertinent here is the observation (10) that the source
of infection in 84% of published cases and in 69% of
reported cases may have been animals or their tissues.

BRUCELLOSIS

Undulant fever is well recognized as an occupational
disease affecting primarily those who come in contact
with infected domestic livestock or their tissues.   The
three classic strains responsible for infection in the
cow, the goat and swine--Brucella abortus, Brucella
melitensis and Brucella suis--are notorious for the
number of infections they have produced among laboratory
personnel working with the organisms or exposed to
infected tissues; of the 190 cases of human brucellosis
reported in the United States in 1971, 3 occurred in
laboratory workers (12).

Rather than discuss and further document such
incidents, I should like instead to call attention to
Brucella canis as a pathogen for man.   This Brucella
species was first incriminated by Carmichael in 1966 as
a cause of canine abortion (13) and subsequent studies
(14) showed infection with B. canis to be widespread in
the canine population of the United States.   Although
infected dogs are usually asymptomatic, overt evidence
of infection may be present; in the female this consists

of generalized lymphadenitis and a long-persisting,
post-abortion vaginal discharge, and in the male of an
epididymitis, scrotal dermatitis and testicular atrophy.
In the annual summary for brucellosis in 1971 (12)
attention is called to 9 reported instances of human
infection with B. canis over the period 1966-1970 (no
cases reported in 1971). Laboratory accidents caused 6
of the 9 infections; exposure to infected dogs resulted
in 2; and the source of infection in one was undetermined. Seven of these individuals had a clinical
illness characterized by headache, fever, malaise, chills,
night sweats, muscular aches and pains in the extremities,
pharyngitis and lymphadenopathy. Lewis (cited in 12),
in a serologic survey of military recruits, found that
5 of 1208, or 0.4%, had agglutination titers to B. canis
of 1-100 or greater. The CDC report suggests that the
incidence of human infections with B. canis may be
higher than the reported morbidity figures indicate, and
in this connection a reminder seems worthwhile, i.e.,
a specific antigen prepared from B. canis must be used
in agglutination tests, since antibody to this species
does not react with the standard brucella antigen
generally employed.

The mode of transmission to man is unknown but it
seems possible that infection may be acquired via
mycoorganisms shed in the vaginal discharge or excreted
in the urine, or by contact with fomites contaminated
by infected animals.

LEPTOSPIROSIS

Leptospirosis is to a considerable extent an
occupational hazard. Man is susceptible to infection
with most members of the genis Leptospira, but from
the standpoint of the laboratory worker and the laboratory
animal caretaker, exposure to canines which harbor L.
canicola constitutes the greatest potential for infection.
The extent to which such infections occur is unknown
since leptospirosis, like syphilis, is a great imitator,
its clinical expression taking the form of any of a
variety of conditions. Cases which escape detection
might be recognized if serologic tests for leptospirosis
were included in the present battery of tests done for
"fevers of unknown origin."

SHIGELLOSIS AND SALMONELLOSIS

The literature contains many examples of human
infection with Shigella acquired from simians (15,16).
Several recent instances might be given as examples.

A recent CDC report (17) summarizes pertinent data on monkey-associated gastroenteritis in man. All three episodes occurred in the state of Washington in May 1972.

In the first outbreak, 4 of 6 family members (2 adults and 2 young children) developed fever, nausea, abdominal pain and diarrhea. Cultures of stool specimens from the 4 patients yielded Shigella flexneri 6 from 3 of the 4 patients. The family had been exposed to a woolly monkey (Lagothrix lagothrica) which became ill with a severe diarrhea shortly after purchase and died shortly thereafter, at the time the illnesses appeared in the family. Cultures of fecal material from the dead monkey yielded, as in the case of the family members, S. flexneri 6.

In the second episode a child and an adult household contact developed fever and vomiting with diarrhea and bloody stools. S. flexneri 2 was isolated from stool specimens from both patients. Both had been exposed to a recently purchased pet "stumptailed" macaque (Macaca arctoides), which had had a diarrhea 2-3 weeks before the human cases developed. Serial bacteriological examinations of stools from this animal failed to yield Shigella.

In the third episode an adult and 2 children in a family developed mild diarrhea. A woolly monkey recently purchased as a pet had diarrhea over this same interval and subsequently died. Stools from this animal were not examined and stool specimens obtained from all three patients after recovery were negative for Shigella.

As concerns laboratory exposures, two animal handlers at one institution developed a diarrhea in July 1971 which was diagnosed clinically as shigellosis (3), but bacteriological examination of stools was not done in either instance. However, just prior to their illness, both individuals had been exposed to monkeys known to harbor infections with Shigella flexneri.

Although salmonellosis in monkeys is common, any potential hazard is chiefly to the highly susceptible young child exposed to pet monkeys. The studies of Rowe (18) indicate that laboratory personnel and animal handlers exposed to nonhuman primates harboring a Salmonella infection run little risk of acquiring it.

MYCOPLASMOSIS

Mycoplasmas enjoy a wide distribution in nature. They appear to possess a high degree of species

specificity, colonizing solely or primarily those
species in which they were first encountered. Ability
to cross species barriers has been considered to be
present in low degree, if at all, and in many cases the
organisms are now regarded as part of the normal microbial
flora, essentially non-pathogenic. Thus of the 7 species
of mycoplasmas found in man, only one, Mycoplasma
pneumoniae, is pathogenic. Recent studies (19,20) of
mycoplasmas encountered in sub-human primates show them
to be separable into three antigenic groups. Two groups
are comprised of strains antigenically related to human
strains, suggesting that at least six of the seven
mycoplasmas may be found in nonhuman primates. The group
3 isolates from nonhuman primates have no demonstrable
antigenic relationship to any of the known human strains,
but ostensibly bear some relationship to M. canis.

Dogs carry at least four well characterized species
of mycoplasma, of which M. canis is one, and perhaps
several additional species as well. Unlike the mycoplasmas of nonhuman primates and of the guinea pig,
which appear not to be incitants of disease in their
natural hosts (20,21), canine mycoplasmas may cause a
pneumonia in the dog (22). It is of interest, therefore,
that Armstrong et al. (22) found M. canis colonizing the
throats of 4 family members in close contact with a
family dog that developed a respiratory illness characterized by wheezing and a productive cough.
Colonization apparently occurred in all 4 members of the
family during the course of their caring for the pet.
In addition to harboring the mycoplasma, all 4 developed
an antibody response. However, only one family member,
who was on immunosupressant therapy for a malignancy,
showed notable symptoms of illness, including a cough
productive of small amounts of sputum. The other 3
family members experienced a mild illness associated
with malaise and occasional loose stools; nasal conges-
tion was present in two and a non-productive cough in
the third. However, additional studies will be required
before M. canis can be definitely incriminated as an
incitant of illness in man.

That mycoplasma species normally found in only a
single animal host and apparently non-transmissible
experimentally to other animals, e.g., Mycoplasma caviae,
can under certain circumstances become pathogenic for
another host was recently reported by Hill (21). A
laboratory worker accidentally inoculated his thumb
with a small amount of M. caviae suspension. Forty-eight
hours later the terminal joint was swollen and painful;
at 72 hours antibiotic therapy was initiated and by 96
hours the entire hand was swollen. By the sixth day the
swelling began to regress and the hand returned to normal

within the next several weeks.   The victim developed
antibodies to M. caviae.

CHLAMYDIAL INFECTIONS

       Psittacosis, or ornithosis, a preferable epithet
inasmuch as the infection is transmitted by a number
of avian species in addition to psittacine birds, is
well recognized as a laboratory hazard by those who
work with the agents or with birds carrying an expressed
infection or a potentially latent one; it is also known
to at least a certain segment of the public, aware of
the connection between birds and "parrot fever."   It is
generally recognized by all concerned that infection is
acquired by the respiratory route and takes the form of
a pneumonia or pneumonitis.

       However, the source of infection need not invariably
be a bird, nor need the route of infection necessarily
be the respiratory.   Meyer and Eddie (23) describe the
development of a pneumonia in an animal caretaker exposed
to guinea pigs inoculated with bovine encephalomyelitis
virus.   (This agent, characterized by large elementary
bodies and antigenic relationship to the psittacosis-
lymphogranuloma venereum group, is a bacterium, not a
virus, and taxonomically falls within the genus
Chlamydia.)   Complement-fixation tests with a psittacosis
antigen revealed a significant rise (8-fold) in antibody
titer over the course of the illness.   In another situation, an elderly man developed an intermittent low-grade
fever and an enlarged axillary lymph node which
eventuated in an abcess.   Bacteriologic cultures of pus
removed by aspiration gave no growth and a diagnosis of
cat scratch disease was considered.   However, a skin
test with Hanger-Rose (cat scratch) antigen and with Frei
(lymphogranuloma venereum) antigen were both negative,
whereas a very high complement-fixing antibody titer
with a psittacosis-lymphogranuloma venereum antigen
suggested infection with a member of this group, a
diagnosis supported by the patient's response to therapy
with tetracycline.   On requestioning, he recalled that
about two months or so before the onset of illness he
had been bitten on the right thumb by a parrot.

       How often infection may be transmitted by this or
similar means is unknown, but it would be of interest
to consider this possibility in the differential
diagnosis of cat scratch disease (to be discussed below).

RICKETTSIOSES

Like the viruses, the rickettsiae have exacted a
high toll from laboratory personnel working with them
or exposed to infected animals or their tissues. Dr. S.
Edward Sulkin informs me that, as of the time of the
writing of this report (December 1971), his data on
overt laboratory-acquired infections total 536 cases
with 22 deaths. Since this number of cases exceeds only
slightly the number reported in 1969 (24), the reader is
referred to that publication for a breakdown of cases
and deaths by agent. By far the greatest totals are for
typhus fever and Q fever; with respect to Q fever,
notorious for the large number of infections it has
produced among laboratory personnel working with the
agent itself, or exposed to infected animals in the
field or in the laboratory, a few comments may bring
into focus the potentialities of this organism. Two
recent reports deal with the occurrence of Q fever
infection or disease in personnel conducting experimental
work on sheep or exposed to the experimental animals.
No work on rickettsial agents was being done by the two
groups involved.

The report by Schachter et al. (25) mentions a case
of pneumonitis diagnosed as Q fever serologically, and
a case of atypical hepatitis given a presumptive diagnosis
of Q fever on the basis of the high stationary antibody
titer to Coxiella burnett in personnel exposed to
presumably healthy sheep. Serologic tests conducted on
other potentially exposed personnel showed that 55 of
95 individuals, or 16%, possessed significant titers of
complement-fixing antibody to Q fever (as against 0.3%
in a control population). Serologic tests on sheep in
the experimental flock showed 14 of 47, or 30%, to
possess complement-fixing antibody. A somewhat similar
experience has just been reported by Curet and Paust
(26). Serologic tests for Q fever were done on 20
individuals with varying degrees of contact with a flock
of experimental sheep. Of the 20 individuals tested,
15 had antibody and 8 of these had had a clinically
evident illness diagnosed as pneumonitis or bronchitis;
3 of the remaining 7 had had biochemical evidence of
hepatic involvement. One individual in the latter group
underwent a laparotomy which revealed a markedly enlarged
liver. Cultures of tissue obtained by liver biopsy
were negative, but histologic examination revealed a
granulomatous hepatitis, a lesion associated with
rickettsial damage (27). Serologic examination of the
ewes in the flock showed 42 of 87 (or 50%) possessed
antibody to Coxiella burneti.

Although Q fever is generally looked upon as a

respiratory disease, it is actually a systemic illness
that may manifest itself in a variety of ways. Hepatic
involvement, sometimes severe, is not rare (28).
Although the frequency is unknown, chronic infections do
occur and may give rise to such serious conditions as
subacute endocarditis (29).

CAT SCRATCH DISEASE

Discussion of this condition has been left to the
end because the etiologic agent is unknown, and because
of uncertainties as to whether we are dealing with one
disease or many (30). It is included here as a possible
occupational hazard of the laboratory worker and animal
caretaker because of a similar condition mentioned
earlier, a regional lymphadenopathy and axillary abcess
caused by a Chlamydial agent. Although approximately
one-half of the patients with cat scratch disease
ostensibly have antibodies to the psittacosis-
lymphogranuloma venereum group, the significance of
such cross reactivity is unclear, and clinical differential diagnosis is often based on differences in
reactivity to the Hanger-Rose skin test and the Frei
test (31). (The antigen for the Hanger-Rose test is
prepared from material aspirated from affected lymph
nodes of patients with cat scratch disease; the antigen
for the Frei test is prepared from similar material from
patients with lymphogranuloma venereum.)

Cat scratch disease is a self-limited disease
characterized by a sub-acute regional lymphadenitis.
There is generally a history of antecedent injury to
the skin peripheral to the involved regional node. The
pathologic process is generally localized and the disease
is mild, although occasional fatal cases have been
reported.

The disease acquired its designation from the fact
that Foshay in the United States and Debre in France,
who first described the condition, were impressed by
the number of instances in which a cat bite or a cat
scratch preceded the illness. However, epidemiologic
studies in recent years have suggested that cats may
serve as a vector rather than a reservoir of the disease.
The disease has followed breaks in the skin inflicted by
penetrating instruments other than a cat's claws or
teeth (e.g., wood splinters, fish hooks, porcupine quills,
pins, rabbit claws), which suggests that "the role of
the cat in this disease may, therefore, be only that of
a peculiarly effective inoculator and it seems likely
that the emphasis should be on the scratch rather than
on the cat in the cat scratch syndrome" (30). Attempts

to isolate a specific causal organism have been un-
rewarding and a few reports of success have not been
confirmed by other workers.

CONCLUSION

It is obvious that some laboratory-acquired
infections may be introduced by trauma, such as a bite
or scratch inflicted by an animal, by inhalation of
infective aerosols, by fecal-oral transmission or by
exposure to fomites. Whatever safety devices we may
install, such as special hoods, air locks, negative
pressure rooms, etc., or safety precautions we may take,
as the use of gowns, masks, gloves, boots, etc., we
will still be left with one very important factor, the
human element. Even if all the physical safety measures
of proved efficacy are utilized and good laboratory
practices are adhered to, accidents will occur and not
the least of these is inadvertent self-inoculation.

To afford protection against such accidents, as
well as against other possible modes of infection, it is
recommended that persons working with a known agent,
or whose work may expose them to contact with a known
agent, be vaccinated against that agent. Vaccines,
however, are not infallible and cannot always be depended
upon to give complete protection even when administered
in the recommended dosages and followed by requisite
booster doses. Where vaccination is utilized as a
protective measure, it should not be considered a
substitute for good laboratory practices or as a license
to take chances or short-cuts.

References

1. Stones, P. B. 1969. Incidence of tuberculosis in
    Macaca mulatta and Cercopithecus aethiops monkeys
    with special reference to the tuberculin test. In
    Hazards of handling simians, Lab. Animal Handbook
    4:11. Lab. Animals Ltd., London.
2. Heywood, R. and P. H. Hague. 1969. Tuberculosis in
    baboons. In Hazards of handling simians, Lab.
    Animal Handbook 4:43.
3. Primate Zoonoses surveillance. 1972. Report No. 8,
    July 1972, pp. 10-11. Center for Disease Control,
    USPHS, Atlanta, Georgia.
4. Garibaldi, R. A., R. E. Drusin, S. H. Ferebee and
    M. B. Gregg. 1972. Isoniazid-associated hepatitis.
    Report of an outbreak. Amer. Rev. Resp. Dis. 106:357.
5. Tribe, G. W. 1969. Clinical aspects of the
    detection and control of tuberculosis in newly

imported monkeys. In Hazards of handling simians,
Lab. Animal Handbook 4:19.

6. Feldman, W. H. and A. G. Karlson. 1963.
Pseudotuberculosis. In Diseases transmitted from
animals to man, ed. T. G. Hull. Charles C. Thomas,
Springfield, Ill.

7. Hubbert, W. T. 1972. Yersiniosis in mammals and
birds in the United States. Amer. J. Trop. Med.
21:458.

8. Mortelmans, J. and P. Kageruka. 1969. Pasteurella
pseudotuberculosis infection in monkeys and man.
In Hazards of handling simians, Lab. Animal Handbook
4:95.

9. Hubbert, W. T. and M. N. Rosen. 1970. I. Pasteurella multocida infection due to animal bite. Amer.
J. Pub. Health 60:1103.

10. Hubbert, W. T. and M. N. Rosen. 1970. II.
Pasteurella multocida infection in man unrelated
to animal bite. Amer. J. Pub. Health 60:1109.

11. Tindall, J. P. and C. M. Harrison. 1972. Pasteurella multocida infections following animal injuries,
especially cat bites. Arch. Derm. 105:412.

12. Brucellosis surveillance. 1972. Annual Summary
1971, October 1972, Center for Disease Control,
USPHS, Atlanta, Georgia.

13. Carmichael, L. E. 1966. Abortion in 200 beagles.
J. Amer. Vet. Med. Ass. 149:1126.

14. Carmichael, L. E. and R. M. Kenney. 1968. Canine
abortion caused by Brucella canis. J. Amer. Vet.
Med. Ass. 152:605.

15. Carpenter, K. P. 1968. In Some diseases of animals
communicable to man in Britain, ed. O. Graham-Jones.
Pergamon Press, Oxford.

16. Mulder, J. B. 1971. Shigellosis in nonhuman primates.
Lab. Animal Sciences 21:734.

17. Primate Zoonoses Surveillance. 1972. Report No. 9,
October 1972. Center for Disease Control, USPHS,
Atlanta, Georgia.

18. Rowe, B. 1969. Salmonellosis in simian and other
nonhuman primates. In Hazards of handling simians,
Lab. Animal Handbook 4:63.

19. Martinez-Lahoz, A., S. S. Kalter, M. E. Pinkerton
and L. Hayflick. 1970. Similarities of mycoplasma
species isolated from man and nonhuman primates.
Ann. N. Y. Acad. Sci. 174:820.

20. Kalter, S. S. and R. L. Heberling. 1972. In Primate
zoonoses surveillance 1972, Report No. 9.

21. Hill, A. 1971. Accidental infection of man with
Mycoplasma caviae. Brit. Med. J. 2:711.

22. Armstrong, D., B. H. Yu, A. Yagoda and M. F. Kagnoff.
1971. Colonization of humans by Mycoplasma canis.
J. Inf. Dis. 124:607.

23. Meyer, K. F. and B. Eddie. 1956. The influence of

tetracycline compounds on the development of antibodies in psittacosis. Amer. Rev. Tuberculosis
& Pulmonary Dis. 74:566.

24. Sulkin, S. E. and R. M. Pike. 1969. Prevention of
laboratory infections. In Diagnostic procedures for
viral and rickettsial infections, 4th ed. Amer.
Pub. Health Ass., New York.

25. Schachter, J., M. Sung and K. F. Meyer. 1971.
Potential danger of Q fever in a university hospital
environment. J. Inf. Dis. 123:301.

26. Curet, L. B. and J. C. Paust. 1972. Transmission of
Q fever from experimental sheep to laboratory personnel. Amer. J. Obst. Gyn. 114:566.

27. Dupont, H. L., R. B. Hornick, H. S. Levin, M. I.
Rappaport and T. E. Woodward. 1971. Q fever
hepatitis. Ann. Intern. Med. 74:198.

28. Clark, W. H., E. H. Lennette, O. C. Railsback and
M. Romer. 1951. Q fever in California. VII.
Clinical features in 180 cases. Arch. Int. Med.
88:155.

29. Ormsbee, R. A. 1965. Q fever rickettsia. In Viral
and rickettsial infections of man, 4th ed., p. 1144.
J. B. Lippincott, Philadelphia.

30. Warwick, W. J. 1967. The cat scratch syndrome,
many diseases or one disease. Progr. Med. Virol.
9:256.

31. News and Notes, Epidemiology, 1972. Psittacosis-
lymphogranuloma venereum. Brit. Med. J. 2:711.

Addendum

After this paper had been completed, the following
articles, of interest to those concerned with biohazards,
came to the attention of the author:

Wedum, A. G., W. E. Barkley and A. Hellman. 1972.
Handling of infectious agents. J. Amer. Vet. Med. Ass.
161:1557.

Brayton, J. B. 1972. Personnel health programs in
biomedical research institutions. Ibid. 161:1568.

DISCUSSION

FRIEDMAN: We have some monkey cells, such as CV-1,
which we are comfortable with and others which we are
not. How sure are we that any are free of herpes B or
other potentially hazardous agents? Can we be comfortable with any monkey cells?

LENNETTE: Maybe you shouldn't be. Nothing in this world
is sure but death and taxes and perhaps your being
comfortable with the use of a specific cell line may lull
you into such a sense of security that you fail to heed
the dictum I mentioned earlier, namely, that all host
systems, as well as agents being passaged, should be
checked from time to time to see what extraneous agents
may be present or are being carried.

HILLEMAN: The point I make in my paper is simple. You
can't give an absolute guarantee of freedom from cryptic
virus in serially passaged cells and you can't give
such guarantee for primary cells of embryo origin
either. However when proper precautions are taken, the
hazards are remote. By contrast, use of tissues of wild-
caught animals is just asking for trouble because of the
lack of control and the known high probability for viral
contamination. Monkeys are too expensive to be grown in
specific pathogen-free colonies and, hence, the simple
solution to the monkey problem is to eliminate the
monkey.

PANEL I
PREVENTIVE MEASURES:
THEORETICAL AND PRACTICAL
CONSIDERATIONS

Robert N. Hull
The Lilly Research Laboratories

Robert Kissling
Center for Disease Control

Edwin H. Lennette
California State Department of Public Health

John C. Parker
Microbiological Associates

Wallace P. Rowe
National Institute of Allergy and Infectious Diseases

DISCUSSION OF INDIGENOUS MURINE VIRUS INFECTIONS
AND EPIDEMIOLOGY OF AN LCM EPIZOOTIC

John C. Parker

Time does not permit a detailed discussion of all
of the potential problems associated with indigenous
rodent viruses, however, I would like to discuss two of
the more important consequences which can and often do
occur. First, I would like to show a slide (Table 1)
illustrating the incidence of indigenous virus infections
in the three most commonly used laboratory rodent
species: the mouse, rat, and hamster. For the most
part the colonies are commercial rodent breeder colonies
from which 25 to, in some instances, over 500 retired
breeders were purchased and tested for virus antibody.
Mouse colonies were heavily contaminated with minute
virus of mice (MVM), mouse hepatitis virus (MHV), mouse
encephalomyelitis virus (GDVII), reovirus type 3, Sendai
virus and pneumonia virus of mice (PVM). More than half
of the rat colonies were infected with Kilham rat virus
(KRV), rat coronavirus (RCV), PVM, Sendai, and H-1
(Toolan) viruses. While an inhibitor to MVM virus was
commonly found in sera from rats, it is not clear whether
it is a nonspecific inhibitor of hemagglutination.
Fewer virus infections are known to occur in the hamster,
with PVM and Sendai viruses the most prevalent. A parainfluenza virus related serologically to SV5 virus also
commonly occurs in hamsters. We have isolated this virus
from hamsters and are presently working up its identity
and biologic characteristics. While it is disappointing
to see the high incidence of virus infection in the
colonies, there is some encouragement since many commer-

65

Table 1
INCIDENCE OF VIRUS INFECTIONS IN RODENT BREEDER COLONIES

I. MICE

COLONIES	VIRUS DETECTION BY ANTIBODY[+]								
	MVM	MHV	GDVII	REO 3	SENDAI	PVM	POLYOMA	M.AD.	K
NUMBER POSITIVE	45	55	55	49	47	44	19	17	4
NUMBER TESTED	59	73	73	73	73	73	73	73	73
PERCENT POSITIVE	76.3	75.3	75.3	67.1	64.4	62.6	26.0	23.3	5.4
ANTIBODY INCIDENCE IN POSITIVE COLONIES (PERCENT)	40.8	26.6	44.2	19.8	42.4	19.6	30.5	9.5	21.4
RANGE OF ANTIBODY INCIDENCE IN POSITIVE COLONIES (PERCENT)	7.8 ↓ 100	1.2 ↓ 100	1.9 ↓ 100	1.4 ↓ 86.8	1.9 ↓ 100	0.6 ↓ 85.7	0.9 ↓ 95.2	1.1 ↓ 64.0	8.7 ↓ 22.7

II. RATS

	MVM	KRV	RCV	PVM	SENDAI	H-1	REO 3	GDVII	M.AD.
NUMBER POSITIVE	23	20	18	17	14	14	11	11	10
NUMBER TESTED	26	26	25	25	25	26	25	25	25
PERCENT POSITIVE	88.4	76.9	72.0	68.0	56.0	53.8	44.0	44.0	40.0
ANTIBODY INCIDENCE IN POSITIVE COLONIES (PERCENT)	20.1	28.5	52.3	61.7	63.7	15.0	13.9	11.4	7.1
RANGE OF ANTIBODY INCIDENCE IN POSITIVE COLONIES (PERCENT)	5.4 ↓ 100	3.1 ↓ 87.5	15.7 ↓ 100	7.1 ↓ 94.4	8.3 ↓ 100	11.1 ↓ 91.7	4.6 ↓ 66.7	1.5 ↓ 42.3	2.9 ↓ 21.4

III. HAMSTERS

	PVM	SENDAI	PARAINF.	REO 3	GDVII
NUMBER POSITIVE	17	9	7	7	3
NUMBER TESTED	18	18	18	18	18
PERCENT POSITIVE	94.4	50.0	38.9	38.9	16.7
ANTIBODY INCIDENCE IN POSITIVE COLONIES (PERCENT)	45.3	64.0	26.2	17.1	10.8
RANGE OF ANTIBODY INCIDENCE IN POSITIVE COLONIES (PERCENT)	4.5 ↓ 91.6	4.3 ↓ 100	4.1 ↓ 84.0	1.3 ↓ 88.1	2.7 ↓ 41.7

Sera were tested for viral hemagglutination-inhibiting antibodies to minute virus of mice (MVM), mouse encephalomyelitis (GDVII), reovirus type 3 (Reo3), Sendai, pneumonia virus of mice (PVM), polyoma, mouse pneumonitis (K), Kilham rat (KRV), H-1, parainfluenza (using SV5 virus) and ectromelia (using vaccinia virus), and for viral complement-fixing antibodies to mouse hepatitis (MHV), mouse adenovirus (M.Ad.), and rat coronavirus (RCV).

cial producers are beginning to realize the need for
virus-free animals and are making sincere efforts to
eliminate virus infections in their animals by cesarian
derivation, followed by establishment of a virus antibody
surveillance or monitoring program.

While there are a total of 13 viruses shown in this
Table, some are more commonly and repeatedly involved in
epizootics within the breeder colonies or laboratories
using the animals or are involved in inadvertent contamination of materials passaged through the infected
animals. In my experience the most important viruses
infecting mice are: polyoma, Sendai, MHV, MVM, Reo3 and
GDVII; for rats, KRV, H-1, Sendai and perhaps RCV; and
for hamsters, Sendai, and perhaps the parainfluenza virus.
Lymphocytic choriomeningitis was not tested for serologically because of the insensitivity of the complement
fixation text; however, it does occur naturally in the
mouse and hamster and poses a serious problem because it
is the only indigenous rodent virus which can be transmitted to and cause serious disease in man.

One major problem precipitated by these indigenous
viruses is their ability to become passengers or
contaminants of materials passaged through the infected
animals. Sixty-three percent of 540 specimens we have
examined were found to have at least one virus contaminant. Many specimens were contaminated with more than
one virus and one specimen was contaminated with 4
different viruses. The more frequently a specimen is
passaged through contaminated animals, the greater the
probability that the specimen will become contaminated.
These serially passaged specimens are analogous to vacuum
cleaners, picking up viruses which infect the host
species and passing them onto the next host. In
subsequent passages these newly acquired virus contaminants may infect susceptible species causing disease in
or even death of the host, or more subtly and unknow-
ingly alter the outcome and interpretation of experimental
results. Thus in order to circumvent these problems, it
becomes more and more important to insist on using animals
which are free of virus infection or at least to use those
animals which have been tested and the viral flora
identified. The experimental materials passaged through
the animals should also be periodically tested to insure
that a contaminant has not been acquired.

In order to illustrate a second major consequence
of indigenous virus infection, I would like to return to
LCM virus in order to show by an example, how a virus
can be accidently introduced into a laboratory and then
move silently to contaminate several different kinds of
biological materials ultimately infecting laboratory

personnel. This outbreak resulted in 23 known cases of human LCM infection which occurred in 7 different laboratories. Numerous other biological systems also were contaminated, including at least 35 serially trans- planted mouse tumor lines, 3 mouse breeder colonies, 3 hamster breeder colonies, one primary cell culture, and one virus reagent pool. The epidemiology of this out- break is illustrated in Fig. 1.

The origin of the contamination in mouse colony #1 was likely the infected wild mice which lived in close proximity to the colony. In colony #1 approximately 5% of the DBA_2 mice were vertically transmitting the infection, and six animal caretakers seroconverted with- out any recallable illness. In at least one instance mouse kidney cell cultures prepared from the DBA_2 mice were contaminated with LCM, which in turn contaminated a mouse adenovirus reagent pool that was prepared in these cultures. Mice from colony #1 were shipped to at least two other colonies, which in turn became con- taminated, and at colony #2 six cases of human infection resulted. However the most serious consequence resulted when hamster colony #1 was contaminated through contact with mice from mouse colony #1. Several animal caretakers in contact with the infected hamsters seroconverted, but they could not recall any unusually severe illness over the period. Subsequently a serially transplanted hamster fibrosarcoma tumor became contaminated with LCM virus during a single passage in these LCM-infected hamsters.

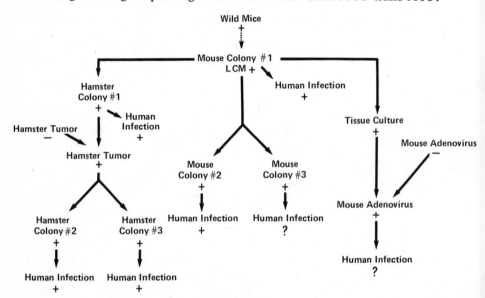

FIGURE 1

Fragments of the LCM-contaminated fibrosarcoma tumor were
then shipped to investigators in other laboratories,
where the infection became widespread in their hamster
colonies. Ten persons who acquired LCM infection in
these laboratories developed a nonmeningitic grippe-like
illness, and one person developed a nonfatal meningitis.
In addition, but not shown in Fig. 1, several mouse
tumor lines, including many lines of the popular L1210
tumor, also became contaminated with LCM during passage
through mice from colony #1. The contaminated tumors
were placed in one of the NCI tumor repositories from
which they were shipped to many laboratories. While
human diseases such as Marburg or LCM, accidently
acquired through contact with infected laboratory
animals, are some of the most dramatic examples of the
consequences that result from inadvertent contamination,
the infectious agents indigenous only to the laboratory
animals are far less dramatic but equally as devastating
in terms of dollars and manhours of effort lost.

Thus the important points to be drawn from this
example, which become applicable to almost any laboratory
working with these types of materials, relate to the
steps which could and should be taken to prevent similar
epidemics from occurring in other laboratories. In this
example contamination of several different types of
biological materials occurred, including wild mice,
laboratory animals, serially transplanted tumor lines,
virus reagents, and cell cultures. Each one of these
materials might represent a vehicle for the introduction
of hazardous agents into any laboratory, and therefore
steps should be taken to monitor all biological specimens
coming into the laboratory in order to certify that these
materials do not inadvertently contain as a contaminant
any undesirable agents.

LABORATORY ACQUIRED INFECTIONS

R. E. Kissling

From 1946 through 1959 the virology laboratories of the Center for Disease Control (CDC) were located in a group of temporary buildings in Montgomery, Ala. These laboratories were generally not equipped with biological safety cabinets, virtually all pipetting was done by mouth, and individual laboratory workers generally supervised animal care. Laboratory safety was an individual responsibility and there were no specific guidelines. Immunizations were given to personnel working with certain viruses (rabies, eastern and western equine encephalitis, and yellow fever).

During that 14-year period there were at least 13 laboratory-acquired infections among a staff of approximately 75 individuals, a rate of 0.012 case/man-year. These 75 people included laboratory workers, animal caretakers, maintenance personnel, and members of the administrative and clerical staff. The specific infections included one case of Q fever, two cases of murine typhus, six of psittacosis, three infections with Venezuelan equine encephalitis (VEE) virus, and one with Rio Bravo virus. The latter infection was contracted through the bite of an experimentally infected hamster. The three VEE cases occurred simultaneously, following accidental spillage of infected guinea pig blood. The remainder appeared to be aerosol-transmitted infections for which no specific accident or incident could be recalled.

    In 1960 the virology laboratories were moved into
new quarters in Atlanta, with controlled air flow and
equipped with biological safety cabinets. However no
uniform safety guidelines were available from 1960
through 1967. During this period four known cases of
laboratory-acquired infections were observed among a
staff of approximately 100 workers, a rate of 0.006 case/
man-year. Almost all of these 100 individuals were
laboratory workers; the group did not include animal
technicians, maintenance personnel, etc. The specific
infections included one case each of Omsk hemorrhagic
fever, VEE, murine typhus, and adenovirus type 2. The
Omsk hemorrhagic fever case was that of a visiting
scientist who stuck his finger with a contaminated
hypodermic needle while inoculating mice. The VEE case
probably resulted from aerosols generated during animal
inoculation procedures. The modes of infection for the
adenovirus and murine typhus cases were not known, but
both individuals were handling large volumes of infectious material in the process of antigen production.

    Since a CDC Biohazards Officer was appointed and
specific laboratory safety guidelines were developed,
two laboratory-acquired infections have been recognized
among the staff of the Virology Branch, a rate of 0.004
case/man-year. One case was a vaccinia lesion on the
hand of a visiting scientist. The other was an Ossa
virus infection resulting from the bite of an experimentally infected hamster.

    The respiratory route appears to be the most common
route by which laboratory infections are acquired.
Often the specific accident or incident leading to these
infections is not recognized. Approximately 79% of the
laboratory-acquired infections were of this type.

    In our experience accidental inoculation via needle
or broken glassware has only rarely led to infection.
Only one such case has been observed.

    Infections acquired directly from laboratory animals
are also rare. Two cases (10.5% of the total cases)
were both the result of hamster bites. No cases of
laboratory-acquired infections have been observed among
the animal technicians caring for the infected animals
belonging to CDC's Virology Branch.

    Seroconversions to St. Louis encephalitis virus
without a recognized illness was relatively common during
the early history of our virology laboratories. In 1954
during the height of poliovirus research in our laboratories, there were three cases of hepatitis. These
occurred while 40-50 young chimpanzees were being used

experimentally in the laboratory and before chimpanzee-
associated hepatitis was recognized as an occupational
hazard. However, 1954-55 was also a peak epidemic year
for hepatitis in the country at large, and these cases
received no epidemiologic follow-up.

In addition to the documented laboratory-acquired
infections, other illnesses occur which cannot be
definitely traced to the work situation. However there
is an impression that respiratory illnesses, especially
influenza and adenovirus infections, occur first and most
frequently among staff members who actually work with
these agents.

Some of the factors that seem to keep laboratory
infections at a minimum include:

1.   Maintenance of correct airflow in laboratories
and corridors.
2.   Education of laboratory workers in safe
techniques and introduction of equipment with special
safety features. Perhaps the single most important move
is the elimination of mouth pipetting.
3.   Classification of viruses according to hazard,
and guidelines for handling each class.
4.   Development of protocols for reporting and
handling accidents, spills, etc.
5.   Erection of biohazard warning signs and
maintenance of controlled access to laboratory areas.
6.   The presence of an individual with authority to
demand safe methods and working conditions.
7.   Establishment of policies regarding the housing
and handling of experimental animals, including centralized responsibility for training and supervision of
animal technicians.

The properties of a virus group (resistance to
drying, heat, etc.) determine, to a large extent, the
hazards involved in handling them. The enteroviruses
are notably labile upon drying. Although we have
worked with these agents for the past 27 years, we have
not had a single proven case of laboratory-acquired
enterovirus infection. The herpesvirus group also appears
to present minimal risk, barring gross accidents. Because
of the serious nature of Herpesvirus simiae infections,
however rare they may be, this virus is handled in a
special laboratory.

The type of work also contributes to the risk of
laboratory infections. The handling of large volumes of
materials, such as are used in antigen preparation or
antigenic analysis, appears to carry considerable risk.
Obviously procedures which create aerosols are hazardous

(homogenizers, centrifuges, sonicators, etc.).

The relative risk to personnel is also directly related to the environment in which infectious material is handled. I am not aware of any 100-percent-foolproof system. The Virology Branch has four types of laboratories in which viruses are handled:

1. General laboratories, where much of the work is done in the open and live agents are handled in biological safety cabinets. Airflow is controlled and flows from corridors into the laboratories. Access is not restricted except that all persons entering the laboratory must have certain immunizations.

2. A high security laboratory, which has an independent ventilation system but is otherwise designed much like the general laboratories. Access is limited, and showering and changing of clothes are required upon exit. Immunization of personnel is required. Agents handled in this laboratory include those known to cause laboratory infections rather commonly and exotic viruses (VEE, Q fever, RSSE, etc.). Vaccines must be available, since no unusual safety features exist in the laboratory. The prime purpose of this laboratory is to restrict geographically these agents to avoid unnecessary immunization of large numbers of personnel not actually working with the agents.

3. A smallpox laboratory, which is quite similar to the high security laboratory. The smallpox laboratory is supplied with a series of glove boxes for handling embryonated eggs inoculated with variola virus or variola suspect specimens.

4. Our maximum security laboratory is a separate building. Clothes changes and showers are required upon exit. Entry is restricted, and all work is performed in Class III cabinets (completely sealed gloveboxes designed for absolute containment). This laboratory is used for handling highly virulent agents for which no immunizations are available. These agents include Marburg, Lassa, and monkey B viruses.

ENDOGENOUS VIRUSES IN CELL CULTURES PREPARED FROM APPARENTLY HEALTHY ANIMALS

G. D. Hsiung & N. S. Swack

Monkey cell cultures harbor a variety of simian viruses. Our interest in latent virus infections of cell cultures was brought about by the recognition of the vacuolating virus of monkeys (SV40) in normal patas monkey kidney cells. During the past 10 years, we have conducted several longitudinal studies of endogenous virus infection in primary monkey kidney cell cultures. We have recently extended these studies to cell cultures prepared from other apparently healthy laboratory and domestic animals. Kidney cell cultures prepared from monkeys, especially rhesus monkeys, showed a higher prevalence of viral agents than cultures prepared from domestic animals. Of the 171 lots of rhesus monkey cells studied, 54% contained one or more viruses, including SV40, SV5, and foamy viruses. Of the 107 lots of green monkey cell cultures examined, 38% contained one or more viral agents. These were mostly cytomegaloviruses and foamy viruses.

Herpesviruses were detected in a small proportion of cell cultures derived from horses and rabbits, but not in bovine cell cultures (Table 1). Furthermore, while only 4% of cell cultures from Hartley strain guinea pigs yielded a herpesvirus, herpesviruses were isolated from 88% of cultures derived from inbred strain 2 guinea pigs (Table 1). This herpesvirus persists for the lifetime of experimentally infected guinea pigs and transplacental transmission has been demonstrated. Thus it seems likely

74

Table 1.   Prevalence of Virus Isolation in Cell Cultures
Prepared from Normal Healthy Animals
(1968-1972)

Animal species	Number studied	Culture lots showing virus infection	
		No.	%
Monkeys			
Rhesus	171	93	54
African green	107	41	38
Domestic animals			
Cattle	64	0	0
Horses	10	2	20
Laboratory animals			
Rabbits	241	6*	9*
Guinea pigs			
Hartley	332	8	4
Strain 2	84	74	88

* Detected by fluorescent antibody technique using a
rabbit herpes-like virus antiserum; 6 of 70 cultures
examined were positive.

that all strain 2 guinea pigs harbor this herpesvirus.

Cell cultures from normal monkeys have on many
occasions been found to harbor two or more different
viruses. Recently we have observed an unusual mixed
infection in a normal strain 2 guinea pig. When a
monolayer culture of spleen tissue from a healthy strain
2 guinea pig was maintained in a medium containing 40
µg/ml 5-bromo-2'-deoxyuridine (BrdU), CPE was not
observed. However electron microscopic examination
revealed numerous C-type virus particles. In the absence
of BrdU, CPE was observed after 20 days in culture, and
electron microscopic studies revealed herpesvirus
particles. Both the guinea pig C-type virus and the
guinea pig herpesvirus would have gone unrecognized had
not special laboratory manipulations been performed.

Large numbers of guinea pigs have been used in
cancer research and immunological studies. Yet prior to
our report, guinea pig C-type virus was observed only in
leukemic strain 2 guinea pigs, and this virus was
considered to be the etiological agent of guinea pig
leukemia. Since it is now apparent that guinea pigs may
be latently infected with one or more potentially oncogenic virus, it would seem prudent to handle these animals
and their tissues with caution.

DISCUSSION

BERG: Can we be certain that established human diploid cell lines are any more free of cryptic viruses than primary monkey cells? I would think that an established monkey cell line such as CV-1 or BSC-1, which have been in use for such a long time, would be preferable to primary monkey cell cultures.

LENNETTE: I am sure many people here are aware that in many instances cell lines, as for example the WI38 human diploid cell line, tend to die out with passage and hence seed cell lines should be kept in the frozen state as a source for initiating additional passage lines. This assures not only having cells which are as close to the source as possible, but also, for virologic purposes, permits preservation of a line with known susceptibilities and resistances. Similarly it cannot be taken for granted that every diploid cell line that is initiated has the same susceptibilities and resistances possessed by other lines that may have been established in the laboratory. Hence in testing a newly established human diploid cell line for susceptibility, for example, to rhinoviruses, it is necessary that the new line be tested with material known to contain rhinoviruses. And by this I do not mean rhinoviruses which have been passaged repeatedly in the laboratory, but rhinoviruses which are essentially clinical material from the patient or at a very low passage level, let us say one to three passages removed from the patient. This latter warning stems from the fact that repeated passage of an agent in the laboratory may lead to the acquisition or deletion of characteristics or properties not inherent in the original virus. An outstanding example of this is the 17D yellow fever virus. Here again it is necessary to have available a large stock of seed virus from which new passage lines can be started because, if the virus is passaged repeatedly beyond a certain chick embryo passage level, it loses its capacity to induce antibodies and to confer protection in man although it still possesses its intracerebral pathogenicity for the laboratory mouse. Other examples might be cited, but this is one of the earliest and perhaps best known.

LEWIS: I would like to point out that there are differences between primary AGMK cells and continuous AGMK cell lines such as BSC-1, CV-1, and Vero in their ability to support the replication of adenovirus 2 and the nondefective adenovirus 2-SV40 hybrids. Nonhybrid adenovirus 2 and the nondefective hybrids will grow almost as efficiently in Vero cells as in HEK. Nonhybrid

adenovirus type 2 replicates very poorly in both
primary AGMK and in CV-1 cells. Certain of the nondefective adenovirus 2-SV40 hybrids plaque with one-hit
kinetics in primary AGMK cells and with two-hit kinetics
in either BSC-1 or CV-1 cells. The reasons for these
differences are unknown. Consequently to obtain
consistent results in our studies we must use primary
AGMK cells in spite of the problems associated with their
use.

TODARO: Many commonly used cell lines have been well
characterized by the American Type Culture Collection.
As a suggestion, I would recommend either obtaining lines
from the ATCC or from the people who originally characterized them. Otherwise, if cells are passed through
several hands, they usually take on different properties.
This certainly occurred with the 3T3 cells, where after
being through several hands it came back with totally
different properties.

LENNETTE: I think this makes for a very good policy,
namely, that if you are passaging an agent, irrespective
of what the host system might be, you check once in a
while to see what you are actually passaging; that is,
are you working with the agent with which you originally
started or have you lost it and are now carrying something that was either latent in the host or picked up
from the laboratory environment.

MANNING: In view of repeated serologic crosses between
herpes simplex and monkey B virus, are animal handlers
who have antibody to herpes simplex protected against
monkey B infection?

HULL: About 50% of adults have B virus antibody
representing a heterologous response to herpes simplex.
This antibody probably is protective. Pre-exposure sera
were available from only a few of the human B virus
cases, and none contained antibody. Herpes simplex alone
is not protective.

FRIEDMAN: Is there any reason to think that any of these
other agents are found in primary or established cell
lines of any sort and are there any examples of suspected
human contamination from tissue culture cells?

LENNETTE: Miss Evelyn Kiresen, my reference librarian, did what I consider to be a fairly comprehensive search of the literature via the computer. Since the computer printout lists articles only by title, author and reference, rather than by content, it becomes virtually necessary to peruse many articles at least superficially if one is to turn up information he is seeking. In the present instance, there were two or perhaps three references mentioning the presence in cell culture of agents of possible human origin and hence these are probably isolated instances. However it is possible that the number of such instances might be somewhat greater than the current literature indicates since, as Dr. Rowe pointed out earlier, if one looked for such agents one might find them. Such searches probably are infrequent, if not rare.

HELLMAN: Mycoplasma are frequently found in animal sera used for tissue culture.

Session II
**Endogenous Agents in Cell Cultures:
Detection, Elimination and Risk**

CELLS, VACCINES, AND SAFETY FOR MAN

Maurice R. Hilleman
Division of Virus and Cell Biology Research
Merck Institute for Therapeutic Research
West Point, Pennsylvania

Never, perhaps, in the history of virology has so much been said about so little that appears to mean so close to nothing. I am referring here to the kinds of cell cultures used to prepare vaccines against viral diseases. The bulk of the deliberations about cells has been in the name of safety--safety steeped in theory with little foundation in established fact.

The preparation of cell cultures for the purpose of vaccine production presents three alternatives:

1. Serially passaged cell cultures that have the possibility for infinite replication: These may be of frankly neoplastic origin or may be derived from cultures in which a heritable transformation has taken place.

2. Serially passaged cells originating from normal tissues: These show the characteristic of a finite limitation in the number of subcultures that can be made and are almost entirely diploid in karyology.

3. Primary cell cultures derived from animal tissues: These may be prepared either from embryonic tissues or from the post-natal animal, usually in early adult life.

TRANSFORMED AND NEOPLASTIC CELLS

The evolution of preferences for cell types for preparing vaccines is best understood in its historic perspective. In 1954 (1) the Armed Forces Epidemiologic Board met to decide whether to permit studies in man of the then newly discovered adenoviruses grown in cell culture. Based on the background of knowledge of tumor cell transplantation, it was declared that cell cultures derived from frankly neoplastic tissues or possessing attributes of neoplasia were definitely ruled out, but that viruses grown in "normal" cells and inactivated by formaldehyde would be acceptable for limited immunization trials. At a second meeting held in 1959 (1), it was made clear that the possible risk attending use of continuous cell lines would be regarded as too great for vaccines against diseases of only mediocre importance.

This view eventually became rule (2) clearly prohibiting, at the time, the use of continuous cell lines for manufacture of live poliovirus vaccine and for live and killed measles virus vaccines. In its definition, the rule excluded the use of the diploid cell strains that were yet to be evolved. This became the subject of controversy which is familiar to most workers in the field and persists to the present time (1,3-10).

Frankly neoplastic cells, such as Hep II or HeLa, and cells transformed in culture provide a desirable technology for cultivating viruses for use in vaccines because the cells can be grown in abundance in submerged culture in tanks and because they generally yield large amounts of virus on inoculation. It is clear that the failure to employ such cells had its origin in the fear of tumor cell transplantation, but this phobia should not necessarily be extended to the production of virus from which intact viable cells can be excluded, and even more so, to vaccines prepared from viral subunits that have been freed of all genetic materials. Perhaps the most difficult problem in defining the safety or non-safety of neoplastic cell cultures for making vaccines is whether there is a problem in the first place, and it may be hoped that the newer technologies in virus induction and in biochemical genetics will eventually provide answers to the problems that now appear in chaotic technical disarray. The objective should be that of finding a cell or cells that grow vaccine viruses well, are free of occult and covert oncogenic virus, and are free of any transmissible oncogenic viral or host genetic material that would be hazardous to man.

Live adenovirus vaccines grown inadvertently in HeLa cell cultures have been tested in human subjects without untoward effect (1). Rats and mice injected with live or killed viruses grown in cancer cells of the same species of origin did not develop tumors (11). Finally, frankly viable cancer cells were shown not to be capable of sustaining growth when inoculated into normal human subjects but to be able to do so only in patients with far advanced cancer, in whom immunologic surveillance and cell-mediated immune mechanisms are characteristically depressed (12).

DIPLOID CELLS

The first effort to evolve serially propagable cells for vaccine production was made by Hayflick and Moorhead (13), who developed the WI38 diploid line of human embryonic lung cells. These workers demonstrated contact inhibition in individual cell cultures, finite limits in the replicability of the cells on serial passage, and diploidy with only minor variations in karyotype. They equated these properties with normalcy. In further support of WI38 cells for vaccine use, Hayflick claimed the advantage of using cells from a source shown to be free of virus and not subject to indigenous viruses that are commonly present in certain animal tissues (e.g., monkey kidney). In due course the WI38 cell found its rightful place as a means for growing poliovirus free of monkey virus contaminants and for the propagation of other vaccine viruses in countries where virus-free primary tissue sources were not readily available.

Though clearly of advantage in some respects, diploid cells do possess their own disadvantages and theoretical problems for safety. Perhaps the greatest problem is a practical one, that of bringing the required number of cultures through 27 or so serial passages to the point of sufficient number and size for vaccine production. The problems along the way include loss through bacterial or fungal contamination, defective media, improperly prepared glassware, etc. Freedom from viral agents, so vigorously claimed by WI38 cell proponents, does not apply beyond cells at the 7th to 9th passage level, at which point the cells are distributed; beyond this are an additional 20 or more passages made in individual laboratories. During this time, detectable or non-detectable viruses, mycoplasma, and even theoretical oncogenic determinants may be introduced into the cells from the medium (which contains serum), from the trypsin, from the air, from the laboratory workers, and from other sources, thereby erasing most of

the claimed advantages. The concept of a "seed cell
system" implies far more than it actually means since
so many serial passages are required after the cells
leave the point of "seed cell" certification.

The strict adherence to defined limits in karyology
for monitoring WI38 cells is probably founded more in
fetish than in fact since the significance of chromosomal
changes in cancer is not understood. Cellular expression
as cancer seems most likely the expression of a single
gene and the gross alteration in chromosomal constitution
commonly seen in cancer seems a more likely result of
neoplastic change rather than its cause. In fact,
frankly neoplastic cells, especially those transformed
by viruses, may display no visible changes in karyotype
(14-17) and, conversely, normal cells may show chromosomal
aberrations (18). If, indeed, chromosomal aberration is
a significant deterrent for use of a particular cell
for vaccine, then no cell should be used for vaccine
production since all cultures--primary, diploid and
neoplastic--show chromosomal aberration to some degree.
Furthermore, contact inhibition displayed by human
diploid cells and regarded to be a strong criterion for
normalcy is also of questionable significance since
contact inhibition can be reduced with resulting continued
cell proliferation after confluence is achieved by
employing such simple procedures as the substitution of
galactose for glucose in the growth medium (19) or by
the continuous perfusion of the culture with fresh
medium (20).

The most recent cause for concern in the use of
diploid cells for vaccine has been based on the oncogene
theory of Todaro and Huebner (21), which states essentially that nearly every animal cell carries in its
genome certain genes (virogenes) which upon derepression
may produce infectious oncogenic virus. One of these
genes, called an oncogene, is capable of inducing
neoplastic transformation. It has been suggested that
diploid cells may present a special hazard since they
might carry an oncogene of human origin that could be
recombined with or otherwise introduced genetically into
the vaccine virus during its growth and carried to the
recipient human host. The basis for the concern seems
nebulous in the absence of the demonstration of the
theoretical oncogene and in view of the failure, to
date, of diploid cell vaccines to cause a detectable
neoplastic change in man. Furthermore, this theory tends
to be self-defeating since, if everyone already carries
the oncogene, it might not be dangerous to add a few
more copies of it.

PRIMARY CELLS

As already noted, primary cell cultures may be made using tissues from post-natal animals, usually young adult, or from prenatal or embryonic tissue. The first cell culture vaccine, poliovaccine, was prepared using virus grown in the kidney cells of adult monkeys. Interestingly, no other cells have been so plagued with problems of indigenous viruses as have monkey cells. Early in its application, monkey kidney cell cultures were shown to be infected with dozens of indigenous viruses, which were carefully catalogued and numbered as SV or SA viruses (22,23). Only one of these viruses (B) initially stood out as a hazard to man. With increasing time, however, a new virus of alarming nature was found, called SV40 virus (24,25), which is oncogenic for rodents and for cells in culture. Later the oncogenic potential of certain of the indigenous simian adenoviruses was also demonstrated. Then there were the cytomegalo-viruses (26) and the adenovirus-associated viruses (27). Finally, there was the so-called Marburg agent (28) that caused a number of deaths in persons who were in direct or indirect contact with grivet monkeys (or their tissues) of a particular geographic origin.

Any vaccine free of contaminating viruses should be safe for human use, but the problem of producing lots of virus free of extraneous agents, the nagging apprehensions of the possible presence of still undetectable viruses of significant clinical hazard to man, the scare caused by the Marburg virus, and the availability of the WI38 diploid cell all combined to bring about the nearly total disuse of primary monkey kidney cells for vaccine production. It should be noted, however, that to date no adverse effect due to simian viruses has been observed in any human recipient of monkey kidney vaccine, including those who received live SV40 virus by oral feeding or injection of poliovaccines (29,30). However, the monkey cell experience did amply demonstrate the possible hazard of using wild-caught animals as a tissue source and hastened the development of the isolated colonies of specific pathogen-free dogs, rabbits, chickens and ducks that are now being used successfully to prepare virus-free primary cell cultures for vaccines.

Primary cell cultures produced from virus-free tissue sources possess a theoretical advantage over serially passaged cells in that there is only a single passage from tissue to vaccine. This rules out the multiplicity of changes for introducing exogenous agents that would be present in all subsequent passage progeny. Primary cells provide a margin of safety also in that a particular batch of tissues is used only once. Thus,

if a hazard is introduced, it will not be perpetuated
beyond the single passage and the single batch of
vaccine in which it is used.

RUBELLA VACCINE. In seeking licensure for the routine
distribution of any vaccine, it bears upon the producer
to demonstrate the safety of the vaccine virus and the
safety of the cells and media used to grow the virus.
Such requirement is made and rigidly enforced by the
Bureau of Biologics of the Food and Drug Administration.

In attempting to develop rubella vaccine in our
laboratories, it was soon found that the virus failed
to proliferate well in cell cultures of chick embryo.
Duck embryo cells, however, did propagate the virus and
caused a rapid attenuation of its virulence for man.

Duck embryo cell culture represented a new substrate
for virus propagation insofar as live virus vaccine for
man was concerned, the only previous application of
duck tissues being with killed rabies virus vaccine
produced in whole duck embryos. Hence it was necessary
to provide data that would establish safety for the
duck cell system. For this vaccine, the HPV-77 virus
modified by passage in duck cells was used. Details are
given in an earlier report (31).

Duck viruses. The domestic duck is remarkably free of
infectious diseases and neoplasia (32-34). No duck
viruses are known to be present in the embryo. Viruses
that infect ducks are shown in Table 1 together with
procedures for their detection.

Avian leukemia. Even though this complex has not been
found in ducks, RIF tests and COFAL tests for chicken
leukemia were applied, with negative results.

Induction of neoplasia in newborn animals. Tests for
oncogenicity of viruses in vivo commonly employ inoculation into newborn hamsters. Six different rubella virus
preparations and their corresponding control materials
were tested in newborn hamsters (see Table 2) with
negative findings. Additionally, newborn ducks
inoculated with the virus failed to develop neoplasia
during a one-year period of observation.

Electron microscopy and immunofluorescence studies.
Representative duck cell cultures infected with HPV-77

Table 1. Methods for Detecting Viruses Infecting Ducks in Nature

Agent	Host	Route of Inoculation	Pathology
Duck plague	Duck embryo Duckling	Allantoic & yolk sac Intramuscular	Death, 7 days Death, 7 days
Duck hepatitis	Duckling Chick embryo	Intramuscular Allantoic	Death, 4 days Death; stunted & edematous embryo
Newcastle disease virus	Chick embryo	Allantoic	Death; hemagglutinins in allantoic fluid
Arboviruses	Mouse-adult and suckling Chick embryo Monkey kidney	Cerebral Cell culture Cell culture	CNS disease; death Cytopathology (EEE, WEE, VEE)[a] Cytopathology (EEE, WEE, VEE)
Ornithosis (non-viral)	Chick embryo	Yolk sac	Death; microscopic observation of agent in yolk sac
Duck influenza virus	Chick embryo	Allantoic	Hemagglutinins in allantoic fluid

[a]EEE = eastern equine encephalomyelitis.
WEE = western equine encephalomyelitis.
VEE = Venezuelan equine encephalomyelitis.

Table 2. Examination of Rubella Virus Vaccines and of Duck Embryo Cell Culture System for Oncogenic Potential in Tests in Newborn Hamsters

Vaccine Lot or Control Fluids	Strain	Passage History[b]				Infectivity Titer/ml in GMK	No. Inoculated	No. Weaned	No. of Survivors (mo)							No. Developed Tumors
		GMK	DE	RK	EAM				1	3	6	9	12	18	25	
Lot 214	Merck-Benoit	19	20		1	$10^{2.7}$	72	64	64	62	35	20	8	5	1	0
Control							48	47	47	45	21	13	6	1	0	0
Rubella	Merck-3429	26				$10^{4.3}$	88	68	68	61	48	22	8	3	2	0
Control							48	47	47	44	33	18	9	1	0	0
Rubella	Merck-3429	21		16		$10^{3.5}$	64	55	55	54	42	21	11	3	2	0
Control							48	34	34	30	21	15	8	2	0	0
Lot 215	Merck-Benoit	11	20		1	$10^{2.5}$	68	64	64	62	61	59	54	23	3	0
Control							55	52	52	51	51	47	44	11	3	0
Lot 267	HPV-77	77	5			$10^{4.3}$	96	80	80	80	69	41	26	4	1	0
Control							56	51	51	51	39	29	20	3	1	0
Lot 295	HPV-77	77	5			$10^{3.9}$	96	68	68	67	55	38	23	5	1	0
Control							56	46	46	45	36	24	13	8	2	0

[a] 0.2 ml, subcutaneous.

[b] GMK = Grivet monkey kidney cell culture; DE = Duck embryo cell culture; RK = Rabbit kidney cell culture; EAM = Embryonated hen's egg intra-amnionic route.

and other rubella virus strains or retained as uninoculated controls were examined by electron microscopy at
various periods of incubation for the presence of
morphological entities resembling viruses. All
together, 350 sections from 168 cultures were photographed
and studied. There was no evidence of the presence of
extraneous agents in the cultures. Additionally, cell
cultures of duck embryo were tested for presence of
antigens (viruses) that would react with pooled sera
from adult ducks and that would be detectable by
immunofluorescence microscopy. No such antigens were
found.

Routine safety tests. Routine tests for safety were
applied (Table 3). This included inoculation into a
variety of animals, embryos, cell cultures, and
artificial media. No extraneous agent was found.

These tests served as the basis for developing the
regulatory standards (35) required by the Bureau of
Biologics for rubella virus vaccine produced in duck
embryo cell culture.

In addition to tests of the vaccine and cell
cultures, the isolated specific pathogen-free duck
flocks used by our laboratories are monitored for
presence of antibody against certain microbial agents
according to the procedures of the Central Veterinary
Laboratory, Ministry of Agriculture, Fisheries and Food,
Weybridge, Surrey, England. These tests are employed
for ruling out the presence in the flocks of infection
with chicken agents, including avian adenovirus, avian
encephalomyelitis virus, avian leukosis virus (COFAL and
Rous antibody tests), fowl pox virus, infectious bronchitis virus, infectious laryngotracheitis virus, Newcastle
disease virus, influenza A virus, Marek's disease virus,
Mycoplasma gallisepticum, Mycoplasma synoviae, and
Salmonella pullorum. It may be added that surveillance
and post-mortem examination of all animals in all
producing flocks provide one of the best measures for
monitoring to assure freedom from disease.

Duck embryo-propagated rubella vaccine (36) was
licensed on June 9, 1969, following completion of tests
in 18,367 human subjects.

SUMMARY

In closing, it may be noted that there are three
kinds of cells that may be applied to vaccine preparation
in the U. S. A   Primary cells in culture that are used

Table 3. Tests for Safety of Live Rubella Virus Vaccine Prepared in Duck Embryo Cell Cultures

Host	Inoculation	Subculture	Observation
	Vaccine prior to clarification		
Cell culture[a]			
Primary			
Chick embryo, whole	Flasks	Yes	Cytopathology
Chick embryo, whole	Flask-plate	Yes	Rous sarcoma inhibition (RIF)
Chick embryo, liver	Flasks	Yes	Cytopathology
Chick embryo, kidney	Flasks	Yes	Cytopathology
Duck embryo, whole	Flasks	Yes	Cytopathology
Duck embryo, liver	Flasks	Yes	Cytopathology
Duck embryo, kidney	Flasks	Yes	Cytopathology
Rhesus monkey kidney	Bottle-tube	Yes	Cytopathology
Grivet monkey kidney	Bottle-tube	Yes	Cytopathology
Rabbit kidney	Bottle-tube	Yes	Cytopathology
Stable			
HeLa[a]	Flasks	Yes	Cytopathology
Embryos			
Chick embryo, 11 day	Allantoic	Yes	Hemagglutinin, embryo death
Chick embryo, 6 day	Yolk sac	Yes	Embryo death or morbidity
Duck embryo, 14 day	Allantoic	Yes	Hemagglutinin, embryo death
Duck embryo, 14 day	Yolk sac	Yes	Embryo death or morbidity
Animals			
Mouse			
Newborn & adult	Intracerebral & intraperitoneal	Yes	Mortality, morbidity, gross pathology
Microbial media			
Thioglycollate	32°C		
Sabouraud	24°C		
PPLO medium	Aerobic & anaerobic, 35°C		
Lowenstein-Jensen	37°C		

Table 3. Continued

Host	Inoculation	Subculture	Observation
	Vaccine following clarification		
Rhesus monkey (CMS)	Intrathalamus, intraspinal & intramuscular		Histopathology of CNS, gross pathology, morbidity
Grivet monkey (neuro)	Intrathalamus & intracisternal		Histopathology of CNS, gross pathology, morbidity
Final container			
Microbial sterility			
Thioglycollate	32°C		
Safety test			
Guinea pigs	Subcutaneous		Mortality, morbidity, gross pathology
Mice	Subcutaneous		Mortality, morbidity, gross pathology
Infectivity titration	Grivet monkey kidney		
Identity	Grivet monkey kidney		
Moisture			
Antibiotic			

aTested as neutralized and non-neutralized mixtures.
Additionally, the cell culture tests were performed on fluids from uninoculated cultures.

routinely to prepare vaccine include chick, duck, rabbit, dog and monkey cells. The problems with indigenous viruses in wild-caught animals now make the monkey kidney an unlikely cell for future vaccine development. Diploid human embryo cell cultures, WI38 in particular, now has partial sanction for use in preparing oral poliovirus vaccine but not for the preparation of vaccines to be injected. In certain other countries, it has sanction for preparing vaccines for injection as well as for oral administration. Neoplastic and transformed cells are clearly forbidden, though such cultures would provide the means for mass production of virus in submerged tank cultivation.

The signs of our times clearly point to increased application of preventive medicine, for economic necessity, if for no other reason. Vaccines present the greatest return in health benefit for the least effort and are likely the first preventive measure to be broadly applied to the world's population. It is indeed fortunate that so many proved and acceptable kinds of cells for vaccine production exist and it is important that more be developed. The question of the relative advantages and disadvantages of primary vs. diploid cells, so much in debate, loses much of its relevance when viewed dispassionately, because all kinds of cells, with the exception of those from wild-caught monkeys, possess most of the same practical and theoretical problems. Furthermore, most of the concepts, upon which judgements of relative safety are based, are founded in myth. Hopefully a future generation, if not our own, will untangle the meaning of neoplasia and chromosomal alteration and provide for the routine use of infinitely replicating cells in submerged culture. Here indeed lies the challenge for the economical production of vaccine and the challenge, in turn, for the biological scientist.

References

1. Hilleman, M. R. 1968. Cells, vaccines, and the pursuit of precedent. Nat. Cancer Inst. Monogr. 29:463.
2. US PHS Regulations for the Manufacture of Biological Products, Title 42, Part 73. Dept. HEW, Publ. 437 (revised Jan. 1967).
3. Report of a Committee on Tissue Culture Viruses and Vaccines. 1963. Continuously cultured tissue cells and viral vaccines. Science 139:15.
4. Hayflick, L., P. S. Moorhead, C. M. Pomerat and T. C. Hsu. 1963. Choice of a cell system for vaccine production. Science 140:766.

5. Perkins, F. T. and L. Hayflick. 1967. Meetings. Cell Cultures. Science 155:723.
6. Sabin, A. 1969. Discussion. Session V of the International Conference on Rubella Immunization. Amer. J. Dis. Child. 118:378.
7. Petricciani, J. C., H. E. Hopps and D. E. Lorenz. 1971. Subhuman primate diploid cells: Possible substrates for production of virus vaccines. Science 174:1025.
8. Hayflick, L. 1972. Human virus vaccines: Why monkey cells? Science 176:813.
9. Petricciani, J. C., H. E. Hopps and D. E. Lorenz. 1972. Human virus vaccines: Why monkey cells? Science 176:814.
10. Andreopoulos, S. 1972. Human vaccines and bureaucracy: Stanford professor's odyssey. Stanford M.D. 11:18.
11. Hull, R. N. 1968. Immunization of experimental animals with vaccines produced in known oncogenic cell lines. Nat. Cancer Inst. Monogr. 29:503.
12. Moore, A. E. 1968. Evidence for oncogenic properties of heteroploid cell lines--discussion. Nat. Cancer Inst. Monogr. 29:291.
13. Hayflick, L. and P. S. Moorhead. 1961. The serial cultivation of human diploid cell strains. Exp. Cell Res. 25:585.
14. Koller, P. C. 1972. The role of chromosomes in cancer biology. Recent Results in Cancer Research, No. 38. Springer-Verlag, New York.
15. Nowell, P. C. and D. A. Hungerford. 1964. Chromosome changes in human leukemia and a tentative assessment of their significance. Ann. N. Y. Acad. Sci. 113:654.
16. Miles, C. P., F. O'Neill, D. Armstrong, B. Clarkson and J. Keane. 1968. Chromosome patterns of human leukocyte established cell lines. Cancer Res. 28:481.
17. Huang, C. C. and G. E. Moore. 1969. Chromosomes of 14 hematopoietic cell lines derived from peripheral blood of persons with and without chromosome anomalies. J. Nat. Cancer Inst. 43:1119.
18. Kleinfeld, R. and J. L. Melnick. 1958. Cytological aberrations in cultures of "normal" monkey kidney epithelial cells. J. Exp. Med. 107:599.
19. Baugh, C. L. and A. A. Tytell. 1967. Propagation of the human diploid cell WI38 in galactose medium. Life Sci. 6:371.
20. Kruse, P. F., Jr. and E. Miedema. 1965. Production and characterization of multiple-layered populations of animal cells. J. Cell Biol. 27:273.
21. Todaro, G. J. and R. J. Huebner. 1972. The viral oncogene hypothesis: New evidence. Proc. Nat. Acad. Sci. 69:1009.
22. Hull, R. N. 1968. Viral flora of primate tissues-- discussion. Nat. Cancer Inst. Monogr. 29:173.

23. Malherbe, H. and R. Harwin. 1957. Seven viruses isolated from the vervet monkey. Brit. J. Exp. Pathol. 38:539.
24. Sweet, B. H. and M. R. Hilleman. 1960. The vacuolating virus, SV40. Proc. Soc. Exp. Biol. Med. 105:420.
25. Girardi, A. J., B. H. Sweet, V. B. Slotnick and M. R. Hilleman. 1962. Development of tumors in hamsters inoculated in the neonatal period with vacuolating virus SV40. Proc. Soc. Exp. Biol. Med. 109:649.
26. Black, P. H., J. W. Hartley and W. P. Rowe. 1963. Isolation of a cytomegalovirus from African green monkey. Proc. Soc. Exp. Biol. Med. 112:601.
27. Melnick, J. L. 1968. Latent viral infections in donor tissues and in recipients of vaccines. Nat. Cancer Inst. Monogr. 29:337.
28. Hennessen, W. 1968. A hemorrhagic disease transmitted from monkeys to man. Nat. Cancer Inst. Monogr. 29:161.
29. Miller, R. W. 1968. Monitoring vaccines for human oncogenicity. Nat. Cancer Inst. Monogr. 29:453.
30. Robbins, F. C. 1968. Monitoring vaccines for human oncogenicity--discussion. Nat. Cancer Inst. Monogr. 29:457.
31. Buynak, E. B., V. M. Larson, W. J. McAleer, C. C. Mascoli and M. R. Hilleman. 1969. Preparation and testing of duck embryo cell culture rubella vaccine. Amer. J. Dis. Child. 118:347.
32. Dougherty, E. 1956. Diseases of ducks. The Yearbook of Agriculture 1956, Animal Diseases, p. 496. U. S. Dept. of Agriculture, U. S. Govt. Printing Office.
33. Rigdon, R. H. 1967. Neoplasms in sterile hybrid ducks--A melanoma and two teratomas. Avian Dis. 11:79.
34. Luginbuhl, R. E. 1968. Viral flora of chick and duck tissue sources. Nat. Cancer Inst. Monogr. 29:109.
35. US PHS Regulations for the Manufacture of Biological Products, Title 42, Part 73. Dept. HEW (revised June 1, 1971).
36. Weibel, R. E., J. Stokes, Jr., E. B. Buynak, J. E. Whitman, Jr., M. B. Leagus and M. R. Hilleman. 1968. Live attenuated rubella virus vaccines prepared in duck embryo cell culture. II. Clinical tests in families and in an institution. J. Amer. Med. Ass. 205:554.

DISCUSSION

CLARK: Is there a higher incidence of cancer in children who were vaccinated with polio vaccine contaminated with SV40? Were any vaccinated children immunosuppressed?

HILLEMAN: Very large numbers of children evidently received live SV40 virus both in killed Salk poliovaccine and in live Sabin poliovaccine. There is no evidence to date to suggest that any of these persons suffered any adverse short- or long-term effect. Long-term follow-up will be continued. Certainly some of the recipients of these vaccines must have been immunosuppressed; we don't know specifically, however.

DIXON: In vaccine follow-up studies is there careful search made for central nervous system disease or chronic glomerulonephritis?

HILLEMAN: Long-term follow-up, to my knowledge, has been focused primarily on cancer, taking a look at poliovaccine background in reported cancer cases. In view of the possible relationship of papovaviruses to CNS disease, purposeful expansion into this area seems indicated.

MILLER: From 1956 to 1958 about 10,000 men in the army received adenovirus vaccine and follow-up studies in theory can be done. It may be difficult to make the study because unlike other veterans, these were peacetime veterans. The follow-up agency (NAS-NRC) is experienced in the record-linkage systems used in studies of wartime veterans. The same system may not apply to peacetime veterans, and the effort of following them may be beyond the capabilities of the agency. The feasibility is being evaluated at present.

EXPERIENCE WITH SV40
AND ADENOVIRUS-SV40 HYBRIDS

Andrew M. Lewis, Jr.
Laboratory of Viral Disease
National Institute of Allergy and Infectious Diseases
Bethesda, Maryland

The recent, rapid advance in the genetic analysis of SV40 suggests that this virus will play an important role in the study of animal cell genetics and in elucidating the mechanism by which viruses transform cells in tissue culture and produce tumors in animals (1-15). A number of these studies have resulted in the discovery or production of a variety of SV40 mutants or recombinants (16-20,7,11). Detailed understanding of the events which lead to the development of neoplastic cells following SV40 infection will likely require the production and study of many more such SV40 mutants. In addition, large quantities of highly concentrated wild-type and mutant SV40 viruses are required for current studies and will certainly be necessary for the biochemical studies which will define at the molecular level the process of virus-induced transformation.

It is necessary to ask whether laboratory manipulations of this oncogenic DNA virus will pose significant risks to either laboratory workers or to the general public. The purpose of this report is to examine the consequences of the inadvertent contamination of poliomyelitis and adenovirus vaccines by SV40, summarize the available data on SV40 infection of humans, and provide a source of information which may be used by anyone wishing to consider the risks associated with the laboratory manipulation of SV40.

VACCINE CONTAMINATION AND HUMAN INFECTION WITH SV40

In 1960 Sweet and Hilleman (21) isolated an
unidentified virus from 7 of 10 lots of primary rhesus
monkey kidney cells that were to be used for poliovirus
vaccine production. This virus produced no noticeable
changes in the infected rhesus cells and was detected
only by the cytopathic effects it produced in primary
African green monkey kidney (AGMK) cells. Due to this
characteristic cytopathology, the new agent was called
vacuolating virus and was officially designated SV40
by Hull's criteria for classifying agents isolated from
monkey kidney tissues (22,23).

Some 20 strains of SV40 were recovered by Sweet and
Hilleman from primary rhesus and cynomologous monkey
kidney cells, as well as from a diversity of viruses
which had been propagated in rhesus kidney cells. Ten
of these strains represented SV40 isolates from types 1,
2, and 3 live, attenuated, poliomyelitis vaccines and
from the seed stocks of the vaccine strains of human
adenoviruses types 1-7 (24). In addition, Sweet and
Hilleman reported the presence of SV40 neutralizing
antibody in the sera of humans who had received
formalin-killed adenovirus or poliomyelitis vaccines
prepared in rhesus monkey kidney cells. Shortly after
this report, Eddy (25) described the development of
sarcomatous tumors in hamsters that had been injected
as newborns with extracts of rhesus kidney cultures.
The agent responsible for the induction of these tumors
was later identified as SV40 (26-28). Soon after these
initial reports, a number of studies confirmed the
presence of infectious SV40 in both live and inactivated
poliomyelitis vaccines and demonstrated the development
of SV40 neutralizing antibody in persons inoculated with
the formaldehyde-inactivated (Salk) vaccine (29-32).
The presence of live SV40 in the inactivated vaccine
was explained by the failure to inactivate a proportion
of the contaminating SV40 virions by the formaldehyde
treatment used to inactivate the polioviruses (33,30,29).
No SV40 neutralizing antibody was detected in children
who received contaminated live attenuated (Sabin) poliomyelitis vaccines (21,33,29,31,34), although some of the
vaccinated children received 100-1000 pfu of SV40 and
excreted detectable virus in their stools for 3-5 weeks
after ingestion of the vaccine. These findings suggested
limited replication of SV40 in the gastrointestinal
tract of these individuals (31,34).

In 1961 Morris (35) conclusively demonstrated that
humans could be infected by SV40. These authors reported
that adult volunteers infected intranasally with SV40-
contaminated respiratory syncytial virus stocks developed

subclinical infection with SV40 as determined by the isolation of virus from throat swabs and the presence of SV40 neutralizing antibody in convalescent serum specimens.

In 1962 Shein and Enders (36,37) showed that SV40 would replicate in human renal cells without producing detectable damage to the cultures; these chronically infected cultures were subsequently transformed as a result of the SV40 infection. Koprowski et al. (38) confirmed the finding that SV40 would transform human cells, and Jensen et al. (39) demonstrated the development of subcutaneous tumor nodules in human subjects who had received injections of tissue culture lines of human cells transformed in vitro by SV40. Horvath (40,41) and Shah (42-44) demonstrated a high incidence (30-37%) of SV40 neutralizing antibodies in persons handling rhesus monkeys or rhesus monkey kidney cell cultures. Both authors found that the longer workers were associated with animals or cell cultures, the more likely they were to have SV40 antibodies.

By 1963 SV40-contaminated poliomyelitis vaccines had been in general use for approximately 8 years; SV40-contaminated adenovirus vaccines had been extensively used in military populations and to a lesser degree in civilian populations for 4-5 years. Thus the individuals and agencies responsible for the development and safety of these vaccines were confronted with the very problem they had tried to avoid by using sub-human primate tissues to produce this product, i.e., the exposure of millions of persons to an occult tumor virus by inoculation or ingestion of inadvertently contaminated vaccines.

FOLLOWUP STUDIES ON HUMANS INFECTED WITH SV40

No illnesses attributable to SV40 were detected during the initial studies of patients receiving the contaminated poliovirus or adenovirus vaccines (45,46, 31). The eight patients studied by Morris et al. (35), who were infected by SV40-contaminated respiratory syncytial virus stocks mixed with respiratory syncytial virus antiserum and administered by intranasal nebulization, remained symptom free during a one-month observation period. Although no data were presented, Fraumeni et al. (47) in a 2-year followup reported that no illnesses which could be associated with the subclinical SV40 infection were found in any of the 35 individuals in Morris' study. Thus SV40 infection appears to produce no discernible illness in humans within 1-2 years after infection.

Considering the longer term effects of SV40 infection, Fraumeni et al. in the same report (47) discussed a preliminary examination of the cancer mortality rates in the United States between 1950-1959 for persons less than 25 years old. This 10-year period included the initial introduction and widespread use of poliomyelitis vaccine in the United States. The only finding was a slight rise in the leukemia mortality rates for children 5-9 years old and 10-14 years old (3.5-3.8 and 2.2-2.5 per 100,000 respectively). After analyzing data from the Salk program for 6-8 year olds vaccinated in 1955, it was concluded that this rise could not be attributed to SV40 infection.

Stewart and Hewett (48) in England examined the immunization records of 99 cases of leukemia in children under 10 years of age and 1108 cases of other types of malignant diseases in children. When these data were compared with the immunization records of matched controls, these authors found no differences in the types of vaccines received between the controls and the children with malignancies (Table 1). Innis (49) in Australia described a similar study (Table 1). In contrast to Stewart and Hewett, he noted, during a comparison of the vaccination records of 706 children over 1 year old with malignant diseases with appropriate controls, that a larger number of children with cancer had received poliomyelitis vaccines. The difference was statistically significant with a X^2 of 12.2;P < 0.0005.

Fraumeni et al. (50) followed up 1077 newborn infants who had received both oral and parenteral poliomyelitis vaccines between 1960 and 1962 at a Cleveland medical center. Each of these vaccine lots was shown to be contaminated with viable SV40, with virion concentrations varying between 1 particle per 5 ml to $10^{4,5}$ $TCID_{50}$/ml. Of the 918 children who had been located by December 1968 there had been 11 deaths, none of which were due to neoplastic diseases. The specific causes of death in these 11 children encompassed infections, accidents, and heritable disorders and could not be attributed to SV40 contamination of the vaccines which these infants had received.

The data from these last three studies indicate that of the millions of persons who received poliomyelitis vaccines contaminated by a simian virus known to produce neoplastic changes in human cells, the records of only 6764 persons have been examined sporadically during the past 15 years. One of these studies (49) suggested the need for more careful and continued surveillance. These findings indicate that there has been no systematic, in-depth follow up study of persons receiving SV40-

Table 1. Studies of Malignant Diseases in Children Following Vaccination

Study	Vaccine	Incidence of Vaccination in Patients and Controls			
		Leukemia[a]		Other childhood malignancies[b]	
		Cases	Controls	Cases	Controls
Stewart and Hewett Oxford, England 1965	poliomyelitis[c]	270	256	259	265
	diphtheria	753	758	768	781
	pertussis	536	561	527	571
	tetanus	176	222	204	227
	smallpox	486	458	455	488

		Malignant disease in children			
		Over 1 year old[e]		Under 1 year old[f]	
		Cases	Controls	Cases	Controls
Innis Sidney & Brisbane Australia 1968	poliomyelitis[d]	618	569	29	28
	diphtheria	646	640	49	52
	pertussis	624	626	48	53
	tetanus	616	609	49	51
	smallpox	29	20		

a 999 matched pairs of children up to 9 years old studied.
b 1108 matched pairs studied; patients' age not recorded.
c Route of vaccination not given.
d Vaccine administered parenterally.
e 706 matched pairs studied.
f 110 matched pairs studied.

contaminated poliomyelitis vaccines and I have been
unable to find any follow up study of military or
civilian personnel who received SV40-contaminated
adenovirus vaccines. Thus while there appears to be no
untoward short-term effects from human infection with
SV40, proper studies evaluating the long-term effects
have not been undertaken. As it is now about 15-20
years since the general use of these contaminated
vaccines, the time for the appearance of any untoward
long term effects may soon be at hand.

INFECTION OF HUMANS WITH SV40-RELATED VIRUSES

     Shah et al. (42) reported the presence of SV40
antibodies in 9% of the sera they obtained from hospital
patients in South India, a region outside the natural
habitat of rhesus monkeys. None of these patients had
been exposed to rhesus monkeys and none had received
poliomyelitis vaccines. In other reports Shah et al.
(43,44) demonstrated SV40 neutralizing antibodies in
patients without a history of poliovirus immunization
and in 2-3% of sera obtained from residents of Maryland
or Ohio before SV40-contaminated vaccines were used and
after such vaccines were rendered SV40 free. These
authors concluded that man may become infected with
viruses antigenically related to SV40.

     In 1965 Zu Rhein and Chou (51) found particles
resembling papovaviruses in brain sections from patients
suffering from progressive multifocal leukoencephalopathy.
Early in 1972 Weiner et al. (52) reported the isolation
of viruses which were related to SV40 from cell cultures
established from the brains of two patients suffering
with this disease. Each of these patients had received
oral poliomyelitis vaccines in the early 1960's and one
of them had titers of SV40 neutralizing antibody ranging
between 320-1280 in multiple serum specimens obtained
over a 4-month period. Their original observation was
further documented by the direct isolation of the SV40-
like virus in human fetal brain cultures infected with
brain homogenates and by the use of electron microscopy
to visualize specific agglutination of the virions in
the brain extracts by SV40 antiserum (53,54). Additional
reports have appeared describing the isolation of other
papovaviruses from humans. Padgett et al. (55) isolated
the JC virus from the brain of another patient with
progressive multifocal leukoencephalopathy. This agent
differed from the virus isolated by Weiner et al. in
that it failed to induce immunofluorescent antigens
reacting with SV40 antisera in infected cells and failed
to grow in primary or continuous lines of AGMK cells.
Other laboratories have also identified papovaviruses

unrelated to SV40 in the brains of patients with
progressive multifocal leukoencephalopathy (56,57).
Gardner et al. (58) isolated a papovavirus from the
urine of a patient following renal transplantation.
This agent was isolated and propagated in a continuous
line of AGMK cells (Vero) and contained antigens which
reacted weakly with SV40 immune sera by hemagglutination
inhibition and immune electron microscopy. Serum from
the patient from whom this virus was isolated contained
low levels of antibody reacting with SV40 by immune
electron microscopy; tests for SV40 neutralizing antibody in this serum were not reported. Collectively,
these studies indicate human infection with several
distinct groups of papovaviruses that appear to be quite
variable in their antigenic relationship with SV40.
It should be emphasized that viruses which seem to be
both serologically identical and serologically unrelated
to SV40 are intimately associated with the same subacute
human demyelinating disease.

Given the diversity of the papovaviruses which have
been isolated thus far from humans, the emergence of
SV40 mutants in individuals infected by contaminated
vaccines must be considered. In this regard recent
reports (59,60) proposed that the emergence of mutant
viruses during acute or chronic viral infection in
certain individuals could possibly be responsible for a
number of chronic diseases that appear to be viral in
origin. This concept was strengthened by Fields'
discovery that a conditional lethal mutant of reovirus
type 3 is capable of producing degenerative brain
disease in rats.

As a consequence of the recovery of SV40-related
viruses from the lesions of progressive multifocal
leukoencephalopathy, future studies evaluating the effects
of human infection with SV40 must consider demyelinating
diseases of the central nervous system as well as
malignancies. Furthermore until definitive studies
clarify the relationship between SV40 and SV40-related
viruses to chronic degenerative central nervous system
disease in humans, it appears to me that the laboratory
manipulation of SV40 involves some risks.

GENETIC RECOMBINATION BETWEEN SV40 AND HUMAN ADENOVIRUSES

The adaptation of human adenoviruses to rhesus monkey cells for vaccine preparation (24) led to the
discovery of a new class of recombinant viruses, the
adenovirus-SV40 (Ad-SV40) hybrids. Following the
discovery of SV40 contamination of adenovirus types 3,
4, and 7 vaccine stocks, extensive efforts were under-

taken by several manufacturers to eliminate the
contaminant by passage with SV40 antiserum or by cloning
in AGMK cells (61-63). Infectious SV40 virions were
eliminated from the vaccine stock of the LL strain of
adenovirus 7 and the JF strain of adenovirus 3 (61,63);
however, exhaustive procedures were unsuccessful in
eliminating SV40 from the RN strain of adenovirus 4 (62).
In 1964 Huebner et al. (61), during an extensive evaluation of the 31 serotypes of human adenoviruses for their
oncogenic effects in hamsters, discovered that the
SV40-free vaccine strain (LL) of adenovirus 7 induced
tumors containing SV40 antigens in hamsters. These
studies suggested that the adenovirus 7 vaccine strain
was a hybrid containing both adenovirus 7 and SV40
genetic information within adenovirus 7 capsids. These
impressions were quickly confirmed (64,65) and other
studies demonstrated that the adenovirus vaccine strains
types 1-5 were all Ad-SV40 hybrids containing SV40
genetic information within adenovirus capsids (66-68,63).
Further studies showed that adenovirus vaccine strains
1, 2, 4, and 5 contained the infectious SV40 genome
within adenovirus capsids in contrast to the adenovirus
3 and 7 vaccine strains, which appeared to contain only
a portion of the SV40 genome (66-68). Since these
initial reports, a number of studies have described
defective and nondefective Ad-SV40 virions (69,70,17)
and have conclusively demonstrated that the hybrid
particles contain a recombinant genome consisting of
various portions of covalently linked adenovirus and
SV40 DNA (71-74). A summary of the properties of the
defective and nondefective Ad2-SV40 and Ad7-SV40 hybrids
is presented in Table 2. These data are shown to
illustrate the variety of recombinations that have been
found between the genomes of these two unrelated viruses.

The defective Ad7-SV40 hybrid population (Table 2)
contains hybrid particles which produce SV40 T and U
antigens (64,65,75) as well as tumor-specific transplantation antigen (TSTA) (76). These Ad7-SV40 hybrid
particles failed to yield SV40 virions (61) and contain
by electron microscopy only 75% of the SV40 genome
inserted approximately 5% from one end of the hybrid DNA
molecule (73). This population contrasts with the
defective Ad2^{++}HEY variant, which contains hybrid
particles that produce SV40, T, U, TSTA and V antigens
as well as nonhybrid SV40 virions in high titer (77).
These hybrid particles appear to contain two types of
hybrid DNA molecules, one containing 1.4 unit lengths of
SV40 DNA, the other containing 2.2 unit lengths of SV40
DNA (Siegel et al., in prep.). Both these recombinant
genomes are missing about 35% of the Ad2 genome, and
these SV40 DNA segments are inserted approximately 28%
from one end of the hybrid DNA molecule. The comparison

Table 2. Properties of Ad-SV40 Hybrid Viruses

	Host Range[a]	SV40 Antigen Induction[b]				Production of SV40 Progeny (pfu/ml)	Size of SV40 DNA Segment in Recombinant Genome[c]
		T	U	TSTA	V		
Defective							
Ad.7[+] (E46[+]/PARA)	HEK/AGMK	+	+	+	–	0	0.75
Ad.2[++]	HEK/AGMK	+	NT	NT	+	$10^{7.7}$	NT
Ad.2[++] HEY[d]	HEK/AGMK	+	+	+	+	$10^{7.0}$	1.4 & 2.2
Ad.2[++] LEY[d]	HEK/AGMK	+	+	–	–	$10^{3.0}$	1.0
Nondefective							
Ad.2[+] ND$_1$	HEK/AGMK	–	+	–	–	0	0.18
Ad.2[+] ND$_2$	HEK/AGMK	–	+	+	–	0	0.32
Ad.2[+] ND$_3$	HEK	–	–	–	–	0	0.06
Ad.2[+] ND$_4$	HEK/AGMK	+	+	+	–	0	0.43
Ad.2[+] ND$_5$	HEK	–	–	–	–	0	0.28

a   Defective hybrid particles require coinfection with nonhybrid adenoviruses to replicate in any host. Nondefective hybrid virions are capable of independent replication without the assistance of nonhybrid adenovirions.

b   T and V antigen induction determined by indirect immunofluorescent technique on infected cells grown on glass coverslips. U antigen induction determined by complement fixation reaction of heated cell pack with SV40 T U[+] hamster sera (75). TSTA antigen induction determined by immunization of weanling hamsters followed by challenge with known tumor inducing dose of SV40-transformed hamster kidney cells.

c   Given as % of one SV40 genome; determined by electron microscopic heteroduplex mapping techniques.

d   HEY = high efficiency yielder; LEY = low efficiency yielder. NT = not tested.

continues with the Ad2^{++} LEY variant which induces SV40
T and U antigen and inefficiently induces trace amounts
of nonhybrid SV40 virions (77). The genome of the
hybrid particles in this population contains 1.0 unit
lengths of SV40 DNA inserted approximately 7% from one
end of the hybrid DNA molecule (Siegel et al., in prep.).
It is interesting to note that these Ad2^{++} LEY hybrid
particles, which contain a full complement of infectious
SV40 genes, fail to induce SV40 TSTA in immunized
hamsters as judged by challenge with SV40-transformed
hamster cells (Siegel et al., in prep.). We speculate
that the break in the SV40 circular chromosome that
preceded its linear insertion into Ad2 DNA occurred in
that region of the SV40 genome that was responsible for
induction of SV40 TSTA.

The properties of the nondefective Ad2-SV40 hybrids
are also listed in Table 2. These hybrid virions differ
from the defective Ad-SV40 hybrid particles in that they
are capable of independent replication without the help
of nonhybrid adenovirus (17,11). Individual viruses in
this group induce either SV40 U (Ad2$^+$ND$_1$), SV40 U and
TSTA (Ad2$^+$ND$_2$), or SV40 U, TSTA, and T (Ad2$^+$ND$_4$)
antigens. These nondefective hybrids contain smaller
segments of the SV40 genome than the defective Ad-SV40
hybrids (12). By electron microscopic heteroduplex
mapping techniques, these SV40 DNA segments vary between
18% of the SV40 genome present in the Ad2$^+$ND$_1$ hybrid to
43% of the SV40 genome present in the Ad2$^+$ND$_4$ hybrid (15).

Two of the nondefective hybrids, Ad2$^+$ND$_3$ and Ad2$^+$ND$_5$,
illustrate the difficulties in identifying such recombinant viruses. In contrast to the other three
nondefective hybrids, these agents induce no detectable
SV40 antigens and have a restricted host range, replicating only in human embryonic kidney (HEK) cells. These
viruses were suspected to contain SV40 DNA sequences
only after finding that they induced SV40-specific RNA
in lytically infected cells. Subsequent studies
demonstrated that the Ad2$^+$ND$_3$ hybrid contained 6% of the
SV40 genome, while the Ad2$^+$ND$_5$ hybrid contained 28% of
the SV40 genome. Thus RNA-DNA hybridization or electron
microscopic heteroduplex mapping techniques are necessary
to detect heterologous DNA sequences within the genome
of this type of Ad-SV40 hybrid.

In addition to carrying segments of SV40 DNA, the
adenovirus genome appears to be able to carry sequences
of genetic information from other sources. Butel et al.
(78) described a monkey-adapted, adenovirus type 7
population which contained defective adenovirus encapsidated particles carrying a monkey adapting component
(MAC). The genome of this MAC particle expressed no

detectable SV40 functions (78,79) and the MAC function could be transferred to adenovirus type 2.

Burlingham and Doerfler (80) have shown that adenovirus types 2 and 12 integrate into host cell DNA during lytic infection. Thus, like SV40, it is possible that under certain conditions DNA sequences from the host cell could become permanently associated with the adenovirus genome. As it is impossible at this time to determine the risks of immunization with recombinant viruses containing any foreign DNA, perhaps viral DNA from any adenovirus strain which is a potential candidate for vaccine use should be tested by biochemical techniques for the presence of recombinant genomes.

SUMMARY

Parenteral and oral poliomyelitis vaccines used in many countries between 1955 and 1962 contained infectious SV40 virions in varying concentrations. Parenteral adenovirus vaccines used extensively in military and to a limited extent in civilian populations in the United States between 1957 and 1960 contained infectious SV40 virions. The adenovirus 3 and 7 vaccines used between 1961 and 1965 contained Ad-SV40 hybrid particles with segments of SV40 DNA enclosed within their recombinant genomes. SV40 infection of humans with as much as 30,000 $TCID_{50}$ appears to produce no initial overt disease in either infants or adults. No information is available on the infection of humans with massive SV40 inoculation (> 10^6 pfu). No thorough, systematic evaluation of the long-term effects of SV40 infection of either civilian or military personnel has been undertaken. Of the four limited studies on the incidence of neoplastic diseases in humans vaccinated with SV40-contaminated vaccines, only one suggests a slightly higher incidence of malignancies in children who received poliomyelitis vaccines. The association of papovaviruses serologically identical to SV40 with degenerative disease of the central nervous system suggests the possibility of chronic human infection with SV40, the existence of human papovaviruses serologically identical to SV40, or the existence of human-adapted SV40 mutants.

The possibility that SV40 mutants emerged during the vaccination of humans with SV40-contaminated vaccines must be considered. Until satisfactory studies evaluate the long-term effects of SV40 infection in humans and clarify the relationship between SV40 and SV40-related agents to chronic degenerative central nervous system disease in humans, it appears to this reviewer that the laboratory manipulation of SV40 involves some risks.

The genomes of SV40 and human adenovirus recombine
during mixed infection in monkey kidney cells. Some
Ad-SV40 hybrids may not produce biological evidence
indicating the recombinant nature of their genome.
Biochemical techniques can detect the presence of
heterologous DNA in such hybrids. Stocks of human
adenoviruses to be used for vaccine production should be
monitored for the presence of recombinant particles by
nucleic acid hybridization and electron microscopic
heteroduplex mapping techniques.

References

1. Danna, K. J. and D. Nathans. 1971. Specific
    cleavage of SV40 DNA by restriction endonuclease of
    Hemophilus influenza. Proc. Nat. Acad. Sci. 68:2913.
2. Danna, K. J. and D. Nathans. 1972. Bidirectional
    replication of SV40 DNA. Proc. Nat. Acad. Sci.
    69:3097.
3. Nathans, D. and K. J. Danna. 1972. Specific origin
    in SV40 DNA replication. Nature New Biol. 236:200.
4. Lindstrom, D. M. and R. Dulbecco. 1972. Strand
    orientation of SV40 transcription in productively
    infected cells. Proc. Nat. Acad. Sci. 69:1517.
5. Khoury, G. and M. A. Martin. 1972. Comparison of
    SV40 DNA transcription in vivo and in vitro. Nature
    New Biol. 238:4.
6. Sambrook, J., P. A. Sharp and W. Keller. 1972.
    Transcription of SV40. I. Separation of the strands
    of SV40 DNA and hybridization of the separated
    strands to RNA extracted from lytically infected and
    transformed cells. J. Mol. Biol. 70:57.
7. Tegtmeyer, P. 1972. SV40 DNA synthesis: the viral
    replicon. J. Virol. 10:591.
8. Patch, C. T., A. M. Lewis, Jr. and A. S. Levine.
    1972. Evidence for a transcription-control region
    of SV40 in the adenovirus 2-SV40 hybrid, $Ad2^+ND_1$.
    Proc. Nat. Acad. Sci. 69:3375.
9. Morrow, J. F. and P. Berg. 1972. Cleavage of SV40
    DNA at a unique site by a bacterial restriction
    enzyme. Proc. Nat. Acad. Sci. 69:3365.
10. Mulder, C. and H. Delius. 1972. Specificity of
    the break produced by restricting endonuclease R_1
    in SV40 DNA as revealed by partial denaturation
    mapping. Proc. Nat. Acad. Sci. 69:3215.
11. Lewis, A. M., Jr., A. S. Levine, C. S. Crumpacker,
    M. J. Levin, R. J. Samaha and P. H. Henry. 1973.
    Studies of nondefective adenovirus 2-SV40 hybrid
    viruses. V. Isolation of additional hybrids which
    differ in their SV40-specific biological properties.
    J. Virol. (in press).
12. Henry, P. H., L. E. Schnipper, R. J. Samaha,

C. S. Crumpacker, A. M. Lewis, Jr. and A. S. Levine. 1973. Studies of nondefective adenovirus 2-SV40 hybrid viruses. VI. Characterization of the DNA from five nondefective hybrid viruses. J. Virol. (in press).

13. Levine, A. S., M. J. Levin, M. N. Oxman and A. M. Lewis, Jr. 1973. Studies of nondefective adenovirus 2-SV40 hybrid viruses. VII. Characterization of the SV40 RNA species induced by five nondefective hybrid viruses. J. Virol. (in press).

14. Zain, B. S., R. Dhar, S. M. Weissman, P. Lebowitz and A. M. Lewis, Jr. 1973. A preferred site for initiation of RNA transcription by $\underline{E.\ coli}$ RNA polymerase within the SV40 DNA segment of the non-defective adenovirus-SV40 hybrid viruses, $Ad2^{+}ND_{1}$ and $Ad2^{+}ND_{3}$. J. Virol. (in press).

15. Kelly, T. J., Jr. and A. M. Lewis, Jr. 1973. The use of nondefective adenovirus-SV40 hybrids for mapping the SV40 genome. J. Virol. (in press).

16. Takemoto, K. K., R. L. Kirschstein and K. Habel. 1966. Mutants of SV40 differing in plaque size, oncogenicity, and heat sensitivity. J. Bacteriol. 92:990.

17. Lewis, A. M., Jr., M. J. Levin, W. H. Wiese, C. S. Crumpacker and P. H. Henry. 1969. A nondefective (competent) adenovirus-SV40 hybrid isolated from the Ad2-SV40 hybrid population. Proc. Nat. Acad. Sci. 63:1128.

18. Tegtmeyer, P. and H. L. Ozer. 1971. Temperature-sensitive mutants of SV40: infection of permissive cells. J. Virol. 8:516.

19. Robb, J. A. and R. G. Martin. 1972. Genetic analysis of SV40. III. Characterization of a temperature-sensitive mutant blocked at an early stage of productive infection in monkey cells. J. Virol. 9:956.

20. Jackson, D. A., R. H. Symons and P. Berg. 1972. Biochemical method for inserting new genetic information into DNA of SV40: circular SV40 DNA molecules containing lambda phage genes and the galactose operon of $\underline{E.\ coli}$. Proc. Nat. Acad. Sci. 69:2904.

21. Sweet, B. H. and M. R. Hilleman. 1960. The vacuolating virus, SV40. Proc. Soc. Exp. Biol. Med. 105:420.

22. Hull, R. N. and J. R. Minner. 1957. New viral agents recovered from tissue cultures of monkey kidney cells. II. Problems of isolation and identification. Ann. N. Y. Acad. Sci. 67:413.

23. Hull, R. N., J. R. Minner and C. C. Mascoli. 1958. New viral agents recovered from tissue cultures of monkey kidney cells. III. Recovery of additional agents both from cultures of monkey tissues and

directly from tissues and excreta. Amer. J. Hyg.
68:31.
24. Hartley, J. W., R. J. Huebner and W. P. Rowe. 1956.
Serial propogation of adenoviruses (APC) in monkey
kidney tissue cultures. Proc. Soc. Exp. Biol. Med.
92:667.
25. Eddy, B. E., G. S. Borman, W. H. Berkeley and R. D.
Young. 1961. Tumors induced in hamsters by
injection of rhesus monkey kidney cell extracts.
Proc. Soc. Exp. Biol. Med. 107:191.
26. Eddy, B. E., G. S. Borman, G. E. Grubbs and R. D.
Young. 1962. Identification of the oncogenic
substance in rhesus monkey kidney cell cultures as
SV40. Virology 17:65.
27. Girardi, A. J., B. H. Sweet, V. B. Slotnick and
M. R. Hilleman. 1962. Development of tumors in
hamsters inoculated in neonatal period with
vacuolating virus, SV40. Proc. Soc. Exp. Biol. Med.
109:649.
28. Deichman, G. I. and E. L. Prigozhina. 1962. Development of tumors in hamsters inoculated with preparations from monkey kidney tissue culture. (in
Russian) Vop. Virus 7:277.
29. McGrath, D. I., K. Russell and J. O'H. Tobin. 1961.
Vacuolating agent. Brit. Med. J. 29:287.
30. Gerber, P., G. A. Hottle and R. E. Grubbs. 1961.
Inactivation of vacuolating virus (SV40) by
formaldehyde. Proc. Soc. Exp. Biol. Med. 108:205.
31. Melnick, J. L. and S. Stinebaugh. 1962. Excretion
vacuolating SV40 virus (papova virus group) after
ingestion as a contaminant of oral poliovaccine.
Proc. Soc. Exp. Biol. Med. 109:965.
32. Gerber, P. 1967. Patterns of antibodies to SV40
in children following the last booster with
inactivated poliomyelitis vaccines. Proc. Soc. Exp.
Biol. Med. 125:1284.
33. Goffe, A. P., J. Hale and P. S. Gardner. 1961.
Poliomyelitis vaccines. Lancet 1:612.
34. Horvath, B. L. and F. Fornosi. 1964. Excretion of
SV40 virus after oral administration of contaminated
polio vaccine. Acta Microbiol. Acad. Sci. (Hungary)
11:271.
35. Morris, J. A., K. M. Johnson, C. G. Aulisio, R. M.
Chanock and V. Knight. 1961. Clinical and serologic
Responses in volunteers given vacuolating virus
(SV40) by respiratory route. Proc. Soc. Exp. Biol.
Med. 108:56.
36. Shein, H. M. and J. F. Enders. 1962. Multiplication
and cytopathogenicity of simian vacuolating virus
40 in cultures of human tissues. Proc. Soc. Exp.
Biol. Med. 109:495.
37. Shein, H. M. and J. F. Enders. 1962. Transformation
induced by simian virus 40 in human renal cell

cultures. I. Morphology and growth characteristics.
Proc. Nat. Acad. Sci. 48:1164.

38. Koprowski, H., J. A. Ponten, F. Jensen, R. G. Ravdin,
P. Moorhead and E. Saksela. 1962. Transformation
of cultures of human tissue infected with SV40.
J. Cell. Comp. Physiol. 59:281.

39. Jensen, F., H. Koprowski, J. S. Pagano, J. Ponten
and R. G. Ravdin. 1964. Autologous and homologous
implantation of human cells transformed in vitro by
SV40. J. Nat. Cancer Inst. 32:917.

40. Horvath, L. B. 1965. Incidence of SV40 virus
neutralizing antibodies in sera of laboratory
workers. Acta Microbiol. Acad. Sci. (Hungary)
12:201.

41. Horvath, L. B. 1972. SV40 neutralizing antibodies
in the sera of man and experimental animals. Acta
Virol. 16:141.

42. Shah, K. V., M. K. Goverdhan and H. L. Ozer. 1970.
Neutralizing-antibodies to SV40 in human sera from
south India: Search for additional hosts of SV40.
Amer. J. Epidemiol. 93:291.

43. Shah, K. V., H. L. Ozer, H. S. Pond, L. D. Palma and
G. P. Murphy. 1971. SV40 neutralizing antibodies
in the sera of U. S. residents without history of
polio immunization. Nature 231:448.

44. Shah, K. V., F. R. McCrumb, Jr., R. W. Daniel and
H. L. Ozer. 1972. Serologic evidence for a SV40-
like infection of man. J. Nat. Cancer Inst. 48:557.

45. Huebner, R. J., J. A. Bell, W. P. Rowe, T. G. Ward,
R. G. Suskind, J. W. Hartley and R. S. Paffenbarger,
Jr. 1955. Studies of adenoidal-pharyngeal-conjunct-
ival vaccines in volunteers. J. Amer. Med. Ass.
159:986.

46. Hilleman, M. R., R. A. Stallones, R. L. Gauld, M. S.
Warfield and S. A. Anderson. 1956. Prevention of
acute respiratory illness in recruits by adenovirus
(R1-APC-ARD) vaccine. Proc. Soc. Exp. Biol. Med.
92:377.

47. Fraumeni, J. F., Jr., F. Ederer and R. W. Miller.
1963. An evaluation of the carcinogenicity of SV40
in man. J. Amer. Med. Ass. 185:713.

48. Stewart, A. M. and D. Hewitt. 1965. Aetiology of
childhood leukemia. Lancet 11:789.

49. Innis, M. D. 1968. Oncogenesis and poliomyelitis
vaccine. Nature 219:972.

50. Fraumeni, J. F., Jr., C. R. Stark, E. Gold and M. L.
Lepow. 1970. SV40 in polio vaccine: Follow-up of
new born recipients. Science 167:59.

51. ZuRhein, G.M. and S.M. Chou. 1965. Particles
resembling papova viruses in human cerebral
demyelinating disease. Science 148:1477.

52. Weiner, L. P., R. M. Herndon, O. Narayan, R. T.
Johnson, K. Shah, L. J. Rubinstein, T. J. Preziosi

and F. K. Conley. 1972. Isolation of virus related
to SV40 from patients with progressive multifocal
leukoencephalopathy. New Eng. J. Med. 286:385.
53. Weiner, L. P., R. M. Herndon, O. Narayan and R. T.
Johnson. 1972. Further studies of a SV40-like
virus isolated from human brain. J. Virol. 10:147.
54. Penney, J. B., Jr., L. P. Weiner, R. M. Herndon,
O. Narayan and R. T. Johnson. 1972. Virions from
progressive multifocal leukoencephalopathy: Rapid
serological identification by electron microscopy.
Science 178:60.
55. Padgett, B. L., D. L. Walker, G. M. ZuRhein and R.
J. Eckroade and B. H. Dessel. 1971. Cultivation of
papova-like virus from human brain with progressive
multifocal leucoencephalopathy. Lancet 1:1257.
56. Horta-Barbosa, L., R. Hamilton, D. A. Fucillo, D.
Hogan, J. L. Sever and J. Gerin. 1972. Progressive
multifocal leukoencephalopathy. New Eng. J. Med.
286:1060.
57. Weiner, L. P. 1972. Progressive multifocal
leukoencephalopathy. New Eng. J. Med. 286:1060.
58. Gardner, S. D., A. M. Field, D. V. Coleman and B.
Hulme. 1971. New human papovavirus (B.K.) isolated
from urine after renal transplantation. Lancet
1:1253.
59. Johnson, R. T. 1970. Subacute sclerosing panencephalitis. J. Infect. Dis. 121:227.
60. Fields, B. N. 1972. Genetic manipulation of
reovirus-A model for modification of disease? New
Eng. J. Med. 287:1026.
61. Huebner, R. J., R. M. Chanock, B. A. Rubin and
M. J. Casey. 1964. Induction by adenovirus type
7 of tumors in hamsters having the antigenic
characteristics of SV40 virus. Proc. Nat. Acad. Sci.
52:1333.
62. Beardmore, W. B., M. J. Havlock, A. Serafini and
I. W. McLean, Jr. 1965. Interrelationship of
adenovirus (type 4) and papovavirus (SV40) in
monkey kidney cell cultures. J. Immunol. 95:422.
63. Morris, J. A., M. J. Casey, B. E. Eddy, W. T. Lane
and R. J. Huebner. 1966. Occurrence of SV40
neoplastic and antigenic information in vaccine
strains of adenovirus type 3. Proc. Soc. Exp. Biol.
Med. 122:679.
64. Rowe, W. P. and S. G. Baum. 1964. Evidence for a
possible genetic hybrid between adenovirus type 7
and SV40 viruses. Proc. Nat. Acad. Sci. 52:1340.
65. Rapp, F., J. L. Melnick, J. S. Butel and T.
Kitahara. 1964. The incorporation of SV40 genetic
material into adenovirus 7 as measured by intranuclear synthesis of SV40 tumor antigen. Proc. Nat.
Acad. Sci. 52:1348.
66. Easton, J. M. and C. W. Hiatt. 1965. Possible

incorporation of SV40 genome within capsid proteins of adenovirus 4. Proc. Nat. Acad. Sci. 54:1100.

67. Lewis, A. M., Jr., S. G. Baum, K. O. Prigge and W. P. Rowe. 1966. Occurrence of adenovirus-SV40 hybrids among monkey kidney cell adapted strains of adenovirus. Proc. Soc. Exp. Biol. Med. 122:214.

68. Lewis, A. M., Jr., K. O. Prigge and W. P. Rowe. 1966. Studies of adenovirus-SV40 hybrid viruses. IV. An adenovirus type 2 strain carrying the infectious SV40 genome. Proc. Nat. Acad. Sci. 55:526.

69. Rowe, W. P. and S. G. Baum. 1965. Studies of adenovirus-SV40 hybrid viruses. II. Defectiveness of the hybrid particles. J. Exp. Med. 122:955.

70. Boeye, A., J. L. Melnick and F. Rapp. 1966. SV40-adenovirus hybrids: Presence of two genotypes and the requirement of their complementation for viral replication. Virology 28:56.

71. Baum, S. A., P. R. Reich, C. J. Hybner, W. P. Rowe and S. M. Weissman. 1966. Biophysical evidence for linkage of adenovirus and SV40 DNAs in adenovirus 7-SV40 hybrid particles. Proc. Nat. Acad. Sci. 56:1509.

72. Crumpacker, C. S., M. J. Levin, W. H. Wiese, A. M. Lewis, Jr. and W. P. Rowe. 1970. Adenovirus type 2-SV40 hybrid population: Evidence for a hybrid DNA molecule and the absence of adenovirus en-capsidated circular SV40 DNA. J. Virol. 6:788.

73. Kelly, T. J., Jr. and J. A. Rose. 1971. SV40 integration site in an adenovirus 7-SV40 hybrid DNA molecule. Proc. Nat. Acad. Sci. 68:1037.

74. Levin, M. J., C. S. Crumpacker, A. M. Lewis, Jr., M. N. Oxman, P. H. Henry and W. P. Rowe. 1971. Studies of nondefective adenovirus 2-SV40 hybrid viruses. II. Relationship of adenovirus 2 DNA and SV40 DNA in the Ad2$^+$ND$_1$ genome. J. Virol. 7:343.

75. Lewis, A. M., Jr. and W. P. Rowe. 1971. Studies on nondefective adenovirus-SV40 hybrid viruses. I. A newly characterized SV40 antigen induced by the Ad2$^+$ND$_1$ virus. J. Virol. 7:189.

76. Rapp, F., S. S. Tevethia and J. L. Melnick. 1966. Papovavirus SV40 transplantation immunity conferred by an adenovirus-SV40 hybrid. J. Nat. Cancer Inst. 36:703.

77. Lewis, A. M., Jr. and W. P. Rowe. 1970. Isolation of two plaque variants from the adenovirus type 2-SV40 hybrid population which differ in their efficiency in yielding SV40. J. Virol. 5:413.

78. Butel, J. S., F. Rapp, J. L. Melnick and B. A. Rubin. 1966. Replication of adenovirus type 7 in monkey cells: A new determinant and its transfer to adenovirus type 2. Science 154:671.

79. Rapp, F., J. S. Butel, S. S. Tevethia and J. L.

Melnick.   1967.   Comparison of ability of defective foreign genomes (PARA and MAC) carried by human adenoviruses to induce SV40 transplantation immunity. J. Immunol. $\underline{99}$:386.
80. Burlingham, $\overline{B}$. T. and W. Doerfler.   1971.   Three size classes of intracellular adenovirus DNA. J. Virol. $\underline{7}$:707.

ENDOGENOUS TYPE C VIRUSES IN CELL CULTURES

George J. Todaro
Viral Leukemia and Lymphoma Branch
National Cancer Institute
Bethesda, Maryland

There are basically two classes of cells that grow in tissue culture. First there are the diploid cells taken directly from an animal or person and put into tissue culture. These cells, called cell strains, have a finite lifetime in tissue culture (1), after which they either die out or spontaneously transform and become permanent cell lines (2). Figure 1 shows the pattern of growth of one such cell line (3T6) that originated from random-bred Swiss mouse embryo cells (2). Initially mouse embryo cells grew rapidly in cell culture, with a doubling time of 16-18 hours, but after several subcultures the average doubling time of the population lengthened and for a prolonged period there was virtually no net growth during a 3-day transfer interval. However as shown in Fig. 1, cell lines having properties that were quite different from the original cells eventually emerged. They were generally found to have lost contact inhibition of cell division and were aneuploid, whereas the original cells were diploid and were more subject to growth regulation.

Cell lines are often tumorigenic and have acquired the capacity for unlimited cell replication, so that in many ways they more resemble microorganisms than differentiated vertebrate cells (3). They also have acquired a certain metabolic autonomy, for they can grow as single cells in a very large volume of medium. These properties of continuous cell lines have made them

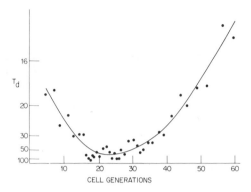

FIGURE 1.   Development of mouse embryo cell line 3T6.
Growth rate of mouse embryo fibroblasts on
successive 3-day transfers at a constant inoculation
density (6 x 10^5 cells/plate).   Each point represents
doubling time (T_d) during a 3-day transfer interval.
Reproduced with permission from Todaro and Green (2).

especially valuable for metabolic and genetic studies
of vertebrate cells because they can be handled in much
the same manner as bacterial cells.   However another
acquired property, which is most relevant to this
conference, is that many established cell lines also
begin to produce type C RNA-containing tumor viruses.
The viruses emerge spontaneously from previously virus-
free mouse cell lines (4,5) or can be induced by chemical
agents, especially 5-bromodeoxyuridine and 5-iododeoxy-
uridine (6-8).   Certain cell lines can be maintained
for many years and through several hundreds of cell
generations without expressing this property, and then,
either spontaneously or as a result of chemical induction,
begin to produce a complete, infectious type C virus
(7,9).

DETECTION OF TYPE C VIRUSES IN CELL CULTURE

The virus can be detected in a variety of ways.   By
electron microscopy typical type C virus can be
visualized.   There is some difficulty, however, in
sampling large enough areas; consequently this method is
not particularly sensitive for detecting low levels of
virus production.   In addition there is some difficulty
in distinguishing type C viruses from closely related
viruses.   Type C viruses can also be recognized by their
biological activity--the ability to form foci or to
transmit sarcoma virus from a nonproducer transformed
sarcoma cell to another cell.   To demonstrate biological
activity requires a cell susceptible to infection by the

virus being tested. As described later the cell from
which the virus emerges is resistant, as a general rule,
to exogenous infection by that same virus.

All mammalian type C viruses have a common interspecies group-specific antigen (gs) that can be detected
by complement fixation tests or, with much greater
sensitivity, by radioimmunoassay (10-12). Similarly all
mammalian type C viruses have an RNA-directed DNA
polymerase (reverse transcriptase) that has interspecies
properties (13). Antiserum made to the reverse transcriptase of one mammalian type C virus efficiently inhibits
the enzymatic activity of the homologous virus polymerase,
but will also neutralize other mammalian type C virus
polymerases without appreciably affecting the normal
cellular DNA polymerases (14). The reverse transcriptase
assay, especially when using synthetic templates, is a
very sensitive assay for detecting low levels of virus
released from cell cultures into the supernatant fluid.
Another approach is to search for viral-specific genetic
information either in the cells themselves (15,16) or
in particles that have properties similar to those of
known viruses (17). The molecular hybridization method
has been extensively exploited because it permits the
detection of partial expression of viral genetic
information. It allows the potential recognition of a
class of cells that have some expression of viral-
specific genetic information in them, but not enough to
make a complete virus (18,12).

NEWLY ISOLATED TYPE C VIRUSES

The RD cell line originally described by McAllister
et al. (19) is a human rhabdomyosarcoma cell line that
propagates readily in culture. When these tumor cells
are inoculated into fetal cats in utero, tumors develop.
From one of these, a type C virus called RD-114 was
isolated (20) having a gs antigen and a reverse transcriptase that differed immunologically from that of the
previous type C viruses isolated from cats (21,22). The
suggestion was therefore made that this virus may have
been derived from the human RD cells or from some
recombinational event between genetic information in the
RD cells and the cat host in which the cells grew.
Recently methods have become available, using immunosuppressed animals treated with antithymocyte serum,
making it possible for human tumor cells to grow in adult
animals of heterologous species (23). The NIH Swiss
mouse was chosen because, although it has gs antigen
and viral information expressed (24,12), it is a strain
with relatively low incidence of leukemia and has never
been reported to produce complete, infectious type C

virus.

The RD cells proved to be transplantable into NIH
Swiss mice treated with antithymocyte (AT) serum (23).
The tumors that developed in these animals could then
either be re-explanted into cell culture or directly
passaged to additional mice. From the third transplant
generation, a cell line called AT-124 was established
that was found to contain a type C virus. This virus has
a host range like that of the RD-114 virus; i.e., it
grows readily in various human and primate cells, but
grows poorly, if at all, in murine cells. However, the
gs antigen and the reverse transcriptase of this virus
are antigenically closely related to the mouse type C
viruses (23). There is a possibility, then, that this
virus may be a recombinant with both mouse and human
type C genetic information. The more likely alternative,
however, is that it represents the isolation of the
endogenous NIH Swiss type C virus and that the RD cell
serves as a permissive host for the replication of the
endogenous NIH Swiss type C virus.

There are naturally occurring mouse and cat type C
viruses that replicate much more efficiently in human
cells than in either mouse or cat cells. In the cat, new
evidence suggests that there are two distinctly different
type C viruses. One of them is clearly an endogenous
virus because it can be induced from virus-negative,
single cell clones (25). The other virus, whose mode of
transmission is not entirely clear but does involve
horizontal spread, is the most commonly found type C
virus in cats and appears to be the etiologic agent in
the production of experimental tumors (26-29). The cat,
then, offers a unique situation because there are two
type C viruses: one is endogenous in the species and
seldom expressed, and the other is quite common in the
species and easily transmitted from animal to animal.
Which, if either, of these is responsible for the
generality of cancer in the species remains to be
determined. However since they are so distinctly different in their antigenic properties and in their nucleic
acids, this becomes a system which might allow us to
distinguish between the role of endogenous virus and
exogenously added viruses in the production of tumors.

ENDOGENOUS MOUSE TYPE C VIRUSES. Balb/c embryo cells
give rise to continuous cell lines. These cell lines
spontaneously transform and acquire the ability to
produce tumors (30). Some of the lines that are spontaneously transformed also begin producing type C viruses,
and the spontaneous production of type C viruses can be
demonstrated from single cell clones that have

spontaneously transformed (9,31). Once a subclone
spontaneously starts to make virus, it continues to
produce that virus indefinitely (Fig. 2) (9). There
must be a mechanism that allows control of the endogenous
virus information, and this mechanism can be disturbed
in such a way that continued production of the virus
results in cells that nevertheless are able to
proliferate. The transformed cells are recognized by
their loss of contact inhibition, their ability to grow
over one another in a random fashion to form multiple
cell layers (Fig. 3), and their ability to produce
tumors upon inoculation into a susceptible host.

While certain of the transformed cells spontaneously

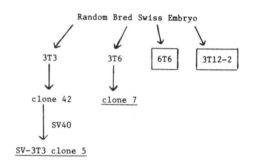

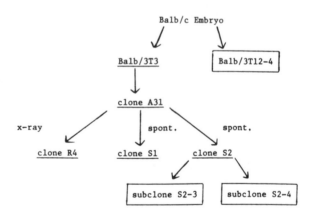

FIGURE 2.   Cell lines with a box around them represent
           cultures that have spontaneously released
type C virus; those underlined have endogenous type C
virus detectable by induction with IdU (30 µg/ml for
24 hr). Subclones S1-6, S2-3 and S2-4 have been through
more than 300 cell generations and more than 2 years in
culture before spontaneously releasing virus. Reproduced
with permission from Todaro (9).

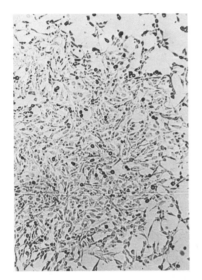

FIGURE 3.   Left, a colony of Balb/c 3T3 clone A31 fixed
            10 days after inoculating a single cell onto
a petri dish.   Right, a colony of S2, a spontaneously
transformed subclone derived from A31.   The individual
cells are epithelioid and readily grow over one another.
The transformed colony is considerably denser and more
compact.   Reproduced with permission from Todaro (9).

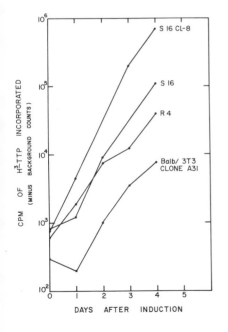

FIGURE 4.   BrdU induction of
            Balb/3T3 and its
transformed derivatives R4,
S16 and S16 Cl-8:   33 µg/ml of
BrdU was applied to just con-
fluent cultures of the cells
in 250-ml plastic tissue
culture flasks for 24 hr.   The
15ml of medium was changed
daily and assayed for reverse
transcriptase activity.   Reproduced with permission from
Lieber and Todaro (31).

produce virus, others do not. However, of the ones that
do not, there is a class which is superinducible for
virus production; these cells will respond much more
rapidly and yield much higher titers of virus production
when treated with inducers such as 5-bromodeoxyuridine
(Fig. 4). Virus can be detected within the first 8 hours
after the addition of the inducer and continues a
logrithmic increase for the first 2 days (Lieber,
Livingston and Todaro, in prep.), after which time it
plateaus and then either slowly declines or continues
to secrete virus indefinitely. With the untransformed
cells, virus production is slower, later and shuts off
more rapidly (9,31,32). With the diploid cell strains,
production of virus is even less on a per cell basis and
lasts for a shorter time. So it appears that the
thymidine analogs can disturb the normal control mechanism, and the effect of this perturbation is most
extreme in the most transformed cells. The diploid
cells recover quickly; the aneuploid but nontumorigenic
cells respond more rapidly but turn off virus production.
The transformed cells respond most rapidly and may or
may not then turn off virus production. Clearly, the
genetic information for production of a complete type C
virus is present in a form that allows it to be rapidly
called for when the proper inducers are added. There are
as yet no examples where virus has been induced by
chemical agents where it has not also been found to
appear spontaneously in the same cell line; the thymidine
analogs increase the probability of an event (secretion
of type C virus) that, nevertheless, occurs spontaneously.

ENDOGENOUS CAT TYPE C VIRUS. The simple postulate that
each species has only one type C virus associated with it
has become somewhat less certain with the finding that a
continuous line of cat cells, CCC (33), produces a
virus with properties very much like the RD-114 virus.
The CCC virus replicates much better in human cells than
it does in cat cells (Table 1). To prove that there was
an endogenous virus with properties like RD-114 in cat
cells, it was necessary to take cat cell clones, select
those clones which were virus negative, and show that
from them one could induce a type C virus. As in the
previously described virus systems (7), the virus appears
and reaches a maximum 2-5 days after treatment with the
inducer and then tends to disappear again. However if
the treated cat CCC cell clones are cocultivated with
primate or human cells, the virus then replicates in the
primate or human cells and grows to high titers (25).
Thus it is possible to obtain from cloned lines of virus-
negative cat cells a virus which has the antigen and the
polymerase properties of the RD-114 virus and has host
range properties quite distinct from those of the most

Table 1. Host Range Properties of CCC and RD-114 Viruses

Cell	Origin	Virus Titer (pmoles [^{3}H]TMP incorporated)		
		CCC	RD-114	FeLV
DBS-1	Rhesus monkey	10.3	16.7	0.20
A204	Human	3.0	29.0	0.58
M413	Human	5.2	24.5	0.13
FEF	Cat	0.08	0.75	35.2
CCC-3A	Cat	0.05	0.12	15.6
NRK	Rat	<0.03	<0.03	<0.03
Balb/c 3T3	Mouse	<0.03	<0.03	<0.03
NIH/3T3	Mouse	<0.03	<0.03	<0.03

In each case 15ml of supernatant were concentrated 70-fold
and tested for reverse polymerase activity. From
Livingston and Todaro (25).

commonly studied cat type C viruses.

Several other cat cell strains have been examined.
None so far have given evidence of induction from virus-
free clones of the typical cat leukemia virus. However,
another cell strain, FFc 2K from normal feline embryonic
kidney, has been found that produces a type C virus with
host range and immunologic properties like the RD-114
and the CCC viruses (Todaro and Livingston, unpubl.).
This suggests that the CCC/RD-114 viruses are the
endogenous viruses of the cat and that the feline
leukemia virus group commonly studied might not be
endogenous (i.e., part of the genetic makeup of somatic
cells) in the cat.

GENERAL PROPERTIES OF ENDOGENOUS VIRUSES

There are now at least five species (chicken, Chinese
hamster, mouse, rat, cat) where it can clearly be shown
that a complete copy of the endogenous type C virus, the
virogene (32,34), is present in the normal cells. Virus
has been shown in normal chick embryo cells that is able
to replicate in Japanese quail cells but not in chicken
cells (35,36). Continuous lines of Chinese hamster,
such as CHO (37), various mouse cells, the rat cell line
NRK (38), and the cat cell line described above all have
single cell clones with the potential to respond rapidly
to BrdU and in so doing produce a complete type C virus
(5). In the avian system, one permissive host for the
endogenous type C viruses is the Japanese quail. In the
mouse system, one permissive host for the endogenous

virus of Balb/c mice is the NIH Swiss strain; the newly
isolated AT-124 virus, a presumed endogenous virus of NIH
Swiss mice, has primate cells as a permissive host; for
the virus from clones of cat lines, CCC, and from normal
cat kidney cells, primate cells (including human) are
again the permissive hosts. The endogenous Chinese
hamster virus and the endogenous rat viruses (Table 2)
so far have not been transmissible to cells of other
species (5). In addition, they are also not infectious
for their own cells. One essential attribute, then, of
endogenous viruses would appear to be an inability to
reinfect the cell lines from which they emerge. This
would suggest that it may well be easier to isolate
endogenous virus of a species using the cells of an
unrelated species as the permissive host. There now
exist several cell lines that secrete endogenous virus
but are not reinfected by it. These represent the only
examples where study of a pure endogenous virus is
possible because with the infectious viruses that are
transmitted horizontally from cell to cell, there is the
very real risk that the virus produced will be a mixture,
either phenotypic or genetic, of the infecting virus and
the endogenous virus of the cell in which the infecting
virus is grown. The more a virus is passed from cell to
cell, the more likely that the virus produced is a
mixture of several genetically different viruses. Even
within a single cell line and from a single cell clone,
there is immunologic evidence that more than one antigenically distinct type C virus can be spontaneously
induced (39).

POSSIBLE BIOLOGIC ROLES OF VIROGENE ACTIVATION. Properties of endogenous virus include their inability to
infect cells that produce them, their presence in many,
perhaps all, species and their greater likelihood to be
produced by transformed rather than by normal cells. The
transformed cells that produce virus appear to be less
tumorigenic than the transformed cells that do not
produce virus (40,41). This, coupled with the fact that
transformed cells are much more likely to release
endogenous virus, suggests that such a system might have
a protective role; virus expression might protect on an
immunologic basis against tumor formation. A system
that would turn on virogene information when the oncogenic
information is already expressed might be of great benefit
to the host. This leads to the possibility that the
endogenous virus under natural circumstances may serve
more to protect against cancer than act as an etiologic
agent. The production of the new virus would result in
a variety of antigenic changes in the cells that in an
immunologically competent animal might facilitate the
rejection of those cells. One of the important biological

Table 2. Some Rat and Chinese Hamster Cell Lines Producing Type C Virus in Long Term Culture

Species	Cell Line	Origin	Special Uses
Rat	NRK	Normal rat kidney	Transformation assays
	LLC-WRC256	Walker rat carcinoma	Tumorigenicity studies
	R2C	Leydig cell testicular tumor	Steroid secreting
	C₆ glial cell	Glial cell tumor	S-100 protein secreting
	RR1022	Schmidt-Ruppin virus induced sarcoma	Rescue of avian sarcoma virus in chickens
Chinese hamster	CHO-K1	Chinese hamster ovary	Nutritional mutants, regulation of gene expression, cell hybridization
	B14-150	Normal peritoneal cells, thymidine kinase resistant	Cell hybridization

All of these cell lines are available from the American Type Culture Collection, Registry of Animal Cell Lines, Rockville, Maryland.

mysteries is the fact that the system which allows a
complete type C virus to be rapidly called forth is
present in so many species and is preserved through
hundreds of generations in cell culture. For such a
system to have been preserved in the evolution of the
species, it must have had, on balance, a selective
advantage for the host. One possibility would be that
virus production serves as an early warning system to
the animal that there are cells in the body that have
already become transformed. However this would not
explain why the system is preserved for hundreds of
cell generations in culture. Over 100 independently
isolated subclones of Balb/3T3 clone A31 have now been
tested, and every one of them has inducible type C
virus. Prior to chemical induction or spontaneous
activation there is partial, but not complete, expression
of the viral type C virus-specific information (12,
Benveniste et al., in prep.).

RELATIONSHIP TO INDUCTION OF DIFFERENTIATED FUNCTIONS.
The system involving the activation of the type C virus
has obvious superficial similarities to the lysogenic
system in bacteria. However in many ways virogene
induction might be considered more analogous to the
switching on of a differentiated function by vertebrate
cells (42). Either spontaneously or after addition of
a small molecule, the cell begins producing proteins
whose genes are normally repressed and assemble a
rather complex package for export from the cell. BrdU
has been known to greatly affect the differentiated
state in culture and to both increase and decrease the
rate of only partial expression of viral genetic information. Upon induction, however, new viral RNA rapidly
appears. Clearly, then, one major control in this system
is at the level of transcription. Whether there will be
additional controls at the level of translation remains
to be resolved. The induction of a lysogenic prophage
involves the excision of that genetic information from
the bacterial chromosome and also the lysis of the cell.
In the induction of type C virus, the cells produce virus
but do not die; whether this, too, involves an excision
mechanism is not clear. It is possible that the system
works entirely by reading off cellular genes. So in a
sense the cell lines that produce endogenous virus
would not be replicating the virus, but would be tran-
scribing and translating information that is part of
their natural genetic makeup.

INTERACTION OF ENDOGENOUS AND EXOGENOUS TYPE C VIRUSES

The results described above suggest the possibility

that the feline leukemia and sarcoma viruses might be
derived from other species. One line of evidence is the
fact that these viruses grow so readily in cat cells and
can be isolated from many tissues of most cats. Such a
virus that spreads readily through the population,
producing a high level of disease, represents an apparently unnatural situation among mammalian species. The
best parallel seems to be in certain inbred mouse strains,
such as AKR and C58, where virus is readily present even
in embryos and continues to be present throughout the
lifetime of the animal. In this case the evidence is
clear that the endogenous virus of the AKR is able to
replicate in AKR cells (43); this may be one major
component in their unusually high susceptibility to
leukemia. It would clearly be of evolutionary benefit
for a species to have developed a mechanism that makes
it resistant to its own endogenous virus. However this
system would not protect it against a type C virus from
a different species. The inability to induce the typical
feline leukemia virus (FeLV) from cat cell clones also
argues against FeLV being an endogenous virus. Another
line of evidence that would support the postulate that
the FeLV type C viruses have been derived from rodents
is the close relationship by polymerase inhibition
studies between the FeLV and the mouse viral polymerases
(13). The mouse viral polymerase antibodies will not
only neutralize the hamster and the rat type C polymerases, but will also neutralize the FeLV polymerase
about as efficiently as the other rodent viruses. On
an evolutionary basis one would not expect this to be
the case since cats are genetically very different from
rodents. However cats and rodents live in close
proximity with one another; in fact, cats subsist
primarily on rodents. This raises the intriguing
possibility that the cat type C virus that is spreading
in domestic cats as an epidemic is derived from a virus
that originated from some rodent, perhaps a rat, at some
recent time, in the last 100-200 years, as the domesti-
cation of cats and of rodents has become intensified. A
rodent virus that would replicate in cat cells could then
spread horizontally through the cat population, and what
we might be witnessing at this moment in time is a
situation where an endogenous virus from one species
can cause extensive disease in another. The alternative
is that there are two completely separate type C viruses
in the cat. These points remain to be resolved; the
antigenic relationship and the relationship by nucleic
acid hybridization between these two very different
viruses and normal cells may help to resolve the cellular
origin of the feline leukemia viruses.

Table 3.   Human Somatic Cell Hybrids with High-titered
           Type C Viruses

Other Species	Hybrid
Hamster	CHO x any human cell
Mouse	L-cell x any human cell
Mouse	RAG x any human cell

TYPE C VIRUS IN SOMATIC CELL HYBRIDS

One final point is that many of the cell lines that
are used to make somatic cell hybrids contain high titers
of type C leukemia viruses (Table 3), and one of the
methods most extensively used involves making thymidine
kinase negative cells by chronic exposure to BrdU.
This procedure can be shown to induce virus from cell
lines of several species that have endogenous type C
virus. Such widely used cell lines as Chinese hamster
ovary (CHO), mouse L and mouse RAG cells secrete high
titers of type C virus. Each of these has been used to
make somatic cell hybrids (37,44) with human cells (45),
and in the cases that we have so far tested, all of the
mouse/human hybrid cells continue to produce type C
virus (Todaro and Ruddle, unpubl.). Because viruses can
alter their host range either by adaptation or selection,
these human hybrid cells would appear to constitute a
potential biohazard since, in this situation, one has an
endogenous virus of a species being produced by cells
which, at least in part, are human. These hybrid cells
are being extensively explored by geneticists all over
the world who do not realize that they contain high
titers of potentially oncogenic viruses.

SUMMARY

That type C viruses are part of our heritage is now
abundantly clear. That type C viral genetic information
(virogenes) is present in somatic cells of a great many
species in a repressed form but having the potential to
become expressed also is clear. What is not clear is
the nature of the relationship between the acquisition
of oncogenic potential by a cell and the expression of
that cell's endogenous type C viral information. Type C
viruses carry oncogenic information and can produce
tumors (leukemias, lymphomas and sarcomas) by exogenous
infection; whether horizontal spread (cell to cell and/or

animal to animal) of exogenous type C virus is responsible for a significant proportion of naturally occurring cancers in vertebrates is uncertain; that they can have oncogenic potential and can produce tumors in a variety of species is firmly established. It follows, then, that these viruses and the cells that produce them must be treated as potentially hazardous agents.

References

1. Hayflick, L. and P. S. Moorhead. 1961. The serial cultivation of human diploid cell strains. Exp. Cell Res. 25:585.
2. Todaro, G. J. and H. Green. 1963. Quantitative studies of the growth of mouse embryo cells in culture and their development into established lines. J. Cell Biol. 17:299.
3. Green, H. and G. J. Todaro. 1967. The mammalian cell as a differentiated microorganism. Annu. Rev. Micro. 21:573.
4. Aaronson, S. A., J. W. Hartley and G. J. Todaro. 1969. Mouse leukemia virus: "Spontaneous" release from mouse embryo cells after long term in vitro cultivation. Proc. Nat. Acad. Sci. 64:87.
5. Todaro, G. J. 1972. Detection and characterization of RNA tumor viruses in normal and transformed cells. Perspectives in virology, p. 81. Academic Press, New York.
6. Lowy, D. R., W. P. Rowe, N. Teich et al. 1971. Murine leukemia virus: High frequency activation in vitro by 5-iododeoxyuridine and 5-bromodeoxyuridine. Science 174:155.
7. Aaronson, S. A., G. J. Todaro and E. M. Scolnick. 1971. Induction of murine C-type viruses from clonal lines of virus-free Balb/3T3 cells. Science 174:157.
8. Klement, V., M. O. Nicolson and R. J. Huebner. 1971. Rescue of the genome of focus forming virus from rat non-productive lines by 5-bromodeoxyuridine. Nature New Biol. 234:12.
9. Todaro, G. J. 1972. "Spontaneous" release of type C viruses from clonal lines of "spontaneously" transformed Balb/3T3 cells. Nature 240:157.
10. Parks, W. P. and E. M. Scolnick. 1972. Radioimmunoassay of mammalian type C viral proteins. II. Interspecies antigenic reactivities of the major internal polypeptides. Proc. Nat. Acad. Sci. 69:1776.
11. Oroszlan, S., M. H. M. White, R. V. Gilden et al. 1972. A rapid direct radioimmunoassay for type C virus group-specific antigen and antibody. Virology 50:294.
12. Parks, W. P., D. M. Livingston, G. J. Todaro et al.

1973. Radioimmunoassay of mammalian type C viral proteins. III. Detection of viral antigen in normal murine cells. J. Exp. Med. (in press).

13. Scolnick, E. M., W. P. Parks and G. J. Todaro. 1972. The reverse transcriptase of primate viruses as immunological markers. Science 117:1119.

14. Ross, J., E. M. Scolnick, G. J. Todaro et al. 1971. Separation of murine cellular and murine leukemia virus DNA polymerases. Nature New Biol. 231:163.

15. Baluda, M. A. and D. P. Nayak. 1970. DNA complementary to viral RNA in leukemic cells induced by avian myeloblastosis virus. Proc. Nat. Acad. Sci. 66:329.

16. Varmus, H. E., R. A. Weiss, R. R. Friis et al. 1972. Detection of avian tumor virus specific nucleotide sequences in avian cell DNA's. Proc. Nat. Acad. Sci. 69:20.

17. Baxt, W. G. and S. Spiegelman. 1972. Nuclear DNA sequences present in human leukemic cells and absent in normal leukocytes. Proc. Nat. Acad. Sci. 69:3737.

18. Tsuchida, N., M. S. Robin and M. Green. 1971. Viral RNA subunits in cells transformed by RNA tumor viruses. Science 176:1418.

19. McAllister, R. M., J. Melnyk, J. Z. Finkelstein et al. 1969. Cultivation in vitro of cells derived from a human rhabdomyosarcoma. Cancer 24:520.

20. McAllister, R. M., M. Nicolson, M. B. Gardner et al. 1972. C-type virus released from cultured human rhabdomyosarcoma cells. Nature New Biol. 235:3.

21. Oroszlan, S., D. Bova, M. H. M. White et al. 1972. Purification and immunological characterization of the major internal protein of the RD-114 virus. Proc. Nat. Acad. Sci. 69:1211.

22. Scolnick, E. M., W. P. Parks, G. J. Todaro et al. 1972. Immunological characterization of primate C-type virus reverse transcriptase. Nature New Biol. 235:35.

23. Todaro, G. J., P. Arnstein, W. P. Parks et al. 1973. A type C virus in human rhabdomyosarcoma cells after inoculation into antithymocyte serum-treated NIH Swiss mice. Proc. Nat. Acad. Sci. (in press).

24. Huebner, R. J., G. J. Kelloff, P. S. Sarma et al. 1970. Group-specific antigen expression during embryogenesis of the genome of the C-type RNA tumor virus: Implications for ontogenesis and oncogenesis. Proc. Nat. Acad. Sci. 67:366.

25. Livingston, D. M. and G. J. Todaro. 1973. Endogenous type C virus from a cat cell clone with properties distinct from previously described feline type C viruses. Virology (in press).

26. Theilen, G. H., T. G. Kawakami, D. L. Dungworth et al. 1968. Current status of transmissible agents in feline leukemia. J. Amer. Vet. Med. Ass. 153:1864.

27. Rickard, D. G., J. E. Post, F. Noronha et al. 1969.

A transmissible virus induced lymphocytic leukemia
of the cat. J. Nat. Cancer Inst. 42:987.

28. Hardy, W. D., G. Geering, L. J. Old et al. 1969.
    Feline leukemia virus: Occurrence of viral antigen
    in the tissues of cats with lymphosarcoma and other
    diseases. Science 166:1019.

29. Gardner, M. B., P. Arnstein, R. W. Rongey et al.
    1970. Experimental transmission of feline fibro-
    sarcoma to cats and dogs. Nature 226:807.

30. Aaronson, S. A. and G. J. Todaro. 1968. Basis for
    the acquisition of malignant potential by mouse
    cells cultivated in vitro. Science 162:1024.

31. Lieber, M. M. and G. J. Todaro. 1973. Spontaneous
    and induced production of endogeneous type C RNA
    virus from a clonal line of spontaneously transformed
    Balb/3T3. Int. J. Cancer (in press).

32. Todaro, G. J. and R. J. Huebner. 1972. The viral
    oncogene hypothesis: New evidence. Proc. Nat. Acad.
    Sci. 69:1009.

33. Lee, L. M., S. Nomura, R. H. Bassin et al. 1972.
    Use of an established cat cell line for investigation
    and quantitation of feline oncornaviruses. J. Nat.
    Cancer Inst. 49:55.

34. Huebner, R. J. and G. J. Todaro. 1969. Oncogenes of
    RNA tumor viruses as determinants of cancer. Proc.
    Nat. Acad. Sci. 64:1087.

35. Hanafusa, T., H. Hanafusa and T. Miyomoto. 1970.
    Recovery of a new virus from apparently normal chick
    cells by infection with avian tumor viruses. Proc.
    Nat. Acad. Sci. 67:1797.

36. Weiss, R. A., R. R. Friis, E. Katz et al. 1971.
    Induction of avian tumor viruses in normal cells by
    physical and chemical carcinogens. Virology 46:920.

37. Kao, F. and T. Puck. 1970. Genetics of somatic
    mammalian cells: Linkage studies with human-Chinese
    hamster cell hybrids. Nature 228:329.

38. Duc-Nugyen, H., E. N. Rosenblum and R. F. Zeigel.
    1966. Persistent infection of a rat kidney cell
    line with Rauscher murine leukemia virus. J.
    Bacteriol. 92:1133.

39. Aoki, T. and G. J. Todaro. 1973. Antigenic proper-
    ties of endogenous type C viruses from spontaneously
    transformed celones of Balb/3T3. Proc. Nat. Acad.
    Sci. (in press).

40. Barbieri, D., J. Belehradek, Jr. and G. Barski. 1971.
    Decrease in tumor-producing capacity of mouse cells
    following infection with mouse leukemia virus. Int.
    J. Cancer 7:364.

41. Stephenson, J. R. and S. A. Aaronson. 1972. Genetic
    factors influencing C-type RNA virus induction. J.
    Exp. Med. 136:175.

42. Stellwagen, R. H. and G. M. Tomkins. 1971.
    Differentiated effect of 5-bromodeoxyuridine on the

concentration of specific enzymes in hepatoma cells in culture. Proc. Nat. Acad. Sci. 68:1147.

43. Rowe, W. P., J. W. Hartley, M. R. Lander et al. 1971. Noninfectious AKR mouse embryo cell lines in which each cell has the capacity to be activated to produce infectious murine leukemia virus. Virology 46:866.

44. Ruddle, F., V. Chapman, F. Riccinit et al. 1971. Linkage relationships of seventeen human gene foci as determined by man-mouse somatic hybrids. Nature New Biol. 232:69.

45. Sell, E. K. and R. S. Krooth. 1972. Tabulation of somatic cell hybrids formed between lines of cultured cells. J. Cell. Physiol. 80:453.

* * * * *

DISCUSSION

OXMAN: Can you, Dr. Todaro, comment briefly on the relative sensitivity of the various methods used to detect latent virus in cells?

TODARO: If you know the permissive host for the virus you are looking for, the biological assay is the most sensitive, because one particle is all you need. Electron microscopy, for example, is less sensitive than the supernatant polymerase assay which, in turn, is less sensitive than hybridization. For our purpose, the reverse transcriptase assay provides rapid, definitive results. However, the determination as to which technique to use depends upon the question being asked, and, of course, the laboratory techniques that are available.

PANEL II
EVALUATION OF THE RESULTS
OF PAST AND FUTURE INTRODUCTIONS
OF SUCH AGENTS TO MAN

Maurice R. Hilleman
Merck Institute for Therapeutic Research

Andrew M. Lewis, Jr.
National Institute of Allergy and Infectious Diseases

Robert W. Miller
National Cancer Institute

George J. Todaro
National Cancer Institute

HUMAN PAPOVAVIRUSES

Kenneth K. Takemoto & Michael F. Mullarkey

The isolation in cell culture of viruses belonging to the papovavirus group has recently been reported from three different laboratories (Table 1). Three of the isolates were from brain in a rare human disease, progressive multifocal leucoencephalopathy (PML), while the fourth was from the urine of a patient who had undergone a renal transplant.

THE VIRUSES.

DAR and EK viruses. Two viruses (DAR and EK) were isolated from cases of PML. They were both indistinguishable from SV40 by neutralization and fluorescent-antibody tests. We have confirmed these results. In addition the DNA of DAR virus could not be distinguished from that of SV40 by hybridization techniques.

JC virus. A papovavirus (JC) was isolated, in human fetal brain cell cultures, from another case of PML. Unlike DAR and EK, JC virus agglutinates human type O and guinea pig erythrocytes. A serologic relationship was not found between JC virus and human wart, SV40 or polyoma viruses.

BK virus. A papovavirus (BK) was isolated from the urine of a patient who had undergone renal transplantation.

Table 1.   Human Papovaviruses

Virus	Isolated From	Cells for Isolation	Investigators
DAR & EK	Human Brain*	AGMK	Weiner et al. (Maryland)
JC	Human Brain*	Human Fetal Brain	Padgett et al. (Wisconsin)
BK	Urine†	Vero	Gardner et al. (England)

* PML cases
† Patient with kidney transplant

This virus was isolated in Vero cells, a continuous line of African green monkey (AGMK) cells. BK virus also agglutinates human type O and guinea pig erythrocytes.

HOST RANGE (SEE TABLE 2). Like SV40, DAR virus grows well in primary, as well as continuous, AGMK cells and poorly in human fibroblasts (WI38 strain). It replicates in fetal brain cells, as does SV40. BK virus grows with CPE in Vero, human fetal brain and WI38 cells, but not in primary AGMK or rhesus monkey kidney cells. To date JC virus has only been reported to grow in human fetal brain cells.

Table 2.   Host Range of SV40 and SV40-like Human Papovaviruses

Virus	Cells			
	AGMK		Human	
	Primary	Continuous	Fibroblast	Fetal Brain
SV40	+	+	±	NT*
DAR	+	+	±	+
BK	−	+	+	+
JC	−	−	−	+

* NT = Not Tested

Table 3.   Summary of Studies on the Antigenic Relationships Between SV40 and Human Papovaviruses

Test	Antibody Against			
	DAR	BK	JC[a]	SV40
Neutralization of SV40 (Plaque Reduction)	+	+	+	+
Hemagglutination-Inhib. of BK virus	-	+	-	-
Fluorescent Antibody (SV40 infected cells)	+	±	±	+
Fluorescent Antibody (BK infected cells)	+	+	NT*	+

* NT = Not tested
a Antiserum to JC was provided by Dr. B. Padgett

ANTIGENIC RELATIONSHIP TO SV40 (SEE TABLE 3).   Rabbit
antisera to DAR, BK, and JC viruses all show some
neutralizing activity for SV40 in plaque reduction tests.
Serologic cross-reactions with SV40 were also noted in
fluorescent antibody tests, using SV40 and BK-infected
cells.

Because these viruses appear to share antigens with
SV40, the question arises as to their origin, e.g.,
whether they were introduced to man by SV40-contaminated
polio vaccines.   The following studies were done to
obtain information concerning this question.

Fifteen newly isolated "wild" strains of SV40
(provided by Drs. Kendall Smith and Bernice Eddy) were
tested for their ability to agglutinate human type O
erythrocytes.   None possessed hemagglutinating activity.
Thus JC and BK viruses, which do agglutinate human type
O erythrocytes, differ from any known SV40 strain in
this respect.   Polyoma, rabbit kidney vacuolating virus
(RKV) and the K mouse pneumonitis virus are other
hemagglutinating members of the papovavirus group.   Antisera against these three agents did not inhibit BK virus
in hemagglutination inhibition (HI) tests or react with
BK-infected cells in fluorescent antibody tests.

Sera from adults collected during the period 1949-
52 (provided by Dr. R. J. Huebner), prior to the widespread use of polio vaccines, were tested for HI antibody
to BK virus.   Seventy percent contained antibody at

dilutions of 1:40 or greater. These results are similar
to those obtained with recently collected sera. Thus
the BK virus does not appear to have been introduced
into man through polio vaccine.

SUMMARY. Viruses belonging to the SV40-polyoma subgroup
of papovaviruses have been isolated from humans. Two
of the viruses, isolated from brains of patients with
PML, were indistinguishable from SV40 by serologic tests;
the DNA from one of these, the DAR virus, was also
identical to that of SV40 by hybridization techniques.
A third isolate from a case of PML, the JC virus,
possessed properties which differed from SV40, namely, a
different host range and hemagglutinating ability.
However rabbit antiserum to JC virus showed cross-
reactivity with SV40 in neutralization tests. A fourth
human papovavirus (BK) was isolated from the urine of a
patient after renal transplantation. It was also shown
to be a hemagglutinating virus and to share antigens
with SV40 as determined by neutralization and fluorescent
antibody tests.

    Antibody studies have provided evidence that BK
virus was present in man prior to large scale poliovirus
immunization, thus ruling out SV40-contaminated polio
vaccine as its source. However the origin and exact
relationships of these different SV40-like viruses remain
to be determined.

DISCUSSION

SHAH: We have tested sera from children from Maryland
for BK virus antibodies and by the age of three about
50% have antibodies.   It is an extremely common infection
of children.

DIMMICK: What is the human ID_{50} for SV40?

LEWIS: I do not believe that this question has ever
been answered.   From the studies I reviewed, as little
as 100 pfu contaminating certain lots of oral poliomyelitis vaccines appeared to produce intestinal
infection in children in that they excreted infectious
SV40 in their stool for 4-5 weeks.   Although some of
these children received 100,000 pfu of SV40 orally in
the contaminated vaccine, none developed detectable
SV40-neutralizing antibody in their serum.   In Morris's
study, adult volunteers received 10,000 $TCID_{50}$ of SV40
contaminating a respiratory syncytial virus pool which
was administered by intranasal nebulization.   Sixty
percent of these individuals developed SV40-neutralizing
antibody in convalescent serum specimens and SV40 was
isolated from throat swabs obtained at 7 days from 2 of
16 of the volunteers.   Thus in humans receiving known
concentrations of infectious SV40, as little as 100 pfu
given orally will produce intestinal infection without
seroconversion, while intranasal inhalation of 10,000
pfu will produce overt infection, seroconversion and
infectious SV40 in nasal secretions.   Since 30-40% of
laboratory workers handling SV40 over a long period of
time develop SV40 antibody, presumably by inhalation of
aerosols, humans can probably be infected intranasally
by less than 10,000 pfu.

ROBB: In 1970 at NIH, I accidently took 10^7 pfu of SV40
by mouth while mouth pipetting.   In spite of repeated
rinses with 70% ethanol, I seroconverted for SV40
neutralization antibody.   In 1972 one of two students
working in my lab (neither had previous cell culture
experience) seroconverted within three weeks in spite of
the recommended SVCP precautions, including Class I
hoods and no mouth pipetting.   Students now must have
previous cell culture training to work in my lab.

MILLER: Dr. Lewis in his excellent papers suggested that
special follow-up of the soldiers who 10 or 11 years ago
received adenovirus vaccine be made, but it wasn't clear
whether he meant the follow-up should concern cancer or

PML or both.

LEWIS: I certainly think there is no question that any follow-up now has to include both. Possibly a number of other diseases that we are not aware of must also be considered.

MILLER: Would you expect, on the basis of experimental observations, healthy adults who are exposed to SV40 to be at high risk of cancer?

LEWIS: I would say that they probably are not, but the data are not available.

POLLACK: In a laboratory where people are working with SV40, would it be rational to exclude pregnant women? What is the epidemiological probability of bringing home the virus to a neonate?

WEDUM: Among 423 laboratory-acquired infections, one was a case of psittacosis contracted by a wife from her convalescent husband. The other was Q fever in the wife of an employee who we believe did not take the required shower before going home.

OXMAN: A worker is more likely to transmit an agent from the laboratory to his family if he is actively infected than if he merely carries it home on his clothing or beard. Unfortunately, we may be unaware that the worker is infected, because an agent which can cause severe disease in some individuals may only produce a mild or subclinical infection in others. For example, herpes simplex infections are usually associated with few, if any, symptoms in adults but may produce a fatal disseminated disease in newborns.

MILLER: Your point was that herpes simplex in infants is very much different from and more severe than a cold sore in adults.

OXMAN: The point is that if, like herpes simplex virus, an agent causes a mild or asymptomatic infection in a laboratory worker, he may not even be aware that he has carried it home and transmitted it to family contacts. Similarly, if an agent produces severe disease (e.g., cancer), but only after a long latent period, it may be silently disseminated before the appearance of symptoms in the carrier.

ROWE: In the course of adenovirus vaccine development, a considerable number of volunteers received monkey kidney-grown adenoviruses, which are now known to have been defective adeno-SV40 hybrids. Is anything known about antibodies to various SV40 antigens in these people?

MORRIS: There are no data on this yet, but these people should be located and tested.

ROWE: I'm not sure everyone in this audience is aware of the natural biology of adenovirus type 2 and why we are so concerned about nondefective Ad2-SV40 hybrid viruses. The basic problem is that adenovirus type 2 establishes itself as a chronic infection (possibly as a lysogenic interaction) in the tonsils and adenoids of young children. Almost certainly the virus persists in these tissues for a period of some years. It maintains itself by sporadic excretion of virus from the chronically infected pharyngeal lymphoid tissues, so that babies or infants come in contact with this virus as part of their normal growing up experience. Serum surveys in all parts of the world give the same pattern; about 60% of children have acquired adeno type 1, 2, and/or 5 by the age of 5 years or so. If comparable viruses carrying SV40 genetic sequences should get into young children, it would be almost like LCM getting into a mouse of the right age. They could become established in the tonsillar chronic infection cycle and become part of the normal human experience for generations and generations to come. So it is not just a matter of a few laboratory personnel being infected and the infection burning itself out. This is a unique situation in that the primary concern is not protection of the laboratory worker, but prevention of the virus from getting into his children.

DARNELL: What one fears is presumably the SV40 component of the hybrid. We jumped to the discussion of what to do about this without thoroughly discussing whether there is anything to be afraid of from SV40 in the first place. I think we should return to that and try to make some sort of sense out of that first before worrying about whether these hybrids are potentially dangerous. My opinion is that since millions of people have gotten SV40 in probably at least as large an amount as they are

ever likely to get from the hybrids and gotten it in
sufficiently large amounts without there being any solid
evidence of SV40 tumors, then this may be a tempest in a
teapot. Perhaps we should work with SV40, and probably
the hybrids also, until we get the answers about transformation that people are striving for.

POLLACK: Infection by SV40 differs in at least two ways
from infection by Ad-SV40. First, a human cell infected
by a hybrid virus would not produce SV40 virions,
whereas SV40-infected human cells would do so. Therefore
hybrid-transformed human cells would not stimulate
antibody to SV40 virions and so would be that much less
antigenic to the host. Second, Ad-SV40 might have the
tissue range of adenoviruses, which includes lymphatic
tissue, while SV40 would not be expected to infect
lymphatic tissue. For these two reasons data on human
infection by SV40 cannot be extended to the hybrid
viruses.

KAPLAN: I'd like to ask whether there is any evidence
as to whether individuals who are immunized against
adenovirus type 2 are also protected against these
nondefective Ad2-SV40 hybrids?

LEWIS: I can't answer the question directly. The only
thing I can say is that all the nondefective hybrids and
all the hybrids I've studied in the adeno 2 category
are completely neutralized by antiserum directed against
the prototype adenovirus 2. Everyone who has ever
worked in my laboratory, with the exception of one
individual, has had adeno 2 neutralizing antibody titers
in their serum in excess of 1:40. Although virus
isolation attempts and paired sera have been obtained on
a number of upper respiratory infections in myself and
other workers, we have never had a documented case of
human infection by either adeno 2 or the nondefective
hybrids in any of our personnel.

ROWE: Fortunately some of the early human volunteer
studies, that is, Huebner's studies of adeno pathogenicity
and vaccine evaluation, were done with viruses of the
adeno 1, 2, 5 subgroup. So we have data on the relation
between serum neutralizing antibody titer and resistance
to challenge with large doses of virus. Antibody titers
of only 1:4 gave quite significant protection, both
against disease and in terms of recovery of virus. I
feel that the screening of persons handling these hybrids
for Ad2 antibody is crucial for containing the hybrids.

Session III
The Tumor Viruses Themselves

AVIAN AND MURINE RNA TUMOR VIRUSES:
MODES OF TRANSMISSION

Murray B. Gardner
Department of Pathology
University of Southern California School of Medicine
Los Angeles, California

The idea that genetic information for RNA tumor
viruses exists as part of the normal cell genome (1-3)
has gained increasingly wide acceptance, largely on the
basis of rapidly accumulating virologic, molecular
hybridization, genetic cross-breeding and seroepidemiologic studies in avian and murine species. Complete
type C virus particles of endogenous origin (Todaro, this
volume) have been generated spontaneously or have been
induced by chemical or physical means from "virus-free,"
single cell clones of four different rodent species,
namely mouse (4-6), rat (7-9), Chinese hamster (10) and
guinea pig (11,12), and from virus-free embryo cell
cultures of chickens and wild jungle fowl (13,14). In
general these endogenous viruses grow poorly or not at
all in their cells of origin and can replicate only in
permissive cells of another species. They have not yet
been shown capable of transforming cells in vitro or of
inducing tumors or leukemia in vivo. Type C virus-
specific DNA or RNA nucleotide sequences have been shown
by various techniques of molecular hybridization to be
present in "virus-free" chicken tumor cells and even in
the chick embryo (15-18) and mice (19,20). Dominant and
recessive autosomal gene loci have been identified by the
segregation pattern of specific genetic markers via
selective breeding programs in chickens (21-23) and in
mice (24-30), independently determining, via Mendelian
inheritance, specific expressions of the endogenous type
C virus genome. These expressions include the presence

of group-specific (gs) antigen in chickens and mice,
helper virus activity for defective Rous sarcoma virus
in chickens, susceptibility to spontaneous or inducible
generation of complete type C virus particles in mice
and the host range properties of the induced virus.
Specific gene loci dominant for resistance in mice (31,
32) and dominant for susceptibility in chickens (33) have
also been identified that determine the degree of
susceptibility of mouse or chicken cells to productive
infection following exposure to different strains of
species-specific exogenous type C viruses. Seroepidemiologic studies have demonstrated that, in aging
populations of laboratory mice (34-36) and feral mice
(37), and following exposure of these strains to chemical
carcinogens (38,39), there occurs a predictable genetically controlled activation of type C virus expression in
the form of gs antigen, infectious virus production and
tumor development. The evidence of gs antigen expression
in early mouse and chicken embryos (40) and the relative
postnatal immunologic tolerance to this antigen (41)
have been well documented.

Similar evidence from genetic breeding programs (42)
and nucleic acid hybridization studies (43) indicates
that the type B RNA tumor viral genome in mice is also
very likely part of the normal cell genome.

Thus it seems very likely from these several lines
of evidence that in chickens and mice the genetic information for type C virus, and in mice also for type B
virus, is integrated into the chromosomal DNA of each
somatic and germ cell, and the perpetuation of these
avian and murine RNA tumor viruses in nature is best
assured by this most intimate type of commensural relationship. Viewed in the context of this viral oncogene
hypothesis (1,2), expression of the latent RNA tumor
virus genetic information as gs or envelope antigens,
reverse transcriptase or complete virus particles
(virogenic expression) and as cell transformation to
malignancy (oncogenic expression) may be controlled
independently by other host cell regulatory genes.
Chemical and physical carcinogens or simple aging may
act to derepress latent RNA tumor virus genes. Hormonal
(44) and immunologic phenomena (45,46), also under
genetic control (47,48), constitute additional factors
influencing the activation of the virus genome, the
subsequent cell to cell spread of activated virus, and
the growth of transformed cells into a clinical cancer.

Only in genetically susceptible strains of mice
(e.g., AKR,C58) and chickens is the activated virus
likely to spread readily in an infectious way within
the animal. In cancer-resistant strains of laboratory

mice and in most strains of wild mice, genetic and
immunologic factors, such as the yet hypothetical intracellular repressor protein molecules (viral genome-coded
?), cell surface receptors and virus neutralizing antibodies probably restrict the number of virus particles
that are produced and limit their capacity to infect
other cells. Activation of the oncogene independently
of the virogene might account for those examples of
neoplasia in which type C particles cannot be seen or
isolated. In these instances the multiplication or
rejection of the initially transformed cells might be
largely determined by the effectiveness of the immunologic
counter-attack mounted against newly expressed or exposed
cell surface proteins, whose appearance is determined by
the oncogenic component of the integrated type C virus
genome. The type of cancer that develops might then be
determined by the specific cell type (e.g., lymphoid,
reticuloendothelial, mesenchymal, epithelial) within
which the oncogene becomes derepressed.

Recent evidence suggests that, in addition to cancer,
indigenous type C virus may also be responsible for an
autoimmune type of renal disease in laboratory mice (49,
50) and a lower motor neuron type of neurogenic paralysis
in wild mice (Officer and Henderson, pers. comm.). Thus
the pathogenetic consequences of type C virus expression
in mice and possibly in other mammalian species may not
be limited to neoplasia, but may well extend to a number
of idiopathic disease processes manifested in diverse
organ systems.

SPREAD OF RNA TUMOR VIRUSES IN MICE

Confirmation of the presence of RNA tumor virus
information as part of the normal host cell genome in
chickens and mice by no means excludes the possibility
of contact transmission of type C virus between members
of the same species. There are three means by which this
may take place: (1) vertical epigenetic spread of
infectious type C virus from the maternal ovary or
genital tract to the developing embryo; (2) transmission
of type C, as well as type B, virus via milk from
maternal to infant mice; and (3) horizontal transmission
of infectious virus via contact fighting or exposure to
virus containing aerosols or body secretions.

By demonstrating the presence of infectious leukemogenic type C virus in AKR mouse embryos, Gross (51)
first showed that vertical transmission of infectious
virus did, indeed, take place in genetically susceptible
strains of mice. However in recent studies (52) infectious virus could not be transmitted to susceptible

C3H mice transplanted in utero in AKR mothers. Furthermore, the prevalence and titer of infectious virus
recoverable from different tissues of embryo and young
AKR mice suggested that the presence of virus could also
be accounted for by a high probability of spontaneous
activation of endogenous virus together with the marked
genetic susceptibility of adjacent cells to infection
with the activated endogenous virus. We have discovered
one and possibly two colonies of wild mice in which a
similar epigenetic vertical transmission of infectious
type C virus is probably taking place. Infectious virus
is recoverable from maternal and embryo wild mice in
these colonies, and the aging postnatal mice are characterized by a remarkably high incidence of gs antigen
expression and type C virus production in various tissues,
including the central nervous system. These mice show a
high incidence of lymphoma (53) and/or hind leg paresis.
In six other populations of wild mice that we have
studied, the incidence of type C virus expression in
embryo and postnatal mice and of spontaneous neoplasia
has been negligible or very low until advanced age.
These findings have seemed most logically interpreted
by the genetic inheritance of latent type C virus information (37).

Without question, murine type C, as well as type B,
RNA tumor viruses can be readily transmitted in the milk
of certain strains of mice to their genetically susceptible offspring (54-56). However even under these
circumstances it has not been possible to convert a
low leukemia strain into a high leukemia strain or vice
versa by foster-nursing. This is in contrast to the
situation with the type B virus and mammary carcinoma,
in which a low tumor line could be changed into a high
tumor line by a single foster-nursing of one generation
of mice (57). Of course it has been known for a long
time that milk-passaged type B virus is an important
determinant of the risk to mammary cancer in mice,
whereas leukemia risk is primarily determined by genotypic
influences and extent of immunological competence. We
have shown that about 60% of pregnant wild mice have
type B virus particles in their lactating breast tissue
and milk regardless of trapping area, whereas type C
particles could be found in the breast tissue and milk
of pregnant wild mice from only one of the trapping
areas examined. These wild mice are remarkably "switched
on" for type C virus activity and exhibit epigenetic
vertical virus transmission (58).

The opportunity for horizontal contact transmission
of type C virus in certain strains of mice certainly
exists since virus has been demonstrated in their saliva,
urine, feces and semen (59,60). Also the experimental

aerosol transmission of mouse type C virus has been demonstrated (61). However, deliberate attempts to demonstrate such horizontal spread by caging together virus-shedding AKR mice with virus-susceptible C3H, Swiss-Webster or DBA mice and assaying for subsequent development of lymphoma, gs antigen, Gross soluble antigen (62) or infectious virus have generally been ineffective and have failed to show any significant degree of contact transmission (63,64,52 and Meier, pers. comm.). Thus although an epigenetic vertical transmission via the genital tract and a horizontal transmission of type C virus via milk may occur in certain genetically susceptible strains of laboratory and wild mice, postnatal intraspecies horizontal spread is probably of little or no consequence in the natural biology of mouse type C viruses.

SPREAD OF RNA AND DNA TUMOR VIRUSES IN CHICKENS

"Naturally occurring" infection with two types of virus capable of causing leukosis, one an RNA tumor virus with several major subgroups responsible for lymphoid leukosis (65) and the other a herpes-type DNA virus responsible for Marek's leukosis (66), are both of widespread prevalence in domestic chicken flocks (67-69). However "naturally occurring" is hardly the proper designation; since Rous' time, man has profoundly altered the biology of chickens and their RNA tumor viruses by means of extensive genetic inbreeding, by housing in close contact many thousands of newborn and young chicks and, until recent years, by the inadvertant inoculation of type C virus as contaminants of avian respiratory disease vaccines prepared in chicken embryo cells. Contact spread of avian type C virus and resultant lymphoid leukosis among susceptible exposed chickens has been well documented (70). Even in wild Malayan jungle fowl (71) the substantial incidence of neutralizing antibody against avian leukosis virus suggests frequent horizontal transmission of this agent.

Avian type C leukosis viruses are also commonly transmitted vertically from hen to egg (72,73) in an epigenetic infectious manner, accounting for lifelong type C viremia and immunologic tolerance to this agent. We have observed one commercial fryer operation in which virtually 100% of the hens and the chicks were infected with type C virus and show immunologic tolerance. Usually the majority of hens in a commercial flock is not previously exposed to avian type C virus and is thus capable, upon exposure, of making neutralizing antibodies which they pass to their embryos, thereby conferring passive protection until 6-8 weeks after hatching.

Subsequently upon exposure these chicks are also capable of an active, acquired antibody response to the avian type C virus. The presence or absence of such maternal or acquired antibody does not clearly relate to the subsequent development of either lymphoid leukosis or Marek's disease (74). In addition a certain number of chicks in each flock often proves genetically resistant to the subgroup A or B leukosis viruses (75), due to the lack of specific cell surface receptors. The endogenous type C virus genome in chicken cells has an envelope type specificity (subgroup E) distinct from the four subgroups (A-D) of exogenous chicken type C virus (76). The endogenous subgroup E virus grows very poorly in chicken cells and thus is not a candidate for horizontal spread among chickens. In addition to inherited endogenous type C virus genomes, intraspecies horizontal spread and vertical transmission of exogenous type C virus strains are of definite importance in the biology of chicken RNA tumor viruses.

By contrast, the Marek's disease herpesvirus is apparently not transmitted vertically (77), either genetically or epigenetically, but is spread horizontally among young chickens by feather follicles (78) in a highly infectious manner. Marek's leukosis now accounts for the major economic loss among commercial chicken flocks throughout the world. Although there is some evidence of possible synergism between Marek's disease DNA herpesvirus and lymphoid leukosis RNA virus in vitro (79), there is no convincing evidence for any such interrelationship in the causation of Marek's leukosis in vivo. A vaccine prepared with an antigenically related but avirulent herpesvirus from turkeys (80) has generally been quite effective in reducing the incidence of Marek's leukosis (81).

EVIDENCE OF TYPE C VIRUSES IN OTHER AVIAN SPECIES. Although lymphoid leukosis and Marek's disease are very common in domestic chickens, they are rarely observed in domestic turkeys (82,83), ducks (84), parakeets (85,86) or wild birds (87,88). We were unable to detect type C virus particles by electron microscopy or chicken gs antigen by complement fixation in a number of spontaneous parakeet tumors, including several lymphomas and sarcomas. The only extant avian type C virus isolated from other than chickens is the reticuloendotheliosis (RE) virus (strain T) (89), derived initially from a turkey. Type C virus-like particles have been seen by electron microscopy in the hearts of young and embryo turkeys afflicted with round heart disease (90), but these have yet to be isolated and characterized. Despite subsequent experimental passage of the RE virus solely in chickens,

it retains a group-specific antigen (91), reverse
transcriptase (92) and envelope antigens (93) immunologically distinct from the chicken leukosis viruses.
Although one study based upon detection of immunofluorescent antibody in eggs suggested that asymptomatic RE
virus infection, presumably the result of contact spread,
might be of fairly common occurrence (16% of eggs with
antibody) in commercial chicken flocks (94), another
study failed to show any antibodies to RE virus in adult
chickens (80).

The lack of antibodies to Rous sarcoma virus in a
sample of healthy turkeys, quails and pigeons (95)
suggests that chicken type C viruses do not spread
naturally to these other avian species. Although type
C virus has not been isolated from avian species other
than chickens and turkeys, the presence of complement-
fixing, group-specific antigen, reactive with hamster
sera prepared against RSV in normal pheasant embryo cells
(96), and the detection of chicken type C virus-related
gene sequences in normal quail embryo cells (17), suggest
that these species also contain some endogenous type C
virus genome information related to that of the chicken.
One would anticipate that these less domesticated and
nonvaccinated species would remain largely repressed for
their endogenous type C virus genome expressions and
would not harbor infectious type C virus of exogenous
origin. However the embryo cells of these nondomesti-
cated fowl are very useful to the researcher because
they are generally more susceptible than chicken embryo
cells to productive infection with the subgroup E
endogenous chicken type C viruses (96,76). Perhaps as
yet undetected helper viruses in these nondomestic fowl
cells are responsible for this phenomenon (17).

Virtually all turkeys after several weeks of age
become infected with the turkey herpesvirus by contact
transmission, but the incidence of tumors is negligible
(97). When given to chickens as Marek's disease vaccine,
the turkey herpesvirus shows very little horizontal
spread to unvaccinated chickens in the same flock (81).
Contact spread of Marek's disease herpesvirus from
chickens to wild birds under experimental conditions has
not been demonstrated (98), nor has this virus been
isolated from two turkey flocks with spontaneous leukosis
(80).

It can be concluded that horizontal spread of avian
RNA or DNA tumor viruses under natural conditions from
chickens to other avian species probably does not occur;
furthermore, these other avian species almost surely do
not represent a reservoir of RNA or DNA tumor viruses
potentially transmittable to humans.

SUSCEPTIBILITY OF HUMAN CELLS TO INFECTION

Certain human cell strains and lines have proven susceptible to infection in vitro with chicken (99-102) and mouse (103-107) type C leukemia and sarcoma viruses. However no permanent virus-productive or nonproductive transformed human cell line has been so derived. Human cell strains from individuals with neoplasia and genetic abnormalities are generally more susceptible to transformation with mouse sarcoma virus (Kirsten strain) than are those derived from "normal" individuals (106), a finding consistent with previous observations using DNA tumor viruses (108). This might serve as a basis for a useful screening test for determining cancer susceptibility in humans. Although host adaptation of mouse or chicken type C viruses grown in human cells may have occurred in several instances (109-111), perhaps by incorporation of human cell surface glycoproteins into the envelope of budding virions, this has been the exception, and usually the chicken and mouse viruses have remained apparently unaltered, even after prolonged growth in human cells (106,107). Despite one such interpretation (111), no positive evidence for genetic alteration of chicken or mouse type C viruses grown in human cells by contribution ("transduction") from the human cell genome has been forthcoming. There has been no evidence that any human "helper" virus may have been activated by passage through human cells of avian or murine sarcoma virus, nor that a human sarcoma virus genome has been "rescued" by growth of chicken and mouse helper leukemia-type viruses in human sarcoma or other cells.

A human glioma tumor cell line (KC) nonproductively infected with Rous sarcoma virus (112) forms syncytia upon exposure specifically to the RD-114 type C virus (113) and could serve as a useful means of quantitating this virus and for searching in human and cat tissues for similar virus (114). Although several other human cell lines may have been nonproductively infected for a short time with Rous sarcoma virus (101), no human cell strains or lines have been reported to be nonproductively infected with and transformed by mouse type C sarcoma virus. By analogy with the bromodeoxyuridine induction of endogenous rat type C virus (9) from the XC cell, a rat cell nonproductively infected with and transformed by RSV (115), a human cell strain or line bearing a mouse type C sarcoma virus genome without virus particle production would be a most promising reagent for attempted induction of a human endogenous type C virus genome.

Human cells (HeLa) were not susceptible to infection with the Marek's disease chicken herpesvirus or the turkey herpesvirus (116).

LACK OF EVIDENCE FOR INFECTION OF HUMANS IN VIVO

Despite the proven susceptibility of some human cell to experimental infection in vitro with chicken and mouse type C virus and the well known ability of these agents to experimentally cross species barrier in vivo, and in the case of Rous sarcoma virus to produce sarcomas in primates (117), several seroepidemiologic surveys failed to give any evidence of in vivo infection or tumorigenic activity of chicken type C virus for humans (95,118-122). Hundreds of sera from humans with and without lymphoma, or following immunization with vaccines containing live avian leukosis virus or from healthy laboratory workers intimately exposed to avian type C viruses were, with one possible exception (123), free of specific complement-fixing, neutralizing or fluorescent antibodies to avian type C virus gs or envelope antigens. Nor could avian type C virus be isolated in vitro from any human leukemic sera (120). World War II veterans given yellow fever vaccines contaminated with viable chicken type C virus have not shown any increased incidence of cancer from 5 to 22 years thereafter (124).

Similarly, no immunologic reactions have been reported indicative of antibodies in human sera, from normal or cancer-bearing patients or exposed laboratory workers, directed specifically to mouse type C virus gs or envelope antigens (118). Recent tests of our laboratory personnel have detected no CF antibodies directed against the lymphomagenic and neurotropic strain of type C virus of wild mouse origin; neutralizing tests have yet to be done. Mouse type C virus gs-1 antigen has not been detected by complement fixation, immuno-fluorescence or radioimmunoassay (125) in human tumors or other tissues (126,127). With the possible exception of the ESP virus (128), which has mouse gs antigenic determinants (129), no murine type C virus has been isolated from primary or cultured human tumor cells. The unique specificity of the gs-1 antigen found in the type C virus isolates of each mammalian species tested thus far (130) strongly suggests that these viruses do not spread between species in an infectious way. This specificity also indicates the value of the gs-1 antigen in identifying the species of origin of any given type C virus. The lack of such precise immunologic specificity is a major limitation in interpretation of earlier studies (131,132) purporting to show mouse type C virus-related envelope antigens in human leukemia.

The detection of nucleotide homologies between the
genome of mouse type C virus and polysomal RNA from
human leukemias (133), sarcomas (134), lymphomas and
ovarian carcinomas (135) suggests that these human
tumors may contain mouse type C virus-related information.
Similarly the homology between the genome of mouse type
B virus and polysomal RNA of human breast carcinomas (136)
suggests the presence in human breast cancer of mouse
type B RNA tumor virus-related information (137-139).
Some recent serologic (140,141,158) and morphologic (142,
143) evidence also suggests the presence in human breast
cancers and milk of type B RNA tumor virus-related
activity. However this evidence certainly does not
indicate spread of mouse type C or B virus to humans,
and the evidence of viral specificity is still not
conclusive as no definitive type C or B virus has yet
been isolated from humans. Determination of the
specificity of various immunologic and molecular genetic
analyses of human tissue for putative human type C or
B virus activity, using reagents prepared from purified
mouse RNA tumor viruses of these types, is likely to be
confounded by the incomplete copying of the viral 70S
genome in the DNA probe, by the inclusion in purified
virus particles of host cell genetic material, and by
the presence in mice and humans of homologous gene
sequences unrelated to virus genes. Also antigenic
moieties on the cell surface or virus envelope, such as
HLA and H-2 alloantigens, may also be shared in common
by mouse and man (144).

RNA TUMOR VIRUSES--SIGNIFICANCE TO MAN

The significance to man of avian and murine RNA
tumor viruses seems to lie not so much in any direct
biohazard that they pose for humans, but rather in what
can be learned from them that may help us understand by
analogy the genetics (145) and natural history of similar
latent, endogenous, inherited virus genes likely to exist
in the human genome. The immunological defenses of
humans will, in all probability, prevent establishment
of any infection in vivo with exogenous type-C viruses
of animal origin. It follows that humans with disorders
of immunologic competence or on immunosuppressive treatment and pregnant women constitute a group at least at
theoretical risk to these agents.

It seems logical that the hypothetical virogene must
be innately cellular and not contracted from a foreign
source, at least not in recent evolutionary times (130).
One major task before us then is to determine how to
reproducibly activate, rescue or genetically complement
this virogene and allow it to reproduce itself ad

infinitum. Many previous attempts in this direction have failed. Powerful genetic repressor mechanisms for type C virus genome expression, an attribute exhibited in general by feral outbred animal species, and/or intrinsic defectiveness or deletion in the endogenous type C viral genome could block the synthesis of complete virus in human cells. Perhaps this difficulty can also be attributed in part to the knowledge that endogenous type C viruses of chicken and murine origin usually grow very poorly in their own cells of origin but often replicate well in cells of other species. Thus perhaps we should do more to expose candidate human cancer material to cells of various nonhuman species, particularly utilizing materials from those individual humans with chromosomal abnormalities, immunologic impairment or known cancer susceptibility. In so doing we may increase our chances of finding a derepressed endogenous human type C virus genome and a permissive cell for its replication, or of producing genetic complementation of a hypothetically defective human type C virus genome, leading in either case to complete virus synthesis. One candidate human type C virus, RD-114, may have been generated in this manner by passage of a human rhabdomyosarcoma (RD) cell line in a fetal cat (146). Recent findings, however, suggest that this virus may be an endogenous cat virus (147,148) different from any of the described cat type C virus isolates.

Another virus, called ATS 124, has been isolated from RD cells transplanted in an immunosuppressed NIH Swiss mouse given antithymocytic sera (149). This virus has mouse-specific gs antigen and reverse transcriptase activity but a host range similar to RD-114 virus, i.e., a tropism for human but not mouse cells. Thus the RD tumor cell genome may have the unique property of complementing the genomes of otherwise defective feline and murine endogenous type C viruses and/or of providing a permissive milieu for the replication of these endogenous viruses. It is quite feasible that the genetic information possibly contributed by the RD cell is part of the human endogenous type C viral genome, in which case RD-114 and ATS 124 would represent hybrid human-cat and human-mouse type C viruses, respectively.

Oncogenic information has not yet been shown to be present in the endogenous chicken or murine type C viruses induced from normal cells, since these have not been proven capable of transforming cells or of inducing lymphoma. However many of these in vivo tests are of very recent vintage. The induction of lymphoma in newborn Balb/c and wild mice following inoculation with type C virus isolated from normal spleens or embryo cells of the isologous strain (2) shows that these indigenous

viruses do bear oncogenic information. It is not yet
known whether the genomes of these type C viruses
indigenous to certain animal populations are identical
to the genome of endogenous type C virus induced from
cells of the isologous strain of animal, nor whether or
not there may exist more than one kind of endogenous
type C viral genome. Nontransforming exogenous type C
viruses added to and replicating in rodent cells can be
shown to be important determinants of chemical transformation (150-153). The activation of latent type C
virus genes during chemical transformation of "virus-
free" hamster (154), rat (155,156) and chicken cells (13),
as well as during spontaneous transformation of mouse
cells (157), suggests that endogenous inherited type C
virus genomes may be basic determinants of chemical and
spontaneous tumors. Inclusion of oncogenic polynucleotide
sequences within any activated endogenous human type C
virus isolate is a necessary prerequisite for eventually
determining those viral gene-directed protein(s) that
are responsible for cell transformation. This knowledge
would lead us toward more specific and definitive means
of cancer therapy and prevention.

References

1. Huebner, R. J. and G. J. Todaro. 1969. Oncogenes
    of RNA tumor viruses as determinants of cancer.
    Proc. Nat. Acad. Sci. 64:1087.
2. Todaro, G. T. and R. J. Huebner. 1972. The viral
    oncogene hypothesis: New evidence. Proc. Nat. Acad.
    Sci. 69:1009.
3. Temin, H. M. 1972. The RNA tumor viruses--back-
    ground and foreground. Proc. Nat. Acad. Sci. 69:
    1016.
4. Rowe, W. P., J. W. Hartley, M. R. Lander, W. E. Pugh
    and N. Teich. 1971. Noninfectious AKR mouse embryo
    cell lines in which each cell has the capacity to
    produce infectious murine leukemia virus. Virology
    46:864.
5. Lowy, D. R., W. P. Rowe, N. Teich and J. W. Hartley.
    1971. Murine leukemia virus: High-frequency
    activation in vitro by 5-iododeoxyuridine and 5-
    bromodeoxyuridine. Science 174:155.
6. Aaronson, S. A., G. J. Todaro and E. M. Scolnick.
    1971. Induction of murine C-type viruses from clonal
    lines of virus free Balb/3T3 cells. Science 174:157.
7. Klement, V., M. O. Nicolson and R. J. Huebner. 1971.
    Rescue of the genome of focus-forming virus from rat
    nonproductive lines by 5-bromodeoxyuridine. Nature
    234:12.
8. Aaronson, S. A. 1971. Chemical induction of focus-
    forming virus from nonproducer cells transformed by

murine sarcoma virus. Proc. Nat. Acad. Sci. 68:3069.
9. Klement, V., M. O. Nicolson, R. V. Gilden, S.
Oroszlan, P. Sarma, R. Rongey and M. B. Gardner.
1972. Induction of rat specific C-type RNA virus in
Rous induced rat sarcoma cell line (XC) by 5-
bromodeoxyuridine. Nature 238:234.
10. Todaro, G. J. 1972. Detection and characterization
of RNA tumor viruses in normal and transformed cells.
Perspectives in virology, ed. M. Pollard, p. 81.
Academic Press, New York.
11. Hsuing, G. D. 1972. Activation of guinea pig C-type
virus in cultured spleen cells by 5-bromodeoxyuridine.
J. Nat. Cancer Inst. 49:567.
12. Nayak, D. P. and P. Murray. Induction of C-type
virus in cultured guinea pig cells. J. Virol. (in
press).
13. Weiss, R. A., R. P. Friis, E. Katz and P. K. Vogt.
1971. Induction of avian tumor viruses in normal
cells by physical and chemical carcinogens. Virology
46:920.
14. Hanafusa, T., H. Hanafusa, T. Miyamoto and E.
Fleissner. 1972. Existance and expression of tumor
virus genes in chick embryo cells. Virology 47:475.
15. Baluda, M. A. 1972. Widespread presence in chickens
of DNA complementary to the RNA genome of avian
leukosis viruses. Proc. Nat. Acad. Sci. 69:576.
16. Rosenthal, P. M., H. L. Robinson, W. S. Robinson, R.
Hanafusa and H. Hanafusa. 1972. DNA in uninfected
and virus-infected cells complimentary to avian tumor
virus RNA. Proc. Nat. Acad. Sci. 68:2336.
17. Varmus, H. E., R. A. Weiss, R. R. Friis, W. Levonson
and J. M. Bishop. 1972. Detection of avian tumor
virus-specific nucleotide sequences in avian cell
DNAs. Proc. Nat. Acad. Sci. 69:20.
18. Neiman, P. E. 1972. Rous sarcoma virus nucleotide
sequences in cellular DNA: Measurement by RNA-DNA
hybridization. Science 178:750.
19. Green, M. 1970. Oncogenic viruses. Annu. Rev.
Biochem. 39:701.
20. Gelb, L. D., S. A. Aaronson and M. A. Martin. 1971.
Heterogeniety of murine leukemia virus in vitro DNA:
Detection of viral DNA in mammalian cells. Science
172:1353.
21. Payne, L. N. and R. C. Chubb. 1968. Studies on the
nature and genetic control of an antigen in normal
chick embryos which reacts in the COFAL test. J.
Gen. Virol. 3:379.
22. Weiss, R. A. and L. N. Payne. 1971. The heritable
nature of the factor in chicken cells which acts as
a helper virus for Rous sarcoma virus. Virology
45:508.
23. Payne, L. N., P. K. Pani and R. A. Weiss. A dominant
epistatic gene which inhibits cellular susceptibility

to RSV (RAV-O).   J. Gen. Virol. (in press).
24. Lilly, F. 1972. Mouse leukemia: A model of a
    multiple-gene disease. J. Nat. Cancer Inst. 49:927.
25. Taylor, B. A., H. Meier and D. D. Myers. 1971. Host
    gene control of C-type RNA tumor virus: Inheritance
    of the group specific antigen of murine leukemia
    virus. Proc. Nat. Acad. Sci. 68:3190.
26. Stephenson, J. R. and S. A. Aaronson. 1972. Genetic
    factors influencing C-type RNA virus induction. J.
    Exp. Med. 136:175.
27. Stephenson, J. R. and S. A. Aaronson. 1972. A
    genetic locus for inducibility of C-type virus in
    Balb/c cells: Effect of a nonlinked regulatory gene
    on detection of virus after chemical activation.
    Proc. Nat. Acad. Sci. 69:2798.
28. Rowe, W. P. 1972. Studies of genetic transmission
    of murine leukemia virus by AKR mice. I. Crosses
    with FV-1^n strains of mice. J. Exp. Med. 136:1272.
29. Rowe, W. P. and J. W. Hartley. 1972. Studies of
    genetic transmissions of murine leukemia virus by
    AKR mice. II. Crosses with FV-1^b strains of mice.
    J. Exp. Med. 136:1286.
30. Rowe, W. P., J. W. Hartley and T. Bremner. 1972.
    Genetic mapping of a murine leukemia virus-inducing
    locus of AKR mice. Science 178:860.
31. Pincus, T., J. W. Hartley and W. P. Rowe. 1971. A
    major genetic locus affecting resistance to infection
    with murine leukemia virus. I. Tissue culture
    studies of naturally occurring viruses. J. Exp.
    Med. 133:1219.
32. Pincus, T., W. P. Rowe and F. Lilly. 1971. A
    major genetic locus affecting resistance to infection
    with murine leukemia viruses. II. Apparent identity
    to a major locus described for resistance to Friend
    murine leukemia virus. J. Exp. Med. 133:1234.
33. Crittenden, L. B., H. A. Stone, R. H. Reamer and W.
    Okazaki. 1967. Two loci controlling genetic
    cellular resistance to avian leukosis-sarcoma
    viruses. J. Virol. 1:898.
34. Peters, R. L., L. S. Rabstein, G. J. Spahn, R. M.
    Madison and R. J. Huebner. 1972. Incidence of
    spontaneous neoplasms in breeding and retired breeder
    Balb/c mice throughout the natural life span. Int.
    J. Cancer 10:273.
35. Peters, R. L., J. W. Hartley, G. J. Spahn, L. S.
    Rabstein, C. E. Whitmire, H. C. Turner and R. J.
    Huebner. 1972. Prevalence of the group specific
    (gs) antigen and infectious virus expressions of the
    murine C-type RNA viruses during the life span.
    Int. J. Cancer 10:283.
36. Peters, R. L., G. J. Spahn, L. S. Rabstein, H. C.
    Turner and R. J. Huebner. 1972. Incidence of group
    specific (gs) antigen of type C tumor viruses in

spontaneous neoplasms of Balb/c mice. Int. J.
Cancer 10:290.

37. Gardner, M. B., B. E. Henderson, R. W. Rongey, J. D.
Estes and R. J. Huebner. Spontaneous tumors of aging
wild house mice. Incidence, pathology and C-type
virus expression. J. Nat. Cancer Inst. (in press).

38. Whitmire, C. E., R. A. Salerno, L. S. Rabstein, R.
J. Huebner and H. C. Turner. 1971. RNA tumor-virus
antigen expression in chemically induced tumors.
Virus-genome specified common antigens detected by
complement fixation in mouse tumors induced by 3-
methylcholanthrene. J. Nat. Cancer Inst. 47:1255.

39. Gardner, M. B., J. E. Officer, R. W. Rongey, J. D.
Estes, H. C. Turner and R. J. Huebner. 1971. C-type
RNA tumor virus genome expression in wild house mice.
Nature 232:617.

40. Huebner, R. J., G. J. Kelloff, P. S. Sarma, W. T.
Lane, H. C. Turner. R. V. Gilden, S. Oroszlan, H.
Meier, D. D. Myers and R. C. Peters. 1970. Group
specific (gs) antigen expression of the C-type RNA
tumor virus genome during embryogenesis: Implications
for ontogenesis and oncogenesis. Proc. Nat. Acad.
Sci. 67:366.

41. Huebner, R. J. and H. J. Igel. 1971. Immunological
tolerance to gs antigens as evidence for vertical
transmission of noninfectious C-type RNA virus
genomes. Perspectives in virology, ed. M. Pollard,
vol. 7, p. 55. Academic Press, New York.

42. Bentvelzen, P., J. H. Daams, P. Hageman and J.
Calafat. 1970. Genetic transmission of viruses that
incite mammary tumors in mice. Proc. Nat. Acad. Sci.
67:377.

43. Varmus, H. E., J. M. Bishop, R. C. Nowinski and N. H.
Sarkar. 1972. Mammary tumor virus specific
nucleotide sequences in mouse DNA. Nature 238:189.

44. Fowler, A. K., C. D. Reed, G. J. Todaro and A.
Hellman. 1972. Activation of C-type RNA virus
markers in mouse uterine tissue. Proc. Nat. Acad.
Sci. 69:2254.

45. Armstrong, M. Y. K., F. L. Black and F. F. Richards.
1972. Tumor induction by cell-free extracts derived
from mice with graft vs. host disease. Nature
235:153.

46. Hirsch, M., S. M. Phillips, C. Solnik, P. H. Black,
R. S. Schwartz and C. B. Carpenter. 1972. Activation of leukemia virus by grafts vs. host and mixed
lymphocyte reactions in vitro. Proc. Nat. Acad. Sci.
69:1069.

47. Tennant, J. R. 1972. Susceptibility and resistance
to viral leukogenesis in the mouse. IV. Mechanisms
of host resistance to leukemia expression. J. Nat.
Cancer Inst. 49:1257.

48. Aoki, T., E. Boyse and L. J. Old. 1968. Wild-type

Gross leukemia virus. II. Influence of immunogenetic factors on natural transmission and on the consequences of infection. J. Nat. Cancer Inst. 41:97.

49. Mellors, R. C., T. Aoki and R. J. Huebner. 1969. Further implications of murine leukemia-like virus in the disorder of NZB mice. J. Exp. Med. 19:1045.

50. Hanna, M. G., R. W. Tennant, J. M. Yuhas, N. H. Clapp, N. K. Batzing and M. J. Snldgrass. 1972. Autogenous immunity to endogenous RNA tumor virus antigens in mice with a low natural incidence of lymphoma. Cancer Res. 32:2226.

51. Gross, L. 1951. Spontaneous leukemia developing in C3H mice following inoculation in infancy with Ak-leukemic extracts or Ak embryos. Proc. Soc. Exp. Biol. Med. 76:27.

52. Rowe, W. P. and T. Pincus. 1972. Quantitative studies of naturally occurring murine leukemia virus infection of AKR mice. J. Exp. Med. 135:429.

53. Gardner, M. B., B. E. Henderson, J. E. Estes, H. Menck and R. J. Huebner. An unusually high incidence of spontaneous lymphomas in a population of wild house mice. J. Nat. Cancer Inst. (in press).

54. Gross, L. 1962. Transmission of mouse leukemia virus through milk of virus-injected C3H female mice. Proc. Soc. Exp. Biol. Med. 109:830.

55. Krischke, W. and A. Graffi. 1963. The transmission of the virus of myeloid leukemia of mice by the milk. Acta Int. Union Against Cancer 19:360.

56. Law, L. W. and J. B. Moloney. 1961. Studies of congenital transmission of a leukemia virus in mice. Proc. Soc. Exp. Biol. Med. 108:715.

57. Gross, L. 1970. Oncogenic viruses, 2nd ed., p. 428. Pergamon Press, New York.

58. Rongey, R. W., A. Hlavackova, S. Lara, J. Estes and M. B. Gardner. Type B and C RNA virus in breast tissue and milk of wild mice (Mus musculus). J. Nat. Cancer Inst. (in press).

59. Mirand, E. A., R. F. Buffett and J. T. Grace, Jr. 1966. Mode of transmission of Friend virus disease. Proc. Soc. Exp. Biol. Med. 121:970.

60. Law, L. W. 1966. Transmission studies of a leukemogenic virus, MLV, in mice. Monogr. Nat. Cancer Inst. 22:267.

61. McKissick, G. E., R. A. Griesemer and R. L. Farrell. 1970. Aerosol transmission of Rauscher murine leukemia virus. J. Nat. Cancer Inst. 45:625.

62. Aoki, T., E. Boyse and L. J. Old. 1968. Wild-type Gross leukemia virus. I. Soluble antigen (GSA) in the plasma and tissues of infected mice. J. Nat. Cancer Inst. 41:89.

63. Gross, L. and Y. Dreyfuss. 1967. How is the mouse leukemia virus transmitted from host to host under natural life conditions? Carcinogenesis: a broad

critique, 20th Ann. Symp. Fundl. Cancer Res., p. 9. Williams & Wilkins, Baltimore.

64. Aoki, T., E. Boyse and L. J. Old. 1968. Wild-type Gross leukemia virus. III. Serological tests as indicators of leukemia risk. J. Nat. Cancer Inst. 41:103.

65. Vogt, P. K. 1965. Avian tumor viruses. Adv. Virus Res. 11:293.

66. Biggs, P. M., A. E. Churchill, D. G. Rootes et al. 1968. The etiology of Marek's disease, an oncogenic herpes-type virus. Perspectives in virology, ed. M. Pollard, vol. 6, p. 211. Academic Press, New York.

67. Calnek, B. W. 1968. Lymphoid leukosis virus: A survey of commercial breeding flocks for genetic resistance and incidence of embryo infection. Avian Dis. 12:104.

68. Cho, B. R., S. G. Kenzy and V. H. Kim. 1968. Case report. Mixed infection with subgroup A and B avian leukosis viruses in commercial white leghorn chickens. Avian Dis. 12:585.

69. Woods, G. N. and J. H. Campbell. 1972. Observations on the epidemiology of classical Marek's disease. In Oncogenesis and herpesviruses, ed. P. M. Biggs et al. Int. Agency Res. Cancer, Lyon.

70. Burmester, B. R. and R. F. Gentry. 1954. The transmission of avian visceral lymphomatosis by contact. Cancer Res. 14:34.

71. Weiss, R. A. and P. M. Biggs. Leukosis and Marek's disease viruses of feral red jungle fowl and domestic fowl in Malaya. J. Nat. Cancer Inst. (in press).

72. Burmester, B. R., R. F. Gentry and N. F. Waters. 1955. The presence of the virus of visceral lymphomatosis in embryonated eggs of normal appearing hens. Poultry Sci. 34:609.

73. Rubin, H., A. Cornelius and L. Fanshier. 1961. The pattern of congenital transmission of avian leukosis virus. Proc. Nat. Acad. Sci. 47:1058.

74. Purchase, H. G., J. J. Solomon and D. C. Johnson. 1969. Avian leukosis-sarcoma viruses and antibody in field flocks and their relationship to "leukosis" mortality. Avian Dis. 13:58.

75. Payne, L. N., L. B. Crittenden and W. Akazaki. 1968. Influence of host genotype on responses to four strains of avian leukosis virus. J. Nat. Cancer Inst. 40:907.

76. Vogt, P. K. and R. R. Friis. 1971. An avian leukosis virus related to RSV(O): Properties and evidence for helper activity. Virology 43:223.

77. Solomon, J. J., R. L. Witter, H. A. Stone and L. R. Champion. 1970. Evidence against embryo transmission of Marek's disease. Avian Dis. 14:752.

78. Calnek, B. W., H. K. Aldinger and D. E. Kahn. 1970. Feather follicle epithelium: A source of enveloped

and infectious cell-free herpesvirus from Marek's disease. Avian Dis. 14:219.

79. Frankel, J. W. and V. Groupe. 1971. Interactions between Marek's disease herpesvirus and avian leukosis virus in tissue culture. Nature 234:125.

80. Witter, R. L., K. Nazerian, H. G. Purchase and G. H. Burmester. 1970. Isolation from turkeys of a cell-associated herpesvirus antigenically related to Marek's disease virus. Amer. J. Vet. Res. 31:525.

81. Purchase, H. G., R. L. Witter, W. Okazaki and B. R. Burmester. 1971. Vaccination against Marek's disease. Perspectives in virology, ed. M. Pollard, vol. 7, p. 91. Academic Press, New York.

82. Simpson, C. E., D. W. Anthony and F. Young. 1957. Visceral lymphomatosis in a flock of turkeys. J. Amer. Vet. Med. Ass. 130:93.

83. Busch, R. H. and L. E. Williams. 1970. Case report. A Marek's disease-like condition in Florida turkeys. Avian Dis. 14:550.

84. Rigdon, R. H. and L. Leibovitz. 1970. Spontaneous occurring tumors in ducks. Review of the literature and report of 3 cases. Avian Dis. 14:431.

85. Blackmore, D. K. 1966. The clinical approach to tumors in cage birds. I. The pathology and incidence of neoplasia in cage birds. J. Small Anim. Pract. 7:217.

86. Petrak, M. L. and C. E. Gilmore. 1969. Neoplasms. In Diseases of cage and aviary birds, ed. M. L. Petrak, p. 459. Lea & Febiger. Philadelphia.

87. Halliwell, W. 1971. Lesions of Marek's disease in a great horned owl. Avian Dis. 15:49.

88. Lombard, L. S. and E. J. Witte. 1959. Frequency and types of tumors in mammals and birds of the Philadelphia Zoological Gardens. Cancer Res. 19:127.

89. Theilen, G. H., R. F. Zeigel and M. J. Twiehaus. 1966. Biological studies with RE virus (strain T) that induces reticuloendotheliosis in turkeys, chickens and Japanese quail. J. Nat. Cancer Inst. 37:731.

90. Staley, N. A., G. R. Noren and E. F. Jankus. 1972. Virus-like particles associated with myocarditis of turkeys. Amer. J. Vet. Res. 33:859.

91. Maldonado, R. L. and H. R. Bose. 1971. Separation of reticuloendotheliosis virus from avian tumor viruses. J. Virol. 8:813.

92. Nowinski, R. C., K. F. Watson, A. Yanif and S. Spiegelman. 1972. Serological analysis of the deoxyribonucleic acid polymerases. J. Virol. 10:959.

93. Bose, H. B. and A. S. Levine. 1967. Replication of the reticuloendotheliosis virus (strain T) in chicken embryo cell culture. J. Virol. 1:1117.

94. Aulisio, C. G. and A. Shelokov. 1969. Prevalence of reticuloendotheliosis in chickens. Immuno-

fluorescence studies. Proc. Soc. Exp. Biol. Med.
130:178.
95. Sarma, P. S., R. J. Huebner and D. Armstrong. 1964.
A simplified tissue culture tube neutralization test
for Rous sarcoma virus antibodies. Proc. Soc. Exp.
Biol. Med. 115:481.
96. Weiss, R. A. 1972. Helper viruses and helper cells.
In RNA viruses and host genome in oncogenesis, ed.
P. Emmelot and P. Bentvelzen, p. 117. North-Holland,
Amsterdam.
97. Witter, R. L. and J. J. Solomon. 1971. Epidemiology
of a herpesvirus of turkeys: Possible sources and
spread of infection in turkey flocks. Infec. Immun.
4:356.
98. Kenzy, S. G. and B. R. Cho. 1969. Transmission of
classical Marek's disease by affected and carrier
birds. Avian Dis. 13:211.
99. Zilber, L. A. and V. J. Shevliaghyn. 1964. Transformation of embryonic human cells by Rous sarcoma
virus. Nature 203:194.
100. Stenkvist, B. and I. Ponten. 1964. Morphological
changes in bovine and human fibroblasts exposed to
two strains of human fibroblasts exposed to two
strains of Rous sarcoma virus in vitro. Acta Path.
Microbiol. Scand. 62:315.
101. Jensen, F. C., A. J. Girardi, R. V. Gilden and H.
Koprowski. 1964. Infection of human and simian
tissue cultures with Rous sarcoma virus. Proc.
Nat. Acad. Sci. 52:53.
102. Shevliaghyn, V. J. and N. V. Karazas. 1970. Transformation of human cells by polyoma and Rous sarcoma
viruses mediated by inactivated Sendai virus. Int.
J. Cancer 6:234.
103. Boiron, M., C. Bernard and J. C. Chuat. 1969.
Replication of mouse sarcoma virus Moloney strain
(MSV-M) in human cells. Proc. Amer. Ass. Cancer
Res. 10:8.
104. Wright, B. S. and W. Korol. 1969. Infection of human embryonic cell cultures with the Rauscher murine
leukemia virus. Cancer Res. 29:1886.
105. Aaronson, S. A. and G. J. Todaro. 1970. Transformation and virus growth by murine sarcoma viruses
in human cells. Nature 225:458.
106. Klement, V., M. H. Freedman, R. M. McAllister, W. A.
Nelson-Rees and R. J. Huebner. 1971. Differences
in susceptibility of human cells to mouse sarcoma
virus. J. Nat. Cancer Inst. 47:65.
107. McAllister, R. M., M. Nicolson, M. B. Gardner, S.
Rasheed, R. W. Rongey, W. D. Hardy and R. V. Gilden.
Comparison of feline and murine type C viruses
released from RD cells with the RD-114 virus.
Nature (in press).
108. Todaro, G. T. and S. A. Aaronson. 1968. Human cell

strains susceptible to focus formation by human adenoviruses type 12. Proc. Nat. Acad. Sci. 61: 1272.

109. Traul, K. A., S. A. Mayyasi, C. E. Garon, G. Schidlovsky and L. M. Bulfone. 1972. Antigenic comparison of Rauscher murine leukemia virus cultivated in human embryo cells and mouse cells. Proc. Soc. Exp. Med. Biol. 139:10.

110. Ablashi, D. V., W. Turner, G. R. Armstrong and L. R. Bass. 1972. Characterization of murine Rauscher leukemia virus propagated in human cells. J. Nat. Cancer Inst. 48:615.

111. Aaronson, S. A. 1971. Common genetic alterations of RNA tumor viruses grown in human cells. Nature 230:445.

112. Macintyre, E. H., R. A. Grimes and A. E. Vatter. 1969. Cytology and growth characteristics of human tumor astrocytes transformed by Rous sarcoma virus. J. Cell Sci. 5:583.

113. McAllister, R. M., M. Nicolson, M. B. Gardner, R. W. Rongey, S. Rasheed, P. S. Sarma, R. J. Huebner, M. Hatanaka, S. Oroszlan, R. V. Gilden, A. Kabigting and L. Vernon. 1972. C-type virus released from cultured human rhabdomyosarcoma cells. Nature 235:3.

114. Rand, K. H. and C. Long. 1972. Syncytial assay for the putative human C-type virus, RD-114, utilizing human cells transformed by Rous sarcoma virus. Nature 240:187.

115. Svoboda, J. 1964. Malignant interaction of Rous virus with mammalian cells in vivo and in vitro. Monogr. Nat. Cancer Inst. 17:277.

116. Purchase, H. G., B. R. Burmester and C. H. Cunningham. 1971. Responses of cell cultures from various avian species to Marek's virus and herpesvirus of turkeys. Amer. J. Vet. Res. 32:1811.

117. Ahlstrom, C. G. 1964. Neoplasms in mammals induced by Rous chicken sarcoma material. Monogr. Nat. Cancer Inst. 17:299.

118. Huebner, R. J. 1967. In vitro methods for detection and assay of leukemia viruses. In Carcinogenesis: a broad critique, 20th Ann. Symp. Fundl. Cancer Res., p. 23. Williams & Wilkins, Baltimore.

119. Piraino, F., E. R. Krumbiegel and H. J. Wisniewski. 1967. Serologic survey of man for avian leukosis virus infection. J. Immunol. 98:702.

120. Solomon, J. J., H. G. Purchase and B. R. Burmester. 1969. A search for avian leukosis virus and antiviral activity in the blood of leukemic and nonleukemic adults and children. J. Nat. Cancer Inst. 42:29.

121. Roth, F. K. and R. M. Dougherty. 1971. Search for

group-specific antibodies of avian leukosis virus
in human leukemic sera. J. Nat. Cancer Inst. 46:
1357.

122. Richman, A. V., C. G. Aulisio, W. G. Jahnes and N.
     M. Tauraso. 1972. Avian leukosis antibody response
     in individuals given chicken embryo derived vaccines.
     Proc. Soc. Exp. Biol. Med. 139:235.

123. Morgan, H. R. and L. A. Wehle. 1967. Antibodies
     to Rous sarcoma virus (Bryan) in fowl, animal, and
     human populations of East Africa. II. Antibodies
     in domestic chickens, wildfowl, primates, and man
     in Kenya, and antibodies for Burkitt lymphoma cells
     in man. J. Nat. Cancer Inst. 39:1229.

124. Waters, T. D., P. S. Anderson, G. W. Beebe and R.
     W. Miller. 1972. Yellow fever vaccination, avian
     leukosis virus, and cancer risk in man. Science
     177:76.

125. Oroszlan, S., M. White, R. V. Gilden and H. P.
     Charman. 1972. A rapid radioimmunoassay for type
     C virus group specific antigen and antibody.
     Virology 50:294.

126. Yohn, D. S., J. S. Horoszewicz, R. R. Ellison, A.
     Mittleman, L. S. Chai and J. T. Grace. 1968.
     Immunofluorescent studies in human leukemia.
     Cancer Res. 28:1692.

127. Levine, P. H., R. B. Herberman, E. B. Rosenberg,
     P. D. McClure, A. Roland, R. J. Pienta and R. C. Y.
     Ting. 1972. Acute leukemia in identical twins:
     Search for viral and leukemia associated antigens.
     J. Nat. Cancer Inst. 49:943.

128. Priori, E. S., L. Dmochowski, B. Myers and J. R.
     Wilbur. 1971. Constant production of type C virus
     particles in a continuous tissue culture derived
     from pleural effusion cells of a lymphoma patient.
     Nature 232:61.

129. Gilden, R. V., W. P. Parks, R. J. Huebner and G. J.
     Todaro. 1971. Murine leukemia virus group-specific
     antigen in the C-type virus-containing human cell
     line ESP-1. Nature 233:102.

130. Gilden, R. V. and S. Oroszlan. 1972. Coevaluation
     of RNA tumor virus genome and vertebrate host genes.
     M. D. Anderson Symp., Houston, Texas. In press.

131. Fink, M. A., R. A. Malmgren, F. J. Rauscher, H. C.
     Orr and M. Karon. 1964. Application of immuno-
     fluorescence to the study of human leukemia. J.
     Nat. Cancer Inst. 33:581.

132. Fink, M. A., M. Karon, F. J. Rauscher, R. A.
     Malmgren and H. C. Orr. 1965. Further observations
     on the immunofluorescence of cells in human leukemia.
     Cancer 18:1317.

133. Hehlmann, R., D. Kufe and S. Spiegelman. 1972. RNA
     in human leukemic cells related to the RNA of a
     mouse leukemia virus. Proc. Nat. Acad. Sci. 69:1727.

134. Kufe, D., R. Hehlmann and S. Spiegelman. 1972. Human sarcomas contain RNA related to the RNA of a mouse leukemia virus. Science 175:182.
135. Green, M. 1972. Molecular hybridization: A powerful approach to the detection of viral nucleic acid sequences in human cancer. J. Nat. Cancer Inst. 48:1559.
136. Axel, R., J. Schlom and S. Spiegelman. 1972. Presence in human breast cancer of RNA homologous to mouse mammary tumor virus RNA. Nature 235:32.
137. Schlom, J. and S. Spiegelman. 1971. Simultaneous detection of reverse transcriptase and high molecular weight RNA unique to oncogenic RNA viruses. Science 174:840.
138. Schlom, J., S. Spiegelman and D. H. Moore. 1972. Detection of high molecular weight RNA in particles from human milk. Science 175:542.
139. Axel, R., S. C. Gulati and S. Spiegelman. 1972. Particles containing RNA-instructed DNA polymerase and virus-related RNA in human breast cancers. Proc. Nat. Acad. Sci. 69:3133.
140. Charney, J. and D. H. Moore. 1971. Neutralization of murine mammary tumor virus by sera of women with breast cancer. Nature 229:627.
141. Muller, M. and H. Grossmann. 1972. An antigen in human breast cancer sera related to the murine mammary tumor virus. Nature 237:116.
142. Moore, D. H., J. Charney, B. Kramarsky, E. Y. Lasfargues, N. H. Sarkar, M. J. Brennan, J. H. Burrows, S. M. Sirsat, J. C. Paymaster and A. B. Vaidya. 1971. Search for human breast cancer virus. Nature 221:611.
143. Sarkar, N. H. and D. H. Moore. 1972. On the possibility of a human breast cancer virus. Nature 236:103.
144. Gotze, D., S. Ferrone and R. A. Reisfeld. 1972. Serologic cross-reactivity between H-2 and HL-A antigens. I. Specific reactivity of rabbit anti-HL-A sera against murine cells. J. Immunol. 109:439.
145. Vogt, P. K. 1972. The emerging genetics of RNA tumor viruses. J. Nat. Cancer Inst. 48:3.
146. McAllister, R. M., W. A. Nelson-Rees, E. Y. Johnson, R. W. Rongey and M. B. Gardner. 1971. Disseminated rhabdomyosarcomas formed in kittens by cultured human rhabdomyosarcoma cells. J. Nat. Cancer Inst. 47:603.
147. Livingston, D. M. and G. J. Todaro. Endogenous type C virus from a cat cell clone with properties distinct from previously described feline C-type virus. Virology (in press).
148. Sarma, P. S., J. Tseng, Y. K. Lee and R. V. Gilden. A covert C-type virus in cat cells similar to RD-114

virus. Nature (in press).
149. Todaro, G. J., P. Arnstein, W. P. Parks, E. H.
     Lennette and R. J. Huebner. A type C virus in
     human rhabdomyosarcoma cells after inoculation
     into antithymocyte serum treated NIH Swiss mice.
     Proc. Nat. Acad. Sci. (in press).
150. Freeman, A. E., P. J. Price, H. J. Igel, J. C.
     Young, J. M. Maryak and R. J. Huebner. 1970.
     Morphological transformation of rat embryo cells
     induced by diethylnitrosoamine and murine leukemia
     viruses. J. Nat. Cancer Inst. 44:65.
151. Price, P. J., A. E. Freeman, W. T. Lane and R. J.
     Huebner. 1971. Morphological transformation of
     rat embryo cells by the combined action of 3-
     methylcholanthrene and Rauscher leukemia virus.
     Nature 230:144.
152. Rhim, J. S., W. Voss and H. Y. Cho. 1971. Malignant
     transformation induced by 7,12-dimethyl-benz(a)-
     anthracene in rat embryo cells infected with Rauscher
     leukemia virus. Int. J. Cancer 7:65.
153. Rhim, J. S., H. Y. Cho, M. H. Joglekar and R. J.
     Huebner. 1972. Comparison of the transforming
     effects of benzo(a)pyrene in mammalian cell lines in
     vitro. J. Nat. Cancer Inst. 48:949.
154. Freeman, A. E., G. J. Kelloff, R. V. Gilden, W. T.
     Lane, A. P. Swain and R. J. Huebner. 1971.
     Activation and isolation of hamster-specific C-type
     RNA viruses from tumors induced by cell cultures
     transformed by chemical carcinogens. Proc. Nat.
     Acad. Sci. 68:2386.
155. Rhim, J. S., F. G. Duh, H. Y. Cho, E. Elder and
     M. L. Vernon. 1973. Activation of a type C RNA
     virus from tumors induced by rat kidney cells
     transformed by a chemical carcinogen. J. Nat.
     Cancer Inst. 50:297.
156. Weinstein, I. B., R. Gebert, U. C. Stadler, J. M.
     Orenstein and R. Axel. 1972. Type C virus from
     cell cultures of chemically induced rat hepatomas.
     Science 178:1100.
157. Todaro, G. T. 1972. Spontaneous release of type
     C viruses from clonal lines of spontaneously trans-
     formed Balb-3T3 cells. Nature 240:157.
158. Priori, E. S., D. E. Anderson, W. C. Williams and
     L. Dmochowski. 1972. Immunological studies on
     human breast carcinoma and mouse mammary tumors.
     J. Nat. Cancer Inst. 48:1131.

FELINE TUMOR VIRUSES

C. G. Rickard, J. E. Post, F. Noronha,
E. Dougherty, III and L. M. Barr
Pathology Department
New York State Veterinary College, Cornell University
Ithaca, New York

Most normal, adolescent or adult cats do not have C-type viruses that can be demonstrated by electron microscopy, transmission, tissue culture inoculation, or immunological methods. On the other hand, such viruses can be demonstrated in the neoplastic tissue and many other organs of most cats with lymphosarcoma or lymphocytic leukemia; such myeloproliferative diseases as myelocytic leukemia, erythremic myelosis, and reticuloendotheliosis; and in fibrosarcomas and occasionally other sarcomas. Morphologically identical virus particles can also be found in many cases of hypoplastic anemia, lymphoid hyperplasia, some neoplasms in which they may be incidental passengers such as mammary tumors (1), in some cat fetuses, in some long-term embryo tissue cultures, and in cat embryo tissue cultures treated with iododeoxyuridine (IdU). The important questions which relate to potential biohazards include the following:

1. What is the host range of these viruses?
2. Are they eliminated in infectious quantities from their hosts into the environment?
3. Is there evidence of horizontal transmission?
4. Is there evidence of human infections or neoplasms induced by the feline viruses?

We have concluded that cats have two kinds of leukemia virus. One is an endogenous virus that is

vertically transmitted in the genetic composition of
the cat and is usually well repressed, but can be
derepressed under various natural and experimental
conditions. The other is a group of exogenous viruses
that are infectious for cat cells and can transmit
horizontally. Three subgroups or serotypes of the latter
have been described (2) and labeled A, B, and C. The
leukemogenic activity of the exogenous (so-called
"infectious") feline C-type viruses is well established
by experimental cell-free transmission. The oncogenic
activity of the endogenous feline viruses has not yet
been demonstrated.

The animal hosts which are suceptible to infection
and oncogenesis by the conventional, exogenous feline
leukemia virus (FeLV) are the cat under natural
conditions, and the cat and dog experimentally. The
host range in the tissue culture includes the cells of
the cat, man, marmoset, dog, and pig. The infectivity
of most strains of exogenous FeLV for human cells, even
on initial isolation or low passage, is equal to or
one or two logs less than for cat cells, although one
isolate of FeLV has been reported (3) not to infect
human cells. On the other hand, the endogenous feline
leukemia virus has very low infectivity for cat cells.
Dr. George Todaro and Dr. Murray Gardner have indicated
it infects human, rhesus, dog, and pig cells.

A young cat with lymphatic neoplasia induced by
injection of an infectious (exogenous) strain of FeLV
is shown in Fig. 1. The lesions closely resemble one of
the common pathological forms of the spontaneous disease
in cats. The widespread distribution of C-type virus
particles in the tissues of such natural and experimental
cases of lymphosarcoma or lymphocytic leukemia is
demonstrated in Table 1. Of eleven such animals, all
had virus demonstrable by electron microscopy. However
the virus particles were more numerous in some organs
than others. Such widespread systemic infection suggests
excretion of virus in the body secretions.

Whereas most clinically normal cats do not have
C-type virus in their blood or tissues, there are a few
exceptions. We have observed 8 such animals over periods
of 1-2 years. Three had infectious (exogenous),
leukemogenic virus and 5 apparently had endogenous virus
since it did not infect cat cells. All of these animals
remained asymptomatic, although Hardy (4) has reported
such animals that later developed lymphosarcoma.

Experimental transmission of FeLV to the dog was
possible (5) with doses similar in magnitude to those
effective in cats and with low-passage FeLV (third passage

Table 1. Distribution of C-type Virus Particles in Eleven Cats with Lymphocytic Leukemia/Lymphoma

	Positive Cats	Positive E.M. Blocks
Bone marrow	11/11*	35/62*
Lymph node	6/8	20/99
Thymus	5/8	11/42
Spleen	3/9	3/26
Pleural fluid cells	1/1	2/4
Buffy coat	3/11	4/24
Intestine	3/6	13/59
Liver	3/8	3/25
Lung	2/7	3/16
Kidney	2/8	19/50
Ovary	1/3	1/8
Tonsil	1/5	1/22
Brain	1/4	1/30
Pancreas	0/5	0/15
Salivary gland	0/5	0/18
Adrenal gland	0/5	0/11
Heart	0/6	0/19
Thyroid gland	0/6	0/18
Urinary bladder	0/3	0/7
Uterus	0/2	0/5
Testis	0/2	0/4
Total, E.M. blocks		116/570

* Number positive/number examined.

Sections were cut from each block to cover one standard grid. All of the visible area of the grid was scanned for virus particles.

Six of the cats had spontaneous, and five had experimental lymphocytic neoplasia. No significant difference was observed between these two groups in the incidence or organ distribution of the virus.

Blocks from neoplastic lymph nodes had a significantly higher incidence of C-type virus particles than normal lymph nodes. C-type viral buds were observed on lymphocytes, eosinophils, and neutrophils in vascular spaces.

in cats from a spontaneous case of feline lymphosarcoma).
The virus was oncogenic, producing lesions very similar
to those in the cat. An example is presented in Fig. 2.
Fetal dogs were more susceptible than puppies inoculated
after birth. As shown in Table 2, fatal lymphosarcomas
were produced in puppies inoculated in utero 14 days or
2 days before birth. The tumors contained virus
demonstrable by electron microscopy and by transmission
with tumor induction to other dogs and cats. Puppies
inoculated on the day of birth developed fatal lymphosarcomas pathologically indistinguishable from those
inoculated by earlier inoculation. However, virus was
not demonstrable in the tumors. Puppies inoculated 2 or
3 days after birth did not develop neoplastic disease.
There was abundant viral group-specific antigen in the
tumors induced by inoculation of fetuses 14 days before
birth, less in those inoculated 2 days before birth, and
little antigen in tumors induced by inoculation on the
day of birth.

The role of endogenous FeLV is less well understood,
inasmuch as it has been possible only recently to distinguish it from the exogenous virus. Endogenous virus
is perhaps potentially subject to expression in all cats.
Let us consider for a moment what constitutes a "normal"
cat. For example, we have an SPF cat colony which was
initiated 6 years ago by taking into a barrier-protected
facility 10 caesarean-derived kittens and which has now
multiplied to about 200 animals. No leukemias or other
neoplastic diseases have appeared. Hundreds of electron
microscopic examinations of bone marrow, thymus, and other
tissues have been conducted on cats of all ages in this
colony, as well as on cat embryo tissue cultures. No
C-type virus has been seen, even in fetuses, with the
exception that C-type virus was seen in three placentas.
On the other hand, typical C-type virus could be induced
by treating some (about a third) of the cat embryo tissue
cultures with IdU. The viruses from the placentas and
IdU-treated tissue cultures did not infect other feline
tissue culture cells upon inoculation. Our interpretation is that some or all of the cats in this colony have
well-repressed endogenous feline leukemia virus.
Tissue cultures from these cats are susceptible to
exogenous FeLV infection and the cats are fully
susceptible to the oncogenic effect of feline leukemia
or sarcoma viruses.

The C-type viruses observed in the anemias, lymphoid
hyperplasias, and other stimulated cells have not been
characterized as yet in terms of host range, gs species
(gs-1) antigen, and antigenic identification of
transcriptases which would permit their classification
as endogenous or exogenous viruses.

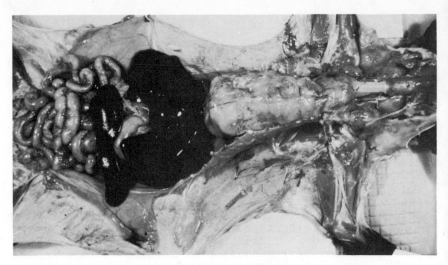

FIGURE 1.   Lymphosarcoma in a cat induced by injection
            of a newborn kitten with feline leukemia
virus. There is massive neoplastic involvement of the
thymus gland (which almost fills the thoracic cavity)
and several cervical and mesenteric lymph nodes.

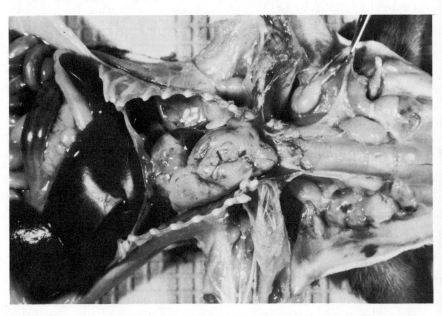

FIGURE 2.   Lymphosarcoma in a dog induced by injection
            with feline leukemia virus. There is neo-
plastic enlargement of the thymus gland, cervical and
prescapular lymph nodes, spleen, and liver. FeLV induced
such lesions when injected into fetal or newborn puppies.

Table 2. Lymphoma/Lymphocytic Leukemia Induced in Dogs by Feline Leukemia Virus

Age at Inoc.	No. Inoc.	No. with Fatal Tumor	Incubation in Days Postinoc. Av. and Range	C-Type Virus in Tumor	GS-Antigen in Tumor, CF Titer*
-14	5	4	54 (48-62)	+	128 (64-256)
- 2	3	1	52	+	64
0	11	7	98 (34-199)	-	4 (0-8)
+ 2	2	0		No Tumor	
+ 3	4	0		No Tumor	

* Reciprocal of titer in complement-fixation test using a rabbit antiserum containing predominantly gs-interspecies (gs 3) antibody.

Puppies were inoculated with FeLV either as fetuses at 14 or 2 days before birth, on the day of birth, or at 2 or 3 days after birth. Electron microscopic examination of the tumors revealed budding C-type virus in all 5 which were induced by inoculation of fetuses, and in none of the 7 induced by inoculation on the day of birth.

It should be emphasized that both the endogenous and the exogenous feline leukemia viruses can infect human cells.

There is abundant evidence that C-type viruses cause fibrosarcomas of the cat. There are associated leukemia helper viruses, as in the chicken and rodents. Sarcomagenic virus is present not only in the tumor, but also in the spleen and presumably in other organs. By electron microscopy C-type particles are found in the bone marrow and salivary glands as well.

As far as host range is concerned, feline sarcoma virus (FeSV) induces tumors in the cat, dog, rabbit, sheep, marmoset, and three Macaca species including the common rhesus monkey. The most progressive, malignant tumors were produced by inoculation of fetal or newborn cats and fetal dogs. Cats injected with the virus after they are a few weeks of age develop small, regressing tumors. Dogs inoculated after birth tend to develop tumors without demonstrable virus which usually regress after 1-2 months. Examples of the lesions induced by feline sarcoma virus in cats and dogs are shown in Figs. 3 and 4.

The evidence for horizontal spread of the conventional (exogenous) feline leukemia viruses comes from several sources. Brodey (6,7), Hardy (8,4) and others have described clusters of cases of lymphosarcoma in unrelated cats in single households and the appearance of such cases after the introduction of an infected animal. W. Jarrett (9) was stimulated to start his work, which led to the first evidence of transmission in this species, by observing 8 cases of lymphosarcoma in a household colony of unrelated cats.

In addition we observed 5 contact controls among 26 such animals that acquired FeLV infection from littermates that had been inoculated intraperitoneally (10). Hoover et al. (11) observed a similar contact animal which developed leukemia. Sibal et al. (12) showed, and others have confirmed, that many normal cats carry neutralizing antibody to the A subgroup of the infectious leukemia viruses. This implies introduction of the antigen after immune competence, and probably postnatal infection by such a virus. Essex et al. (13,14) observed that mother cats developed some immunity after their kittens had been inoculated with FeLV. This was expressed as an increased resistance to challenge by feline sarcoma virus in their subsequent litters. Humeral antibody was also demonstrated by a positive cell-membrane fluorescent antibody test.

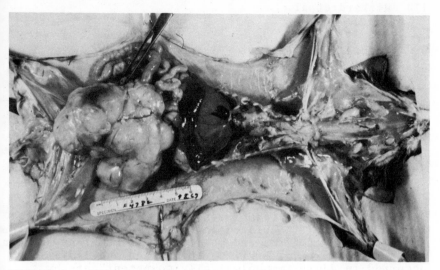

FIGURE 3.   Fibrosarcoma induced in a cat by subcutaneous
            injection of feline sarcoma virus into a
newborn kitten.   The large, multilobulated lesion at the
left of the picture developed at the site of inoculation.
The sarcoma had metastasized to several regional lymph
nodes.

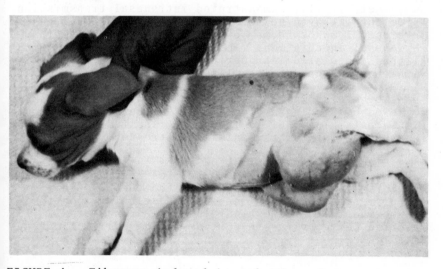

FIGURE 4.   Fibromas induced in a dog by subcutaneous
            injection of a newborn puppy with feline
sarcoma virus.   The two large, globular lesions in the
lower abdomen developed at the two inoculation sites.
Such lesions tend to regress after several weeks.   However
invasive and metastasizing fibrosarcomas can be induced
by injection of feline sarcoma virus into fetal puppies.

Recently, W. Jarrett (9) reported the results of additional experiments on horizontal transmission, which he conducted in collaboration with Essex and Hardy. Both young adult and neonatal cats were infected with purified FeLV and mixed with both littermates and unrelated cats. Control litters were housed at a distance. Evidence of horizontal transmission was based on: (1) the appearance of FeLV gs antigen in the circulating leukocytes of the inoculated cats and later in the contact animals, (2) the finding of replicating FeLV in the platelets of the inoculated cats and later the contact animals, (3) the raising of antibodies demonstrated by the cell membrane antigen test in both inoculated and contact animals, (4) the appearance of leukemia in both inoculated and contact cats, and (5) the demonstration of virus and positive cell membrane antigen tests in both groups. Virus replication and leukemia occurred in adult immunologically competent cats by inoculation or contact. The isolated control animals remained negative.

The route of horizontal infection is not established Experimentally we demonstrated that urine from a cat with renal lymphosarcoma and blood from an asymptomatic carrier were leukemogenic when inoculated into kittens. Hoover et al. (11) demonstrated intranasal transmission. Three out of seven newborn gnotobiotic or SPF kittens developed lymphosarcoma between 170 and 329 days after intranasal inoculation. In the same experiment, 7 of 8 kittens inoculated intraperitoneally developed leukemia with the somewhat shorter latent period of 82-180 days.

We have shown that light scarification of the skin followed by placing a drop of either leukemia or sarcoma virus-containing fluid on the skin resulted in infection in about a tenth of the animals so treated.

Electron microscopic studies of cats with lympho-sarcoma show typical C-type particles budding from renal tubule cells and in the lumens of the tubules, in alveolar spaces in the lung, in intestinal lymphatic tissue and in the intestinal lumen. Gardner et al. (15) reported C-type virus in salivary gland cells of cats with leukemia or sarcoma. Such observations suggest routes of possible excretion of virus.

The most pressing question relating to potential biohazard in man is whether enough of the feline leukemia or sarcoma virus is excreted by cats and whether there are susceptible humans which can get infected. There is no reported evidence that humans have been infected. Many attempts have been made to demonstrate virus or distinctive feline oncornavirus antigens in human tumors

or tissues, as well as serological studies on human sera.
Schneider and Riggs (16) reported that in a serologic
survey employing an indirect immunofluorescence test of
626 veterinarians and 67 other persons for evidence of
FeLV infection, one veterinarian was seropositive but
he became seronegative when retested 8 months later.
Hong (17) tested sera from 41 laboratory and animal
care personnel who had been in potential contact with
FeLV and FeSV, using complement-fixation, passive
hemagglutination, and focus reduction tests, with
negative results. The results of other workers apparently have been negative, and some such work has not
been reported.

     In view of the fact that newborn and fetal cats
and dogs are much more susceptible to the oncogenic
effect of the FeLV and FeSV, perhaps the search for
infected humans should be concentrated on children who
were potentially exposed when they were very young or in
utero. Experimental evidence in the dog, mentioned
above, indicated that slightly older, less susceptible
animals developed fatal lymphatic leukemias or fibrosarcomas which had no demonstrable virus or gs antigens.
Although there was a low titer of humeral antibody, this
was not demonstrable in all animals and the persistence
of this antibody is not well documented. Thus it is
possible that infection of humans under suboptimal
conditions of infective dose or susceptibility might
result in leukemias or sarcomas which would be difficult
to associate with the inciting virus.

     Laboratory personnel who work with FeLV or FeSV
should use appropriate biohazard procedures. The in vitro
virology and tissue culture techniques will be discussed
by others at this conference. The handling of infected
cats requires some special precautions. Proper restraint
should be used to prevent bites and scratches. Cages
should be cleaned by procedures which do not produce
aerosals from fluids possibly contaminated with urine,
saliva, blood, or feces. One satisfactory cage design
is shown in Figs. 5 and 6.

     An experience in our laboratory illustrates the
need for better methods of detection and evaluation of
suspected biohazards. Four years ago, a 23-year-old
medical student became ill with acute myelocytic
leukemia. He had worked during the previous summer,
6-8 months before, in our research project on feline
leukemia. He had participated in the inoculation of
kittens with FeLV, collection of blood and bone marrow
specimens from infected kittens, and other animal work.
An extensive laboratory and epidemiological investigation
of his disease was conducted by the National Cancer

FIGURE 5. Experimental cats confined in modified
Horsfall cages. The cages are of stainless
steel construction with gasketted doors. Air is introduced at negative pressure through a high efficiency
(FG-50) filter, pulled through the cages, and filtered
again before discharge from the roof of the building,
which is at an isolated location.

FIGURE 6. Mother cat and two kittens in a modified
Horsfall cage. Usually a pregnant cat is
placed in a sterile cage and the kittens inoculated on
the day of birth. The corrugated paper floor covering
and the cardboard litter box containing commercial ground
corncob litter are disposed of by incineration. Cages
are autoclaved after use.

Institute with our full cooperation. Several laboratories
examined his blood, plasma concentrates, serum, bone
marrow, blood cells, etc., for evidence of virus, antigens, antibodies, or other components that might
establish an association with feline leukemia virus.
The results of the tests were negative. It might also
be mentioned that the patient's disease was myelocytic,
whereas the strain of FeLV used in our laboratory at
that time produced lymphocytic neoplasia in the cat and
dog. Thus the evidence was negative. However negative
evidence is by definition usually not conclusive.
Specimens from this patient and others like him will
presumably be tested further as new procedures are
developed. Additional epidemiological investigations
will be carried out. In the meantime all laboratories
that work with known or suspected oncogenic agents need
realistic guidelines for operation.

References

1. Feldman, D. G. and L. Gross. 1971. Electron
    microscopic study of spontaneous mammary carcinomas
    in cats and dogs: Virus-like particles in cat
    mammary carcinomas. Cancer Res. 31:1261.
2. Sarma, P. S. 1973. Viral envelope antigens of
    feline leukemia and sarcoma viruses. Unifying
    concepts of leukemia, Bibl. Haemat. No. 39 (in press).
3. Jarrett, O., H. M. Laird and D. Hay. 1972.
    Restricted host range of a feline leukemia virus.
    Nature 238:220.
4. Hardy, W. D., Jr., Y. Hirshaut and P. Hess. 1973.
    Detection of the feline leukemia virus and other
    mammalian oncornaviruses by immunofluorescence.
    Unifying concepts of leukemia, Bibl. Haemat. No. 39
    (in press).
5. Rickard, C. G., J. E. Post, F. Noronha and L. M.
    Barr. 1973. Interspecies infection by feline
    leukemia virus: Serial cell-free transmission in
    dogs of malignant lymphomas induced by feline leu-
    kemia virus. Unifying concepts of leukemia, Bibl.
    Haemat. No. 39 (in press).
6. Brodey, R. S., S. K. McDonough, F. L. Frye and W. D.
    Hardy. 1970. Epidemiology of feline leukemia
    (lymphosarcoma). Comparative leukemia research 1969,
    Biblio. Haemat. No. 36, p. 333. Karger, Basel.
7. Brodey, R. S. 1971. Comments on epidemiological
    aspects of feline leukemia virus. J. Amer. Vet.
    Med. Ass. 158:1123.
8. Hardy, W. D., Jr. 1971. Immunodiffusion studies
    of feline leukemia and sarcoma. J. Amer. Vet. Med.
    Ass. 158:1060.
9. Jarrett, W. F. H. 1972. Feline leukemia. J. Clin.

Path. 25 (suppl. 6):43.
10. Rickard, C. G., J. E. Post, F. Noronha and L. M. Barr. 1969. A transmissible virus-induced lymphocytic leukemia of the cat. J. Nat. Cancer Inst. 42:987.
11. Hoover, E. A., C. B. McCullough and R. A. Griesemer. 1972. Intranasal transmission of feline leukemia. J. Nat. Cancer Inst. 48:973.
12. Sibal, L. R., M. A. Fink, E. J. Plata, B. E. Kohler, F. Noronha and K. M. Lee. 1970. Methods for the detection of viral antigen and antibody to feline leukemia virus. J. Nat. Cancer Inst. 45:607.
13. Essex, M., G. Klein, S. P. Snyder and J. B. Harrold. 1971. Antibody to feline oncorna-associated cell membrane antigen in neonatal cats. Int. J. Cancer 8:384.
14. Essex, M. 1973. Relationship between humeral antibodies and the failure to develop progressive tumors in cats injected with feline sarcoma virus. Unifying concepts of leukemia, Bibl. Haemat. No. 39 (in press).
15. Gardner, M. B., R. W. Rongey, E. Y. Johnson, R. DeJournett and R. J. Huebner. 1971. C-type tumor virus particles in salivary tissue of domestic cats. J. Nat. Cancer Inst. 47:561.
16. Schneider, R. and J. L. Riggs. 1973. A serological survey of veterinarians for antibody to feline leukemia virus. J. Amer. Vet. Med. Ass. 162:217.
17. Hong, C-B. 1973. Feline leukemia-sarcoma viruses: Detection of viral antigens and antibodies and cloning of the virus. Ph. D. thesis, Cornell University.

SIMIAN TUMOR VIRUSES

R. A. Griesemer and J. S. Manning
California Primate Research Center and
Departments of Veterinary Pathology and Veterinary Microbiology
University of California
Davis, California

The discovery since 1968 of five oncogenic viruses
in nonhuman primates has created an exciting new area
for research in viral oncology. Simultaneously it has
raised questions about the possible dangers to man from
exposure to simian tumor viruses. One might presuppose
that because nonhuman primates more closely resemble
man than do other laboratory animals, their viruses are
more likely to infect man. This view was well stated in
the introduction to a recent symposium on viruses of
South American monkeys by Weller (1), who said:

"... it is amazing that the iatrogenic exposure of
man to these simian viruses has been relatively free
of unforeseen consequences.... Yet, until the hazards
to man of the new oncogenic herpesviruses are defined,
distribution of the new agents should be monitored and
limited to investigators competent and equipped to
deal with dangerous viruses. Currently we are forced to
develop policies designed to protect the health of the
public on the basis of an educated guess, rather than
on the rational evidence of scientific facts."

The purpose of this report is to review the data
related to hazards of simian tumor viruses to man and
experimental nonhuman primates and to provide recommena-
tions for biohazard control. Considered particularly
important are (1) production of tumors in man, (2)
seroconversion in man, (3) host range of the viruses,

179

(4) contact or cage-mate transmission, (5) excretion of virus in urine and feces, (6) infection by aerosols, and (7) distribution of the virus in nature.

HERPESVIRUS SAIMIRI

Of the three known DNA oncogenic viruses of nonhuman primates (Herpesvirus saimiri, Herpesvirus ateles, Yaba virus), Herpesvirus saimiri has attracted the most attention because the neoplastic disease produced in experimental animals is rapidly fatal. First reported in 1968 (2), it is nonpathogenic for its natural host, the squirrel monkey (Saimiri sciureus) (3). Of 65 captured squirrel monkeys 88% had serum-neutralizing antibodies to H. saimiri, and virus could be recovered from the blood or circulating lymphocytes of 11 of 13 (4). Melendez et al. (5) have found that 70-75% of squirrel monkeys collected at random have high levels of serum-neutralizing antibodies. Serum neutralizing antibodies also occur naturally in spider and owl monkeys (6). When inoculated with the virus experimentally, malignant lymphoma or reticuloproliferative disease is produced in marmosets, owl monkeys, spider monkeys, cinnamon ringtail monkeys, African green monkeys, and rabbits (7-15,5). The mortality rate in marmosets and owl monkeys is very high, with most dying 20-50 days after injection (16). Viral antigen or viral particles cannot be demonstrated in fresh tumor cells but appear in vitro after these cells are cocultivated with susceptible cells (17,18). Vero, marmoset kidney, or owl monkey kidney cells are most often used, but the virus can also be propagated in African green monkey, squirrel monkey, and rhesus monkey cells (19), in human WI38 cells (20), human embryo, spider monkey kidney, and dog fetal lung cells (5). Talapoin monkeys, Galagos (12), rhesus monkeys, stumptailed monkeys, baboons, chimpanzees (5), chick embryos and mice (3) appear not to be susceptible. Dogs did not develop antibodies or lesions within one year after inoculation (5).

As part of their studies, Wolfe et al. (13) have included experiments related to biohazards. They did not observe horizontal transmission from experimentally infected marmosets to susceptible cage-mates. One marmoset was exposed to infected marmosets for 10 consecutive months without developing antibodies to H. saimiri and when challenged with this virus died with lymphoma. Although the presence of tumor cells in the lacrimal glands, tonsils, salivary glands, intestines, and kidneys would suggest excretion of virus in tears, saliva, feces and urine, Wolfe et al. (13) were unsuccessful in recovering virus from throat and

rectal swabs from infected marmosets. They did succeed
in rescuing H. saimiri from unclarified rectal specimens,
however, indicating that the cellular component of
excreta may contain viral information. Ablashi et al.
(19) recovered H. saimiri from an oral swab of one of
five infected owl monkeys that had a nasal discharge.
Anal swabs were negative. Wolfe et al. (13) have
monitored serum from personnel in contact with infected
monkeys or cultures and as yet have found no serum
antibodies to H. saimiri. Melendez et al. (3) reported
that acute and convalescent sera from two patients with
infectious mononucleosis did not neutralize H. saimiri
in vitro.

HERPESVIRUS ATELES

Herpesvirus ateles was isolated from the kidney
culture of a black spider monkey (Ateles geoffroyi) (21).
It is oncogenic, producing lymphomas in cottontop
marmosets and owl monkeys (21-23). Spider monkeys (the
reservoir host), squirrel monkeys, and rabbits are not
susceptible (5). Six of 12 spider monkeys had naturally
acquired serum-neutralizing antibodies (5). H. ateles
multiplies in owl monkey kidney, whole human embryo,
human embryonic lung, squirrel monkey fetal lung,
spider monkey kidney, goat bursa, rabbit kidney, and
hamster heart cell cultures and forms plaques in squirrel
monkey fetal lung and hamster heart cells (24,5). One
contact-exposed animal developed lymphoma from which the
virus was re-isolated (Hunt, pers. comm.). Unfortunately,
insufficient data are available as yet to evaluate the
biohazard potential of this virus.

YABA VIRUS

A third simian oncogenic DNA virus, Yaba poxvirus,
was originally isolated from tumor material collected
during an epizootic of self-limiting subcutaneous
histiocytomas in a monkey colony at Yaba, Nigeria (25).
Few spontaneous outbreaks have occurred since. African
green monkeys often have serum antibodies (26). Asiatic
monkeys, particularly rhesus and cynomolgus, are the most
susceptible hosts. African monkeys (Cercopithecus sp.)
are resistant, although African monkeys born in the
United States were susceptible (27). After subcutaneous
inoculation, histiocytic tumors become palpable at the
inoculation sites in 5-10 days, attain maximum size of 2-5
cm in three weeks, and regress spontaneously by 6-8
weeks after inoculation (28,20). It is not uncommon for
new tumors to form in the skin while others are in
various stages of regression in the presence of humoral

viral neutralizing antibody. Ultimately all tumors regress spontaneously but this may take several months. The experimental host range includes chick embryos (30) and cultured simian cells (31). Mice, rats, hamsters, guinea pigs, dogs and cats are not susceptible (32).

The mode of transmission of spontaneous Yaba-induced histiocytomas is unknown. In one report (33) infected and uninfected monkeys were kept in the same room and even in the same cage. During a 3-year period, only two contact controls in this colony of 250 monkeys contracted the disease. Wolfe et al. (29) demonstrated aerosol transmission of Yaba virus experimentally. Pulmonary tumors were found in 5 of 14 monkeys killed between 2 and 16 weeks after exposure. Virus was not recovered from nasal or rectal swabs. A sixth monkey developed a large deforming nasal tumor and multiple subcutaneous tumors which regressed 22 weeks after exposure. Virus was recovered from nasal swabs while the nasal tumor was present. Horizontal transmission to three cage-mate contact-exposed monkeys did not occur.

Yaba virus is the only simian tumor virus known to be infectious for man. Man has been infected accidentally (33) and intentionally (34). Humans and monkeys respond to comparable doses. In man, histiocytomas develop at inoculation sites in 5-7 days. The tumors are generally smaller in man than in monkeys and regression occurs more rapidly (in 3-4 weeks).

MASON-PFIZER MONKEY VIRUS

A virus morphologically similar to known oncorna-viruses was isolated from a spontaneous carcinoma of the breast in a rhesus monkey by cocultivating tumor cells with embryonic monkey cells (35-38). The virus replicates in rhesus monkey embryo cells, rhesus monkey lung cells, chimpanzee lung cells, human leukocytes, and human embryonic cells (39) and has a number of biophysical properties in common with RNA tumor viruses (40). Multinucleated cells are formed in vitro in several simian and human cell lines (41). Rhesus foreskin cells are transformed by Mason-Pfizer virus and the transformed cells induce palpable tumors that regress in newborn rhesus monkeys (42). The oncogenic potential of the Mason-Pfizer virus is as yet unknown, as is its causal relation-ship, if any, to rhesus breast cancer. In addition, Nowinski et al. (43) did not find evidence for Mason-Pfizer virus antigens in extracts of human tumors.

SIMIAN SARCOMA VIRUS (LAGOTHRIX)

C-type oncornavirus particles were demonstrated by electron microscopy in a naturally occurring fibro-sarcoma in a 3-year-old woolly monkey (Lagothrix sp.) (44) and isolated in woolly monkey muscle cells (45). The virus, designated simian sarcoma virus type 1, has the morphologic and physicochemical properties of a C-type oncornavirus. The internal gs-1 antigens and reverse transcriptase enzymes are similar to those in the gibbon lymphosarcoma virus but differ from those in murine, feline, and avian oncornaviruses and from simian foamy viruses (46,47). Simian sarcoma virus transforms marmoset lung, marmoset skin, marmoset muscle, human fetal lung, human embryo, and cat embryo cells in vitro (48). Wolfe et al. (48) demonstrated the presence of an excess of nontransforming associated virus in simian sarcoma virus preparations. When inoculated intra-muscularly with simian sarcoma virus, marmosets developed well-differentiated fibrosarcomas which contained virus particles (49). The malignant potential and host range of this virus have yet to be fully determined.

GIBBON LYMPHOSARCOMA VIRUS

C-type oncornavirus particles were demonstrated electron microscopically in the neoplastic tissues of a gibbon (Hylobates lar) with spontaneous lymphosarcoma and subsequently isolated from primary cultures of gibbon tumor cells (50). Normal bovine, human, and simian cells were susceptible to infection in vitro but exhibited neither transformation nor cytopathic effects. Evidence for oncogenicity is lacking, but the virus associated with a spontaneous gibbon lymphosarcoma possesses the other characteristics of known RNA oncogenic viruses.

Case reports of spontaneous lymphosarcoma in gibbons are not uncommon (51-55). Johnsen et al. (54) report an unusually high incidence of four affected gibbons in a colony of 120 over a 16-month period. The affected animals were all over 5 years of age, previously splenectomized, and had been given inoculations of whole blood from common donor animals. More recently, several gibbons in this colony, including animals caged together, have developed myeloid leukemia (56), suggesting contact transmission. The search for possible viral etiology is still in progress.

SIMIAN FOAMY VIRUSES

Although not yet demonstrated to be oncogenic, simian foamy viruses must be mentioned because they share several properties with RNA tumor viruses. The simian foamy viruses are of similar buoyant density, incorporate radioactive uridine but not thymidine, show inhibition of radioactive uridine incorporation in the presence of actinomycin D, show reduction of infectious virus replication in presence of BudR (57), possess particle-bound RNA-dependent DNA polymerase, and show strikingly similar UV inactivation rates (58,57).

Simian foamy viruses are frequently found in simian tissues. Swack et al. (59) could observe infection in 66% of rhesus kidney cell cultures kept more than 8 weeks. In a study of spontaneous lymphomas in rhesus monkeys, Manning (60) isolated simian foamy viruses from tumor tissues of two of five monkeys. Johnston (61) was able to isolate simian foamy viruses from throat swabs of 38 o. 104 monkeys but not from urine or stools. Hooks et al. (62) found foamy viruses in the brains of most chimpanzees studied. Of 88 chimpanzees in one colony, 91% had antibody to foamy viruses. All of 16 chimpanzees in the bush had antibodies to foamy viruses.

No signs of disease were observed in mice, guinea pigs, hamsters, rabbits and chick embryos inoculated with simian foamy viruses (63,62). The foamy viruses recently isolated from human kidney cell cultures (64) and cultures from a human nasopharyngeal carcinoma (65) await comparison with simian foamy viruses. Johnston (61) found no antibody to simian foamy viruses in 100 human sera collected in Taiwan from areas where monkeys are abundant and known to be infected, nor in five pools of human gamma globulin from the United States. Hooks et al. (62) found no antibody to two chimpanzee foamy viruses in nine animal handlers or in 13 other human sera. Swack et al. (59) could not demonstrate antibody in the serum of an animal caretaker.

OTHER POSSIBLE SIMIAN TUMOR VIRUSES

Lapin (66,67) has reported the possible transmission of human leukemia to nonhuman primates. The tumors produced in macaques, green monkeys, and baboons contain oncornavirus particles. Because there is some evidence for subsequent spread from animal to animal, however, the possibility of activation of a simian tumor virus must be considered. It is possible, of course, that nonhuman primates may share tumor viruses with man.

Melendez et al. (23) have mentioned a possible association between human patients with lymphosarcoma and spider monkeys (Ateles geoffroyi) in Guatemala.

In addition to the "outbreaks" of lymphosarcoma in gibbons (54) and macaques (66) already mentioned, an outbreak of spontaneous lymphosarcoma in rhesus monkeys has been reported recently (68). In that outbreak there have now been 39 cases over a 4-year period with an incidence rate for the four years of 936/100,000 rhesus/ year. Epidemiologic studies thus far suggest a pre-cipitating event followed by vertical transmission. Efforts to demonstrate causative oncornaviruses or herpes-viruses have been unsuccessful to date (69).

Additional data on simian tumor viruses will be found in the Proceedings of the Fifth International Symposium on Comparative Leukemia Research (70).

CONCLUSIONS

Relatively little information is available at present upon which to base realistic recommendations to protect personnel or experimental animals from simian tumor viruses. Herpesvirus saimiri and Yaba poxvirus are classified as "moderate-risk" (see NCI, vo. classifi-cation) agents.

Yaba virus is clearly infectious for man and could conceivably be acquired by the airborne route as well as by accidental injection. Spread from animal to animal is slight when good husbandry is practiced. As yet there is little data to suggest Herpesvirus saimiri poses a threat to man. Infected animals excrete little or no virus. The absence of seroconversion in investi-gators working with this virus suggests either that H. saimiri is not a human pathogen or that the present methods for containing viruses classified as a moderate-risk are sufficient to prevent significant exposure. The data on simian herpesviruses indicate that each species of New World monkeys, and especially squirrel monkeys, should be housed separately.

It is essential that investigators working with simian tumor viruses follow the lead of Wolfe et al. (13) and include in their experimental design studies on (1) seroconversion of personnel, (2) the fate of virus inoculated into animals, (3) routes and amounts of excretion of virus by infected animals, (4) host range of susceptibility, (5) spread to contact-exposed animals, and (6) aerosol transmission. New studies are needed on (a) infectivity and oncogenicity of viral nucleic acids,

(b) variation in oncogenicity by virus serotypes, and (c)
rates of inactivation of simian tumor viruses by chemical
and physical means under common laboratory conditions.
Until more information is obtained about the more recently
discovered simian tumor viruses, it would be prudent to
treat them as if hazardous to man. However, it is
likely that indigenous, non-oncogenic simian viruses,
such as Herpesvirus simiae or hepatitis virus, are more
hazardous than simian tumor viruses.

References

1. Weller, T. H. 1972. Herpes-like simian viruses:
    retrospective and prospective considerations. J.
    Nat. Cancer Inst. 49:209.
2. Melendez, L. V. et al. 1968. An apparently new
    herpesvirus from primary kidney cultures of the
    squirrel monkey (Saimiri sciureus). Lab. Animal
    Care 18:374.
3. Melendez, L. V. et al. 1969. Herpesvirus saimiri.
    I. Further characterization studies of a new virus
    from the squirrel monkey. Lab. Animal Care 19:372.
4. Falk, L. A. et al. 1972. Isolation of Herpesvirus
    saimiri from blood of squirrel monkeys (Saimiri
    sciureus). J. Nat. Cancer Inst. 48:1499.
5. Melendez, L. V. et al. 1972. Herpesviruses saimiri
    and ateles--their role in malignant lymphomas of
    monkeys. Fed. Proc. 31:1643.
6. Deinhardt, F. et al. 1972. Induction of neoplasms
    by viruses in marmoset monkeys. J. Med. Primatology
    1:29.
7. Melendez, L. V. et al. 1969. Herpesvirus saimiri.
    II. Experimentally induced malignant lymphoma in
    primates. Lab. Animal Care 19:378.
8. Daniel, M. D. et al. 1970. Malignant lymphoma
    induced in rabbits by Herpesvirus saimiri strains.
    Bact. Proc., p. 195.
9. Melendez, L. V. et al. 1970. Lethal reticulo-
    proliferative disease induced in Cebus albifrons
    monkeys by Herpesvirus saimiri. Int. J. Cancer 6:431.
10. Ablashi, D. V. et al. 1971. Malignant lymphoma
    with lymphocytic leukemia induced in owl monkeys by
    Herpesvirus saimiri. J. Nat. Cancer Inst. 47:837.
11. Melendez, L. V. et al. 1971. Acute lymphocytic
    leukemia in owl monkeys inoculated with Herpesvirus
    saimiri. Science 171:1161.
12. Rabin, H. 1971. Assay and pathogenesis of oncogenic
    viruses in nonhuman primates. Lab. Animal Sci. 21:
    1032.
13. Wolfe, L. G. et al. 1971. Oncogenicity of Herpes-
    virus saimiri in marmoset monkeys. J. Nat. Cancer
    Inst. 47:1145.

14. Hunt, R. D. et al. 1970. Morphology of a disease with features of malignant lymphoma in marmosets and owl monkeys inoculated with Herpes saimiri virus.
15. Hunt, R. D. et al. 1972. Herpesvirus saimiri malignant lymphoma in spider monkeys. J. Med. Prim. 1:114.
16. Hunt, R. D. and L. V. Melendez. 1972. Clinical, epidemiologic, and pathologic features of cytocidal and oncogenic herpesviruses in South American monkeys. J. Nat. Cancer Inst. 49:261.
17. Falk, L. A et al. 1972. Demonstration of Herpesvirus saimiri-associated antigens in peripheral lymphocytes from infected marmosets during in vitro cultivation. J. Nat. Cancer Inst. 48:523.
18. Pearson, G. et al. 1972. Intracellular and membrane immunofluorescence investigations on cells infected with Herpesvirus saimiri. J. Nat. Cancer Inst. 49: 1417.
19. Ablashi, D. V. et al. 1972. Certain characteristics of Herpesvirus saimiri cultured in subhuman primate culture cells. Amer. J. Vet. Res. 33:1689.
20. Ablashi, D. V. et al. 1971. Propagation of Herpesvirus saimiri in human cells. J. Nat. Cancer Inst. 47:241.
21. Melendez, L. V. et al. 1971. New herpesviruses from South American monkeys. Preliminary report. Lab. Animal Sci. 21:1050.
22. Melendez, L. V. et al. 1972. Herpesvirus ateles, a new lymphoma virus of monkeys. Nature New Biol. 235:182.
23. Melendez, L. V. et al. 1972. Two new herpesviruses from spider monkeys (Ateles geoffroyi). J. Nat. Cancer Inst. 49:233.
24. Daniel, M. D. et al. 1972. Plaque characterization of viruses from South American nonhuman primates. J. Nat. Cancer Inst. 49:239.
25. Bearcroft, W. G. C. and M. F. Jamieson. 1958. An outbreak of subcutaneous tumors in rhesus monkeys. Nature 182:195.
26. Tsuchiya, Y. and I. Tagaya. 1971. Sero-epidemiological survey on Yaba and 1211 virus infections among several species of monkeys. J. Hyg. 69:445.
27. Ambrus, J. L. and H. V. Strandstrom. 1966. Susceptibility of Old World monkeys to Yaba virus. Nature 211:876.
28. Sproul, E. E. et al. 1963. The pathogenesis of Yaba virus-induced histiocytomas in primates. Cancer Res. 23:671.
29. Wolfe, L. G. et al. 1968. Experimental aerosol transmission of Yaba virus in monkeys. J. Nat. Cancer Inst. 41:1175.
30. Strandstrom, H. V. et al. 1966. Propagation of Yaba virus in embryonated hen eggs. Virology 28:479.

31. Yohn, D. S. et al.   1964.   A quantitative cell culture assay for Yaba tumor virus.   Nature 202: 881.
32. Feltz, E. T.   1961.   Studies on a virus induced tumor in monkeys.   Proc. Amer. Assoc. Cancer Res. 3:224.
33. Ambrus, J. L. et al.   1963.   A virus-induced tumor in primates.   Nat. Cancer Inst. Monogr. 10:447.
34. Grace, J. T., Jr. and E. A. Mirand.   1963.   Human susceptibility to a simian tumor virus.   Ann. N. Y. Acad. Sci. 108:1123.
35. Chopra, H. C.   1970.   Electron microscopic detection of an oncogenic-type virus in a monkey breast tumor.   Proc. Amer. Ass. Cancer Res. 11:16.
36. Chopra, H. C. and M. M. Mason.   1970.   A new virus in a spontaneous mammary tumor of a rhesus monkey.   Cancer Res. 30:2081.
37. Jensen, E. M. et al.   1970.   Isolation and propagation of a virus from a spontaneous mammary carcinoma of a rhesus monkey.   Cancer Res. 30:2388.
38. Mason, M. M. et al.   1972.   History of a rhesus monkey adenocarcinoma containing viral particles resembling oncogenic RNA viruses.   J. Nat. Cancer Inst. 48:1323.
39. Chopra, H. C. et al.   1971.   Studies on virus particles resembling oncogenic RNA viruses in monkey breast carcinoma.   Cancer 28:1406.
40. Manning, J. S. and A. J. Hackett.   1972.   Morphological and biophysical properties of the Mason-Pfizer monkey virus.   J. Nat. Cancer Inst. 48:417.
41. Fine, D. L. et al.   1971.   Simian tumor virus isolate: demonstration of cytopathic effects in vitro.   Science 174:420.
42. Pienta, R. J. et al.   1972.   Transformation of rhesus foreskin cells by Mason-Pfizer monkey virus.   J. Nat. Cancer Inst. 48:1913.
43. Nowinski, R. C. et al.   1971.   Serological and structural properties of Mason-Pfizer monkey virus isolated from the mammary tumor of a rhesus monkey.   Proc. Nat. Acad. Sci. 68:1608.
44. Theilen, G. H. et al.   1971.   C-type virus in tumor tissue of a woolly monkey (Lagothrix spp) with fibrosarcoma.   J. Nat. Cancer Inst. 47:881.
45. Kawakami, T. G.   1971.   Personel communication.
46. Scolnick, E. M. et al.   1972.   Reverse transcriptases of primate viruses as immunological markers.   Science 177:1119.
47. Scolnick, E. M. et al.   1972.   Immunological characterization of primate C-type virus reverse transcriptases.   Nature New Biol. 235:35.
48. Wolfe, L. G. et al.   1972.   Simian sarcoma virus, type 1 (Lagothrix): focus assay and demonstration of nontransforming associated virus.   J. Nat. Cancer

Inst. 48:1905.
49. Wolfe, L. G. et al. 1971. Induction of tumors in marmoset monkeys by simian sarcoma virus, type 1 (Lagothrix): A preliminary report. J. Nat. Cancer Inst. 47:1115.
50. Kawakami, T. G. et al. 1972. C-type virus associated with gibbon lymphosarcoma. Nature New Biol. 235:170.
51. Newborne, J. and V. B. Robinson. 1960. Spontaneous tumors in primates--a report of two cases with notes on the apparent low incidence of neoplasms in sub-human primates. Amer. J. Vet. Res. 21:150.
52. Di Giacomo, R. F. 1967. Burkitt's lymphoma in a white-handed gibbon (Hylobates lar). Cancer Res. 27:1178.
53. De Paoli, A. and F. M. Garner. 1968. Acute lymphatic leukemia in a white-cheeked gibbon (Hylobates concolor). Cancer Res. 28:2559.
54. Johnsen, D. O. et al. 1971. Malignant lymphoma in the gibbon. J. Amer. Vet. Med. Ass. 159:563.
55. Jones, M. D. et al. 1972. Lymphoblastic lympho-sarcoma in two white-handed gibbons, Hylobates lar. J. Nat. Cancer Inst. 49:599.
56. Kawakami, T. G. 1972. Personal communication.
57. Parks, W. P. and G. J. Todaro. 1972. Biological properties of syncytium-forming ("foamy") viruses. Virology 47:673.
58. Parks, W. P. et al. 1971. RNA-dependent DNA polymerase in primate syncytium-forming ("foamy") viruses. Nature 229:258.
59. Swack, N. S. et al. 1970. Foamy virus infection of rhesus and green monkeys in captivity. Amer. J. Epidemiol. 92:79.
60. Manning, J. S. 1972. Unpublished data.
61. Johnston, P. B. 1961. A second immunologic type of simian foamy virus: monkey throat infections and unmasking by both types. J. Infect. Dis. 109:1.
62. Hooks, J. et al. 1972. Characterization and dis-tribution of two new foamy viruses isolated from chimpanzees. Archiv. fur die Gesamte Virusforschung 38:38.
63. Stiles, G. E. et al. 1964. Comparison of simian foamy virus strains including a new serological type. Nature 201:1350.
64. Hsiung, G. D. 1968. Latent virus infections in primate tissues with special reference to simian viruses. Bact. Rev. 32:185.
65. Anchong, B. G. et al. 1971. An unusual virus in cultures from a human nasopharyngeal carcinoma. J. Nat. Cancer Inst. 46:299.
66. Lapin, B. A. 1969. Experiments in monkeys with human leukemia. Primates in Medicine 3:23.
67. Lapin, B. A. 1972. Personal communication.

68. Stowell, R. E. et al. 1971. Outbreak of malignant
    lymphoma in rhesus monkeys. Lab. Invest. 25:476.
69. Manning, J. S. and R. A. Griesemer. 1972. Unpub-
    lished data.
70. Chieco-Bianchi, L. and R. Dutcher, eds. 1971.
    Unifying concepts of leukemia. Proc. 5th Int.
    Symp. Comp. Leuk. Res., Padova. Bibliotheca
    Hematologica no. 39. Karger.

EBV AND OTHER POTENTIAL PROBLEMS
IN THE HANDLING OF LONG-TERM
HUMAN LYMPHOCYTE CULTURES

Jerry A. Schneider
Department of Pediatrics
University of California School of Medicine
La Jolla, California

It has been known for several years that many
inherited metabolic defects are expressed in fibroblast
cultures established from skin biopsies of affected
individuals (1). There are, however, two major
difficulties in utilizing these fibroblasts: They have
a limited life span (2), and they grow only as monolayers,
making it difficult to obtain large yields of cells.

The development of methods for the propagation of
human lymphocytes in long-term culture (3) promises to
solve both of the above problems and appears to have
heralded another revolution in the study of inborn
errors of metabolism. These cells also express the
metabolic defect of the individual from whom they were
obtained, and in addition they have an infinite life
span. Moreover, it is now even possible to obtain these
cells from commercial sources. Thus we have a situation
in which many clinical investigators and biochemists are
beginning to grow long-term lymphocyte lines to use as a
tissue for biochemical studies. Most of these investigators have had limited, if any training in techniques
of bacteriology or virology and have given little thought
to the possible hazards associated with these cells.

In 1959 Rothfels et al. (4) reported that several
long-term cell lines of supposed mouse, monkey, or
human origin were, in fact, all derivatives of either
mouse L-cells or HeLa cells. These findings were

191

corroborated by several other laboratories (5,6). In
1967 Gartler reported that 18 established human cell
lines of supposed independent origin were identical with
regard to the isozyme type of their glucose-6-phosphate
dehydrogenase and phosphoglucomutase. Since all of these
cell lines contained the A-type glucose-6-phosphate
dehydrogenase, which is often found in Negroes, and since
the HeLa line was derived from a Negro and was the first
established human cell line, he concluded that the 18
human cell lines were most likely overgrown by HeLa cell
contaminants (7). This has been confirmed by many
investigators, and more recently, using the new
quinacrine staining technique, Miller et al. (8) have
shown that several of these established cell lines contain HeLa "marker" chromosomes (see also 9). These
examples of cross-contamination of cell types may have
been caused by pipetting errors and aerosols.

In view of such observations, it is relatively easy
to convince an investigator that if he does not use good
technique he may cross-contaminate his cell lines.
Fortunately if an investigator takes adequate precautions
to prevent cross-contamination of his cell lines, he
will automatically do many of the things that are
necessary to protect himself and his fellow workers.

Unfortunately, there is no hard data available to
let us determine whether or not there is a danger to
laboratory workers involved in handling long-term
lymphocyte cultures. There is, however, one virus
which is believed to be present in all long-term
lymphocyte lines and which might be dangerous to laboratory workers. This is the herpesvirus designated
Epstein-Barr virus (EBV). Epstein et al. (10,11) first
described the presence of a herpes-like virus in
lymphoblast cultures from a Burkitt's lymphoma in 1964.
Since that time particles of similar morphology and antigenicity have been detected in cultures of other Burkitt
lymphomas (12-16), in white blood cells from patients
with leukemia (17) and infectious mononucleosis (18-22),
and in cultures of apparently normal individuals (23,3).
Compelling serologic data has accumulated linking EBV to
Burkitt's lymphoma (24-29,16), infectious mononucleosis
(21,22,28,29) and nasopharyngeal carcinoma (27-33).
Elevated titers of EBV antibodies have also been reported
in Hodgkin's disease (34), sarcoidosis (35,36), leprosy
(37), systemic lupus erythematosus (38-40), viral
hepatitis (41,42) and in patients receiving numerous
blood transfusions during open heart surgery (43).
Considerable controversy exists concerning the etiologic
significance of EBV in these diseases. One view holds
that EBV is the cause of infectious mononucleosis and
perhaps, along with another factor, the cause of Burkitt's

lymphoma, the two conditions with which EBV has been most closely associated. The other view holds that EBV is simply a passenger virus which is attracted to poorly differentiated lymphoblastic cells. The controversy has been extensively discussed in three excellent recent reviews (44-46).

The first question to ask is whether EBV can induce cell transformation. Investigators have shown that this virus converts peripheral lymphocytes in culture into lymphoblastoid cells with long-term growth potential (47-51). Chang (52) has reported that cell-free filtrates from lymphoid cell lines, which were negative for EBV antigens but derived from healthy individuals who were positive for EBV antibody, were able to transform umbilical cord leukocytes. Glade et al. (53) had previously reported establishing long-term lymphocyte cultures from 5 of 15 patients with infectious mononucleosis, whereas the same technique had been unsuccessful in establishing such cultures from blood of 13 patients with clinical disease but without infectious mononucleosis, or from blood of 10 normal controls. Similarly Henle et al. (54) were able to stimulate growth of peripheral leukocytes after they were exposed to lethally irradiated cells that had been cultured from a patient with Burkitt's lymphoma. More recently, two reports have described the transformation of human leukocytes by throat washings from patients with infectious mononucleosis (55,56). Propert and Epstein (57) have recently reported the morphologic transformation in vitro of human fibroblasts treated with EBV under conditions of cell fusion (inactivated Sendai virus). Thus the evidence is very compelling that EB virus can cause the transformation of lymphocytes to a lymphoblastoid cell with long-term growth potential.

The next question is whether EBV is present in all long-term lymphocyte cultures. Serologic tests have revealed the presence of EBV in Burkitt lymphoma cell cultures (58), and McCormick et al. (59) have reported the apparent spontaneous conversion of two lymphoid cell lines from a "virus-free" state to a condition where EBV could be identified both by antigenic studies and by electron microscopy. More recently, Gerber (60) treated three human lymphoid cell lines with 5-bromodeoxyuridine (BrdU) in an attempt to make thymidine kinase negative cells for fusion studies. The cell lines used were Raji cells (from a Burkitt lymphoma), a line of cells established from a patient with pneumonia, and a line of cells established from a normal individual. All three cell lines were negative for EBV antigens, yet following treatment with BrdU, all three developed EBV antigens. In an independent study reported at the same time, Hampar and colleagues (61) also treated a "virus-negative" human

lymphoblastoid line (Raji) with BrdU and were able to
identify particles of EBV. Sugawara et al. (62) have
recently reported similar findings. Glaser and O'Neill
(63) have described similar results following cell
hybridization of Burkitt lymphoblastoid cell lines with
mouse or human cell lines. Continued synthesis of EBV
antigens by the hybrid cells was induced by 5-iododeoxy-
uridine.

Using a DNA-DNA hybridization technique, Zur-Hausen
et al. (64) have demonstrated EB virus DNA in biopsies
of Burkitt tumors and nasopharyngeal carcinomas. With
the same technique Zur-Hausen and Schulte-Holthausen
(65) were also able to demonstrate a small amount of
EBV DNA in a line of Burkitt lymphoma cells (Raji)
which were thought to be virus-free on the basis of
electron microscopy and immunofluorescence. Using E.
coli RNA polymerase in the presence of radioactive
precursors, it has been possible to synthesize [^{3}H]EBV-
specific complementary RNA (cRNA) of high specific
activity (66-69). Utilizing DNA-RNA hybridization
experiments with this material, Nonoyama and Pagano (66)
were able to identify EB viral DNA in Burkitt lymphoma
cells and two established leukocyte cell lines. They
estimated that the EBV DNA in these cells ranged from
0.06-1.6% of the total DNA. More recently the same
investigators (67) have separated EB virus DNA from the
large chromosomal DNA of Raji cells (from a Burkitt
lymphoma) and feel these DNAs are not covalently linked.
Their data also indicated that the EBV DNA in Raji cells
exists in strands of nearly complete size.

Zur-Hausen (69) has tested ten apparently virus-free
lymphoblastoid tissue culture lines with the DNA-RNA
hybridization technique. These lines had been established
from patients with Burkitt's lymphoma, Hodgkin's disease,
anaplastic carcinoma of the nasopharynx, leukemia, and
from healthy donors. EBV-specific DNA was identified in
all ten cell lines. The amount demonstrable varied
considerably from line to line and seemed to be cell-line
specific. To further investigate the persistence of EB
viral genomes in human lymphoblastoid cell lines, Zur-
Hausen (69) retested the cell lines by nucleic acid
hybridization after one year of continuous growth. The
hybridized counts were in the same range, indicating a
stable association of the viral DNA with the host cell
genome. In another experiment, Zur-Hausen (69) infected
these "virus-free" lymphoblastoid lines with EBV concentrates, which led to the temporary synthesis of
viral structural antigens in some of the cells. This
infection could be demonstrated for only a short period
of time, and after approximately two weeks the viral
capsid antigen was no longer detectable. But hybridiza-

tion studies with EBV cRNA revealed that the DNA of
these cells annealed with more counts than DNA of unaffected cells from the same line. This suggested that
lymphoblastoid cells surviving infection with EBV
increased their content of virus-specific nucleic acid.

Thus it is clear that long-term lymphocyte lines
do contain EBV. Consequently laboratory workers who
do not have antibodies to this virus may risk getting
infectious mononucleosis, which can be a devastating
disease.

Another question to be answered is whether EBV is
a human tumor virus. There is some direct evidence
of the oncogenicity of EBV. Deal et al. (70) have shown
that EBV can cause human lymphocytes to grow in an
uncontrolled and highly invasive manner in a heterologous
host. More recently Gerber and Hoyer (71) have presented
evidence that infection of human leukocytes with EBV
induces cellular DNA synthesis. They believe the
activation of DNA synthesis is unusual in cells infected
by DNA viruses and that only oncogenic DNA viruses
stimulate cellular DNA synthesis. The investigators
state that this is the first direct evidence of an EBV
function which is usually found only in oncogenic DNA
viruses. However, it remains to be explained why
evidence of EBV infection may be found in many healthy
individuals, as well as in patients with infectious
mononucleosis. The possibilities would seem to include
(1) that there are different types of EBV which we are
not yet able to differentiate, (3) there are differing
host responses to the same virus, and (3) EBV is simply
a passenger virus which is seen in association with
various conditions but is not the etiologic agent. In
any event, if human lymphocytes can be transformed in
vitro, the possibility exists that they may be oncogenic
in man. Thus these cells should be considered hazardous
and the Minimum Standard of Biological Safety and
Environmental Control for Contractors of the SVCP
(Office of Biohazard and Environmental Control, N.C.I.,
N.I.H., Bethesda, Md.) should be followed by all investigators using them.

References

1. Raivio, K. O. and J. E. Seegmiller. 1972. Genetic
   diseases of metabolism. Ann. Rev. Biochem. 41:543.
2. Hayflick, L. 1965. The limited in vitro lifetime
   of human diploid cell strains. Exp. Cell Res.
   37:614.
3. Moore, G. E., R. E. Gerner and H. A. Franklin. 1967.
   Culture of normal human leukocytes. J. Amer. Med.

Ass. 199:519.

4. Rothfels, K. H., A. A. Axelrad, L. Siminovitch, E. A. McCulloch and R. C. Parker. 1959. The origin of altered cell lines from mouse, monkey, and man, as indicated by chromosome and transplantation studies. Proc. 3rd Canadian Cancer Res. Conf., p. 189. Academic Press, New York.

5. Defendi, V., R. E. Billingham, W. K. Silvers and P. Moorhead. 1960. Immunological and karyological criteria for identification of cell lines. J. Nat. Cancer Inst. 25:359.

6. Brand, K. G. and J. T. Syverton. 1962. Results of species-specific hemagglutination tests on "transformed", nontransformed, and primary cell cultures. J. Nat. Cancer Inst. 28:147.

7. Gartler, S. M. 1967. Genetic markers as tracers in cell culture. Nat. Cancer Inst. Monogr. 26:167.

8. Miller, O. J., D. A. Miller, P. W. Allderdice, V. G. Dev and M. S. Grewal. 1971. Quinacrine fluorescent karyotypes of human diploid and heteroploid cell lines. Cytogenetics 10:338.

9. Francke, U., D. S. Hammond and J. A. Schneider. 1973. The band patterns of twelve D98/AH-2 marker chromosomes and their use for identification of intraspecific cell hybrids. Chromosoma (in press).

10. Epstein, M. A. and Y. M. Barr. 1964. Cultivation in vitro of human lymphoblasts from Burkitt's malignant lymphoma. Lancet 1:252.

11. Epstein, M. A., G. A. Achong and V. M. Barr. 1964. Virus particles in cultured lymphoblasts from Burkitt's lymphoma. Lancet 1:702.

12. Pulvertaft, R. J. V. 1964. Cytology of Burkitt's tumor (African lymphoma). Lancet 1:238.

13. Stewart, S. E., E. Lovelace, J. J. Whang and V. A. Ngu. 1965. Burkitt tumor: Tissue culture, cytogenetic and virus studies. J. Nat. Cancer Inst. 34:319.

14. Rabson, A. S., G. T. O'Connor, S. Baron, J. J. Whang and F. Y. Legallais. 1966. Morphologic, cytogenetic and virologic studies in vitro of a malignant lymphoma from an African child. Int. J. Cancer 1:89.

15. Pope, J. H., B. G. Achong, M. A. Epstein and J. Biddulph. 1967. Burkitt lymphoma in New Guinea: Establishment of a line of lymphoblasts in vitro and description of their fine structure. J. Nat. Cancer Inst. 39:933.

16. Henle, G., W. Henle, P. Clifford, V. Diehl, G. W. Kafuko, B. G. Kirya, G. Klein, R. H. Morrow, G. M. R. Munube, P. Pike, P. M. Tukei and J. L. Ziegler. 1969. Antibodies to Epstein-Barr virus in Burkitt's lymphoma and control groups. J. Nat. Cancer Inst. 43:1147.

17. Iwakata, S. and J. T. Grace, Jr. 1964. Cultivation

in vitro of myeloblasts from human leukemia.   N. Y.
State J. Med. 64:2279.
18. Pope, J. H. 1967. Establishment of cell lines
from peripheral leukocytes in infectious mononucleosis. Nature 216:810.
19. Diehl, V., G. Henle, W. Henle and G. Kohn. 1968.
Demonstration of a herpes group virus in cultures of
peripheral leukocytes from patients with infectious
mononucleosis. J. Virol. 2: 663.
20. Henle, G., W. Henle and V. Diehl. 1968. Relation
of Burkitt's tumor-associated herpes-type virus to
infectious mononucleosis. Proc. Nat. Acad. Sci.
59:94.
21. Niederman, J. C., R. W. McCollum, G. Henle and W.
Henle. 1968. Infectious mononucleosis: Clinical
manifestations in relation to EB virus antibodies.
J. Amer. Med. Ass. 203:205.
22. Niederman, J. C., A. S. Evans, L. Subrahmanyan and
R. W. McCollum. 1970. Prevalence, incidence and
persistence of EB virus antibody in young adults.
New Eng. J. Med. 282:361.
23. Gerber, P. and S. M. Birch. 1967. Herpes-like
particles in cultures of buffy coat from nonleukemic donors. Bacteriol. Proc., p. 153 (abstract).
24. Levy, J. A. and G. Henle. 1966. Indirect immunofluorescence tests with sera from African children
and cultured Burkitt lymphoma cells. J. Bacteriol.
92:275.
25. Klein, G., G. Pearson, G. Henle, W. Henle, G. Gold-
stein and P. Clifford. 1969. Relation between
Epstein-Barr viral and cell membrane immunofluorescence in Burkitt tumor cells. III. Comparison of
blocking of direct membrane immunofluorescence and
anti-EBV reactivities of different sera. J. Exp.
Med. 129:697.
26. Pearson, G., G. Klein, G. Henle, W. Henle and P.
Clifford. 1969. Relation between Epstein-Barr
viral and cell membrane immunofluorescence in
Burkitt tumor cells. IV. Differentiation between
antibodies responsible for membrane and viral
immunofluorescence. J. Exp. Med. 129:707.
27. Gunven, P., G. Klein, G. Henle, W. Henle and P.
Clifford. 1970. Epstein-Barr virus in Burkitt's
lymphoma and nasopharyngeal carcinoma. Nature
228:1053.
28. Emerson, S. U., K. Tokuyasu and M. I. Simon. 1970.
Differential reactivity of human serums with early
antigens induced by Epstein-Barr virus. Science
169:188.
29. Banatvala, J. E., J. M. Best and D. K. Waller. 1972.
Epstein-Barr virus-specific IgM in infectious mononucleosis, Burkitt lymphoma, and nasopharyngeal
carcinoma. Lancet 1:1205.

30. Old, L. J., E. A. Boyse, H. F. Oettgen, E. deHarven, G. Geering, B. Williamson and P. Clifford. 1966. Precipitating antibody in human serum to an antigen present in cultured Burkitt lymphoma cells. Proc. Nat. Acad. Sci. 56:1699.

31. deSchryver, A., S. Friberg, Jr., G. Klein, W. Henle, G. Henle, G. deThe, P. Clifford and H. C. Ho. 1969. Epstein-Barr virus-associated antibody patterns in carcinoma of the post-nasal space. Clin. Exp. Immunol. 5:443.

32. Henle, W., G. Henle, H. C. Ho, P. Burtin, Y. Cachin, P. Clifford, A. deSchryver, G. deThe, V. Diel and G. Klein. 1970. Antibodies to Epstein-Barr virus in nasopharyngeal carcinoma, other head and neck neoplasms and control groups. J. Nat. Cancer Inst. 44:225.

33. Klein, G., G. Geering, L. J. Old, G. Henle, W. Henle and P. Clifford. 1970. Comparison of the anti-EBV titer and the EBV-associated membrane reactive and precipitating antibody levels in the sera of Burkitt lymphoma and nasopharyngeal carcinoma patients and controls. Int. J. Cancer 5:185.

34. Levine, P. H., D. V. Ablashi, C. W. Berard, P. P. Carbone, D. E. Waggoner and L. Malan. 1971. Elevated antibody titers to Epstein-Barr virus in Hodgkins disease. Cancer 27:416.

35. Hirshaut, Y., P. Glade, L. O. B. D. Vieira, E. Ainbender, B. Dvorak and L. E. Siltzbach. 1970. Sarcoidosis, another disease associated with serologic evidence for herpes-like virus infection. New Eng. J. Med. 283:502.

36. Wahren, B., E. Carlens, A. Espmark, H. Lundbeck, S. Lofgren, E. Madar, G. Henle and W. Henle. 1971. Antibodies to various herpesviruses in sera from patients with sarcoidosis. J. Nat. Cancer Inst. 47:747.

37. Papageorgiou, P. S., C. Sorokin, K. Kouzoutzakoglou and P. R. Glade. 1971. Herpes-like Epstein-Barr virus in leprosy. Nature 231:47.

38. Evans, A. S., N. F. Rothfield and J. C. Niederman. 1971. Raised antibody titers to E. B. virus in systemic lupus erythematosus. Lancet 1:167.

39. Evans, A. S. 1971. E. B. virus antibody in systemic lupus erythematosus. Lancet 1:1023.

40. Stevens, D. A., M. B. Stevens, G. R. Newell, P. H. Levine and D. E. Waggoner. 1972. Epstein-Barr virus (herpes-type virus) antibodies in connective tissue diseases. Arch. Int. Med. 130:23.

41. Glade, P. R., Y. Hirshaut, S. D. Douglas and K. Hirschhorn. 1968. Lymphoid suspension cultures from patients with viral hepatitis. Lancet 2:1273.

42. McCollum, R. W., J. C. Niederman, A. S. Evans and J. P. Giles. 1969. Viruses in viral hepatitis.

Lancet 1:778.
43. Gerber, P., J. H. Walsh, E. N. Rosenblum and R. H. Purcell. 1969. Association of EB-virus infection with the post-perfusion syndrome. Lancet 1:593.
44. Epstein, M. A. 1970. Aspects of the EB virus. Adv. Cancer Res. 13:383.
45. Miller. G. 1971. Human lymphoblastoid cell lines and Epstein-Barr virus: A review of their inter-relationships and their relevance to the etiology of leukoproliferative states in man. Yale J. Biol. Med. 43:358.
46. Smith, R. T. and J. C. Bausher. 1972. Epstein-Barr virus infection in relation to infectious mono-nucleosis and Burkitt's lymphoma. Ann. Rev. Med. 23:39.
47. Diehl, V., G. Henle, W. Henle and G. Kohn. 1969. Effect of a herpes group virus (EBV) on growth of peripheral leukocyte cultures. In Vitro 4:92.
48. Pope, J. H., M. K. Horne and W. Scott. 1969. Identification of the filtrable leukocyte-transforming factor of QIMR-WIL cells as herpes-like virus. Int. J. Cancer 4:255.
49. Gerber, P., J. Whang-Peng and J. A. Monroe. 1969. Transformation and chromosome changes induced by Epstein-Barr virus in normal human leukocyte cultures. Proc. Nat. Acad. Sci. 63:740.
50. Miller, G., J. F. Enders, H. Lisco and H. I. Kohn. 1969. Establishment of lines from normal human blood leukocytes by co-cultivation with a leukocyte line derived from a leukemic child. Proc. Soc. Exp. Biol. Med. 132:247.
51. Baumal, R., B. Bloom and M. D. Scharff. 1971. Induction or long-term lymphocyte lines from delayed hypersensitive human donors using specific antigen plus Epstein-Barr virus. Nature New Biol. 230:20.
52. Chang, R. S. 1971. Umbilical cord leukocytes transformed by lymphoid cell filtrates from healthy people. Nature New Biol. 233:124.
53. Glade, P. R., J. A. Kasel, H. L. Moses, J. Whang-Peng, P. F. Hoffman, J. K. Kammermeyer and L. N. Chessin. 1968. Infectious mononucleosis: Contin-uous suspension culture of peripheral blood leuko-cytes. Nature 217:564.
54. Henle, W., V. Diehl, G. Kohn, H. Zur-Hausen and G. Henle. 1967. Herpes-type virus and chromosome marker in normal leukocytes after growth with irradiated Burkitt cells. Science 157:1064.
55. Chang, R. S. and H. D. Golden. 1971. Transformation of human leukocytes by throat washing from infectious mononucleosis patients. Nature 234:359.
56. Pereira, M. S., J. M. Blake, A. M. Field, F. G. Rodgers, L. A. Bailey and J. R. Davies. 1972. Evidence for oral excretion of E. B. virus in infec-

tious mononucleosis. Lancet 1:710.
57. Probert, M. and M. A. Epstein. 1972. Morphological transformation in vitro of human fibroblasts by Epstein-Barr virus: Preliminary observations. Science 175:202.
58. Gerber, P. and D. R. Deal. 1970. Epstein-Barr virus-induced viral and soluble complement-fixing antigens in Burkitt lymphoma cell cultures. Proc. Soc. Exp. Biol. Med. 134:748.
59. McCormick, K. J., D. M. Mumford, W. A. Stenback and J. J. Trentin. 1971. Lymphoid cell lines: Activation of a latent herpes-like virus particle. Nature New Biol. 230:83.
60. Gerber, P. 1972. Activation of Epstein-Barr virus by 5-bromodeoxyuridine in "virus-free" human cells. Proc. Nat. Acad. Sci. 69:83.
61. Hampar, B., J. G. Derge, L. M. Martos and J. L. Walker. 1972. Synthesis of Epstein-Barr virus after activation of the viral genome in a "virus-negative" human lymphoblastoid cell (Raji) made resistant to 5-bromodeoxyuridine. Proc. Nat. Acad. Sci. 69:78.
62. Sugawara, K., F. Mizuno and T. Osato. 1972. Epstein-Barr virus-associated antigens in non-producing clones of human lymphoblastoid cell lines. Nature New Biol. 239:242.
63. Glaser, R. and F. J. O'Neill. 1972. Hybridization of Burkitt lymphoblastoid cells. Science 176:1245.
64. Zur-Hausen, H., H. Schulte-Holthausen, G. Klein, W. Henle, G. Henle, P. Clifford and L. Santesson. 1970. EBV-DNA in biopsies of Burkitt tumors and anaplastic carcinomas of the nasopharynx. Nature 228:1056.
65. Zur-Hausen, H. and H. Schulte-Holthausen. 1970. Presence of EB virus nucleic acid homology in a "virus-free" line of Burkitt tumor cells. Nature 227:245.
66. Nonoyama, M. and J. S. Pagano. 1971. Detection of Epstein-Barr viral genome in nonproductive cells. Nature New Biol. 233:103.
67. Nonoyama, M. and J. S. Pagano. 1972. Separation of Epstein-Barr virus DNA from large chromosomal DNA in non-virus-producing cells. Nature New Biol. 238:169.
68. Zur-Hausen, H., V. Diehl, H. Wolf, H. Schulte-Holthausen and U. Schneider. 1972. Occurrence of Epstein-Barr virus genomes in human lymphoblastoid cell lines. Nature New Biol. 237:189.
69. Zur-Hausen, H. 1972. EB virus-associated macro-molecules in cells derived from human tumors. Molecular studies in viral neoplasia, 25th Ann. Symp. Fund. Cancer Res. Houston, p. 57 (abstract).
70. Deal, D. R., P. Gerber and F. V. Chisari. 1971. Hetero-transplantation of two human lymphoid cell

lines transformed in vitro by Epstein-Barr virus. J. Nat. Cancer Inst. 47:771.

71. Gerber, P. and B. H. Hoyer. 1971. Induction of cellular DNA synthesis in human leukocytes by Epstein-Barr virus. Nature 231:46.

PANEL III
HOW DANGEROUS ARE THESE AGENTS: HAVE THEY ACTUALLY PRODUCED MALIGNANCY IN MAN

Richard A. Griesemer
University of California Primate Research Center

Clarke Heath
Center for Disease Control

Robert J. Hueber
National Cancer Institute

Henry S. Kaplan
Stanford University School of Medicine

Robert W. Miller
National Cancer Institute

Jerry A. Schneider
University of California School of Medicine

CURRENT EPIDEMIOLOGIC EVIDENCE REGARDING KNOWN OR SUSPECTED ONCOGENIC VIRUSES IN RELATION TO HUMAN CANCER

Clark W. Heath, Jr.

From an epidemiologic viewpoint, evidence is sparce that infectious agents cause human cancer, whether one considers all infectious agents or just viruses with known or suspected oncogenic potential. There are no obvious tendencies for cancers to spread by contact from animals to humans or from humans to humans, no clear-cut seasonality or short-term epidemicity, and no strong associations with exposure to specific infectious agents. However, unlike traditional infectious disease models, the role of viruses in cancer may well be one where latent periods are long and variable, where multiple co-factors influence the expression of clinical disease more strongly than infection does itself, or where viruses behave like genes (vertical, rather than horizontal, transmission). Standard epidemiologic approaches are not easily adapted to such settings.

There are, however, fragments of epidemiologic evidence to suggest a role for infection in two situations, one related to a known animal RNA tumor virus, the feline leukemia virus (FeLV), the other related to a suspect human DNA tumor virus, the Epstein-Barr virus (EBV).

FeLV has received particular attention in relation to human malignancy because of features that seem to distinguish it from other animal oncornaviruses: (1) the ease with which it can be isolated from outbred cats, (2) its ability to infect human cells in tissue culture, and (3) the fact that leukemia/lymphoma is a considerably more common tumor in cats than in other species. Despite these features, most epidemiologic studies seeking links between feline and human tumors have found no obvious relationships. Cancer seems no more common in persons occupationally exposed to diseased cats (veterinarians) than in the general population, and in two household surveys human cancer has not seemed unduly frequent in homes containing leukemic cats, or vice versa, feline leukemia in households containing human cancer. One set of data, however, does suggest an interspecies relationship, although the suggestion may not be directly concerned with animal cancer. The data come from the Tristate Survey, a large and complex case-control study conducted in New York, Minnesota, and Maryland 10-15 years ago, long before it was realized that a virus causes cat leukemia. Among many questions asked of cases and controls were ones regarding contact with pets and, in particular, regarding recent contact with sick pets. For childhood cases (300 cases, 831 controls), a suggestive association was found with pet cats, the only category of pets to yield such a result. The association was especially strong when attention was restricted to sick cats (relative risk = 2.24, p = 0.008). Unfortunately the data stop at that point since no information was collected concerning specific diagnoses among sick pets. We therefore don't know whether leukemia was or was not unduly represented among the sick cats of leukemic children. The apparent relationship, if not an artifact, might as easily reflect nonspecific animal illness triggering human disease as the direct spread of tumor-producing virus. The observation's possible importance is underscored, however, by the fact that a similar association was found among adult cases in the same survey. Here, in addition to sick cats, sick birds and to a lesser degree sick dogs seemed associated with human leukemia.

The possible relationship of EBV to human cancer is a complex issue, and it will probably not be resolved until we understand more fully the relationship of other herpesviruses to other tumors, e.g., Marek's disease in chickens and lymphoma in primates. At present it seems possible that EBV serves as one of several interacting factors involved in the etiology of certain lymphoid tumors, other factors perhaps being genetic constitution, RNA tumor viruses, immunologic status, and the action of as yet ill-defined trigger factors. How major a role EBV may play in all of this and precisely what that role may be remain unknown.

The epidemiologic data which bear on the question of EBV as a human oncogenic virus have been thoroughly

reviewed on several occasions. The main points focus on
(1) the frequency and concentration of anti-EBV antibodies
in patients with such malignancies as Burkitt's tumor
(BT), nasopharyngeal carcinoma, and Hodgkin's disease,
(2) the peculiar epidemiologic features of BT itself,
and (3) the etiologic relationship of EBV to infectious
mononucleosis (IM) and of IM possibly to cancer.

For some time there has been clinical speculation
that IM might per se represent a self-limited form of
acute lymphocytic leukemia, or at least might predispose
to later lymphoid malignancy. In neither case has much
supportive evidence appeared. Cases of leukemia and
lymphoma associated with IM have been reported but without any clear patterns that might suggest cause and
effect. (In some instances IM has been seen to follow
leukemia.) More remote IM-cancer links were sought in
a systematic study conducted in Connecticut, where
names of patients with IM were matched with names from
the state's tumor registry. A questionably significant
excess of lymphoma cases, mostly Hodgkin's disease, was
found, but no excesses for other cancers. A similar
study, based on military and Veterans Administration
records, has recently been completed and has likewise
produced no firm evidence of any increase in cancer
among persons with past histories of IM. Actually
studies such as these are more a test of the idea that
the disease IM itself predisposes to cancer than that
EBV is a risk factor, since EBV is ubiquitous and infects
a majority of the population whereas clinically apparent
IM occurs in fewer than 10%. Were a relationship between
IM and cancer found, instead of attributing it to EBV
directly, one might well instead relate it to the lymphocyte transformation which occurs in clinical IM. Such
transformation could perhaps result in the activation of
an endogenous oncogenic C-type RNA virus.

The epidemiologic features of BT include, among
other things, a peculiar geographic distribution in
Africa (concentration of cases in warm, wet, mosquito-
infested areas) and a strong tendency, in parts of
Uganda at least, for cases to appear in clusters. Both
are often cited as evidence of a viral or other infectious
etiology. It is not always stressed, however, that
neither of these features necessarily reflects EBV
infection. In fact, the geographic distribution of BT
in Africa seems to correspond to the geographic distribution of holoendemic malaria, and severe and chronic
malaria may be an important factor underlying BT development.

Discrete time-space clustering, while it clearly
implies underlying infectious etiology, is in theory most

easily linked to infectious agents which spread
episodically or which are related to epidemic disease
patterns. From what little is known at present, EBV
does not seem to fit this mold. As yet there is little
or no evidence to suggest that EBV infections occur
episodically within given populations, and certainly it
is rare for cases of IM to come in epidemics. Without
some evidence of epidemic behavior, it is hard to account
for BT case clustering in terms of EBV infection alone;
one must postulate some cofactor modifying EBV behavior
or consider the possibility that BT is actually caused by
some other unidentified infectious agent. None of this,
of course, rules out the possibility that EBV may act in
some other etiologic role with respect to BT, quite
unrelated to the tumor's geographic distribution or to
its patterns of occurrence in time and space.

COMMENTS BY

William D. Hardy, Jr.

I would like to comment on the horizontal transmission of feline leukemia virus (FeLV) between cats and
the possible danger of this virus to man. We have
studied the infectious spread of FeLV in pet cats under
natural conditions with an indirect fluorescent antibody
test, which detects FeLV group-specific (gs) antigen in
peripheral leukocytes and platelets of infected cats.
Thirty-three percent of normal cats (mostly unrelated)
living in contact with FeLV infected cats were infected
(Table 1). One hundred and forty of these 177 cats were
followed for development of disease. Twenty-four cats
developed lymphosarcoma within 6 months. By contrast,
FeLV was detected in only 2 of 1462 (0.14%) normal cats
from environments with no history of FeLV-associated
diseases.

With this information I feel we should be more concerned about the possible biohazards of this virus.
Examining human serum by complement-fixation for antibodies to FeLV gs antigens is not a suitable method for
detection of FeLV infection. We are examining human
sera, especially from individuals in contact with
infected cats, for the presence of neutralizing antibody
to FeLV in order to resolve this question.

Table 1.   Immunofluorescent Detection of Feline Leukemia Virus
in Normal Cats According to Exposure Environment

Environment of Normal Cats	No. Studied	No. with Normal Cats FeLV +	Normal Cats	
			No. Tested	No. FeLV +
FeLV-associated disease households	103	52	543	177
Normal multiple cat households	47	0	130	0
Single cat households	497	0	497	0
Stray cats	638		638	2*
Experimental colonies	4	0	197	0
			2005	179

* 2 cats used repeatedly as blood donors

COMMENTS BY

Robert J. Huebner

I will try to limit my discussion to the type C
tumor viruses and will not discuss the DNA tumor viruses.
There is no evidence at this point that would convince
anybody with any critical sense that known RNA tumor
viruses have produced a single tumor in man.   This
observation of course must be limited to those viruses
for which we have adequate reagents for detection.   I
think it is important to consider in the beginning the
possible modes of transmission of the RNA tumor viruses.
We and others have postulated genetic transmission of
both A and B type RNA viruses and all of their various
phenotypic expressions.   However, when this virus does
become expressed as an infectious agent, then it can be
in many instances transmitted epigenetically.   Such
transmissions can be vertical through the maternal
reproductive tract, the same sort of transmission
exhibited by LCM and cytomegaloviruses.   That the RNA
tumor virus is transmitted quite readily from mouse to
mouse and chicken to chicken vertically has been
established experimentally.   It must be assumed also
for rat and hamster and cat viruses.

What about the possibility of horizontal spread?
Now there are several types of possible horizontal
spread. One that occurs in the mouse (but of course not
in the chicken) is milk transmission, both C and B type--
not just B as some might think--type C is transmitted
very well in the same way through the milk. Of course
the virus has been demonstrated, as Dr. Gardner may have
mentioned, in saliva and urine of wild mice, but
horizontal spread from this source has not been established in mice. However, there isn't any question that
the various avian type C viruses, particularly the
leukosis viruses of the chicken, are transmitted
horizontally from chicken to chicken.

There is evidence suggesting that chickens, mouse,
and cat viruses are not transmitted to man. Dr. Miller
possibly discussed the evidence related to avian viruses.
Twenty million persons were inoculated with yellow fever
vaccine during World War II. These vaccines invariably
contained avian leukosis virus. No excess of leukemia
or cancer can be traced to this practice. Transmission
of mouse viruses to humans has not been demonstrated,
despite the fact that mouse viruses, like cat viruses
(particularly those that are activated genetically),
often grow better in human cells than they do in mouse
cells. The cat FeLV and RD114 viruses also grow well
in human cells. However tests of humans for virus-
specific antibodies to cat, mouse and chicken tumor
viruses have been uniformly negative; this includes
cancer patients and veterinarians, as well as those
given yellow fever vaccines. The evidence then does not
suggest successful transmission of animal RNA tumor
viruses to humans, even those most intensely exposed.
Obviously serological tests of all those now being
exposed to tumor viruses in laboratories should be
monitored for specific antibodies. Such a group no doubt
has the most intense exposure. Dr. Miller's suggestion
that we look at cancer in the young for evidence of viral
expression is an excellent one since this group is likely
to be most susceptible to infection. This could be done.
We do have specific antiviral antisera available for
doing such studies for the well-known tumor viruses.
Similarly, sera from children with cancer should be
tested for antibodies to specific antigens of the known
animal viruses.

COMMENTS BY

Robert W. Miller

The question before the conference is, Are there
biohazards in cancer research? The answer, it seems to
me, is that there are biohazards in all microbiological
studies, including cancer research. These may now extend
to people who work with tissue cultures. The hazards,
though, in microbiological research concern diseases that
can kill a lot faster than cancer. Whatever preventive
measures are taken for the more lethal known risks of
working in these laboratories will protect against whatever might come from cancer viruses.

Another question: Is cancer induced by viruses in
man? Studies conducted to date have not implicated any
virus. The studies concern peculiarities of tumor
occurrence and tests of a host of hypotheses that have
come from the laboratory. We simply do not have the
epidemiologic methods to demonstrate that there is any
cancer in man induced by a virus. Nevertheless the
possibility cannot be excluded that there is such a
virus. The findings indicate only that we are unable to
demonstrate a viral etiology.

Laboratory workers are a special group, however.
Unlike ordinary people who are exposed to cancer in dogs,
cats or human beings, laboratory workers have a very
concentrated exposure to viruses that are known to be or
potentially are oncogenic. It is a general principal of
epidemiology that in looking for an etiologic agent that
might be causing disease, one should look at the most

211

heavily exposed groups first.   The laboratory worker
surely fits this circumstance.

This principal is exemplified in chemical carcinogenesis.   During World War I U. S. soldiers were exposed
briefly to mustard gas and subsequently there was a trend
toward increased lung cancer, but the differences from
normal were not statistically significant.   In Japan
during the war, there were mustard-gas workers who
manufactured the gas without wearing protective clothing.
Subsequently these workers suffered an epidemic of
respiratory tract cancer: 49 cases among about 500
workers.   So, one must look at the right group, the most
heavily exposed, among whom the influence of this chemical
carcinogen could readily be recognized.

Laboratory workers have not only heavy exposures to
known viruses, but also to the viruses that they invent,
a circumstance that does not pertain to the general
public.   Thus for two reasons laboratory workers are in
a different category of exposure than the rest of the
population.

Another approach would involve study of subnormal
people with ordinary exposures--people whose subnormality
causes them to be at very high risk of certain cancers.
We know, for example, that all categories of persons at
high risk of leukemia are characterized by pre-existent
chromosomal abnormality.   If through laboratory experimentation you find a leukemogenic virus, it would be
appropriate to study these high-risk persons for
exposure to that particular agent.   With respect to
lymphoma, high risk in man is associated not with
chromosomal abnormalities but with immunological
deficiency, either inborn as in Wiskott-Aldrich syndrome
or ataxia-telangiectasia, or iatrogenic as in immunosuppressive therapy for organ transplantation.   When
lymphoma in laboratory animals is found to be caused by
a virus that might be in the general environment, it
would be appropriate to look for it especially in
patients with the above-mentioned pre-existent immunosuppressive disorders.

Finally, exposures during pregnancy may be much more
important than exposures later in life.   Again from
studies in chemical carcinogenesis, we know that DES
taken by the mother during pregnancy has induced a rare
form of cancer in their daughters 14-25 years later.
It is easy to imagine that viruses might cross the
placenta in the same way.   The problem is that one has to
define which viral exposures during pregnancy should be
studied now to determine some years hence if there is
indeed a relationship.   It would be preferable, and

sometimes possible, to identify women with exposures of
interest at some time in the past, and the health of
their offspring can be evaluated in the time that has
elapsed since then.

* * * * *

DISCUSSION

GRIESEMER: Aside from Yaba virus, there is no evidence
that any of the animal tumor viruses produce malignancies
in man. I'm convinced that the risks are much greater
from the nononcogenic indigenous agents in animals and
cells than from the oncogenic viruses.

In regard to infectivity of viral nucleic acids, my
associates and I have succeeded in transmitting SV40 DNA
by the aerosol route to weanling thymectomized hamsters.
The tumors produced contained T antigen. Those people
working with nucleic acids may be at some risk.

Also I believe studies are needed on inactivation of
oncogenic viruses under laboratory conditions. A
comment was made that implied it is relatively easy to
inactivate oncogenic viruses on laboratory surfaces with
chemicals. It would be comforting if supporting experimental data were published.

SCHNEIDER: HeLa cells were, I think, the first
established human cell line and it has been used by
thousands of people, mostly by mouth-pipetting. Most
people who work with it think it is completely safe
simply because everybody else thought so; very few stop
to ask whether there is a danger. The question that I
would like to ask the epidemiologist is, If among people
who worked with HeLa cells--let's say 10 years ago--there
were a 1% increase in the incidence of cancer, would we
know it? If it were a 10% increase? How high would the
incidence have to be before it would become apparent to
people that there was a danger?

MILLER: If the cancer was of a distinctive type which
ordinarily does not occur in that age group, I think
three or four cases would be very suspicious. If,
however, the tumor was a common one for the age group,
for example lymphoma in adolescence, it would be
difficult to recognize a few additional cases.

OZER: How should we evaluate exposure to laboratory
personnel? Serological testing of workers seems to have
varied considerably among laboratories and NIH programs.
From the discussion this morning, it appears that
exposure to murine C particles was evaluated by testing
for antibody to gs antigen using the complement fixation
assay (CF). We know from experience with SV40 that we
can readily detect antibody to SV40 V antigen by neutralization and precipitation of radioactively labeled
virions. However, an analogous approach to that used
with C particles would be negative (i.e., we have never
observed antibody to T antigen). I wonder if Dr. Huebner
would address himself to what might be the most sensitive
assay we can use, particularly radioimmune assays?

HUEBNER: If the RNA viruses are indeed getting into
other species, there should be some immune response and
this should be measurable. I agree that the current
tests for the so-called "tumor" antigens (unlike the gs
antigens of the RNA viruses) would not be adequate for
this purpose. What we should look for are those antibodies which would be directed against the gs antigens
and viral envelopes. This requires very highly concentrated, purified, banded viral antigens. This kind of
effort has not yet been done for the newer RNA tumor
viruses, although rather extensive tests have been made
for infection with avian RNA viruses and the DNA viruses
(polyoma, SV40 and adenovirus). C57BL, Balb/c, Swiss,
NZB, wild mice, and cats do develop natural antibodies
to banded virus, each of which is rather specific.
Tests to date have been done by complement fixation, but
we should be able to increase sensitivity by the radioimmune assay test. In fact, we have been testing sera
from cancer patients and sera from tumored animals
against these banded virus preparations and we find there
is virtually no evidence of antibodies to FeLV and RD114
or the Crandell cell viruses. We would like to have sera
from young infants and children who have leukemia and
who have been exposed to cats; that is one test we
haven't yet done and it is one that can be done. I
think we have to do the tests that we can do now; that
is, talk less about it and just get to work.

DIXON: In searching for antibodies to potential viral
agents, it would probably be wise to check the kidneys
of the subject for deposition of gamma globulin in the
glomeruli. If by fluorescent microscopy considerable
amounts of gamma globulin are found in the glomeruli, the
kidneys can be homogenized and the gamma globulin eluted
at an acid pH. Gamma globulin obtained in this way has
been a rich source for antibodies to a variety of chronic

viral agents of both animals and man.   In situations
where an excess of virus is present in the circulation,
renal eluates may be the only possible source of antibody, as was the case in chronic LCM infections of mice.
As part of any well organized search for antibody, the
kidney should be held in the deep freeze for possible
examination if examination of the blood proves negative.

OXMAN: The case of Dr. Henle's technician, whose
acquisition of infectious mononucleosis led to the
discovery of EBV's role in that disease, demonstrated
the value of well-studied isolated cases of disease in
exposed individuals.   Are there any comparable "anecdotal
cases," for example malignancies in workers exposed to
tumor viruses, which might provide similar leads?

HUEBNER: I have many of them.   I can tell you the latest
one discovered in our animal laboratory.   We had a group
of technicians who suddenly beg250 fighting with each
other, their personalities changed tremendously, three
of them walked out and quit.   So we bled them.   We found
that we had an epidemic of LCM, the virus having been
brought in by some hamsters that looked perfectly normal.
We had one other situation: One of our secretaries (a
Scotch girl), who previously had moved frequently having
been in three different laboratories, was a rather dour
person with no sense of humor whatsoever, but after she
got LCM she was absolutely changed--she smiled all the
time and she was very much in demand.

     Technicians maybe more so than the professional
people no doubt have the most intimate exposures to these
agents.   They're the ones who most often cut or inject
themselves accidentally and have other exposure to
accidents that distribute the viruses into the environment.   I can recall about three cancers in NIAID
laboratory workers; one of these occurred in the baby
son of one of our scientists, who was exposed to our
laboratory mice which were brought home by his father.
The baby died of a brain tumor a couple of months later;
this is most likely coincidental.   I don't know of any
laboratory epidemic or even individual case situations
where we could actually trace the disease to an exact
source, unless of course one was able to make a specific
laboratory diagnosis of viral infection.   For this we
must rely on available technologies; otherwise we're
going to have many situations where we're going to create
fear, and fear in this case, as in any other, will be
directly proportionate to the lack of real knowledge.
The less we know about a problem, the more concern it
generates.   The Davis case is a good example.   Her

lymphoma was in all the newspapers and weekly magazines.
She attributed her lymphoma to exposure to chickens
during most of her professional career. We put her in
the Clinical Center for about a month; she had some
existing subcutaneous tumors which were removed. We
grew her cells, took her sera and did everything that
we could at that time to see if she had any avian virus
in her tissues or neutralizing antibody to the viruses.
We tested for gs antigens in the tumor cells and for
antibody in her sera but found no evidence of any avian
virus infection.

GIRARDI: Where virus expression may be limited in a
cell infected non-productively, a search for surface
antigen expression could be helpful; an example would be
normal cat cells carrying the endogenous RD-like agent.

TODARO: An in-depth look at the few anecdotal cases
could also provide clues. An obvious example is the one
Dr. Rickard cited of the animal caretaker who worked
with leukemic cats and who subsequently developed tumors.
Several people at NCI tested material from the caretaker
for antibody to the gs antigen and for virus neutralization. All tests were negative. Personally, I'm not
satisfied with those results because now more sensitive
tests are available.

KAPLAN: Is there any evidence that the RNA tumor viruses
infect man?

HELLMAN: Dr. Yohn at Ohio State University suggests that
with certain types of malignancy sera there seems to be
some cross reactivity against the interspecies-specific
antigen (gs3) of feline tumor virus prepared in rabbits.

OZER: The question previously asked of SV40 is certainly
appropriate here. Have sera of workers with these
viruses or of those exposed to them, as for example on
farms, been tested?

GARDNER: We found no CF antibodies reacting with whole
banded wild mouse viruses in the sera of our laboratory
group, who work very closely with these virus preparations.

HELLMAN: I think that's true of all sera that have been
tested, in particular the work that was done at Ed

Lennette's laboratory on people that have been associated with leukemic cats.   There seems to be no correlation.

TODARO: I don't think it has been done.   The NZB strain of mice, for example, has a virus that grows in human cells as does the NIH Swiss mouse.   Precautions are certainly necessary to eliminate the chance of human infection by a virus that has human cell lines as a permissive host.   I don't think there is enough to be gained scientifically by deliberately growing viruses in human cells to make those experiments worth doing.

FRIEDMAN: Could you bring us up to date on what attempts have been made statistically to look at correlation of human tumors with pets in the house?

HUEBNER: Epidemiological studies of associations between cats with and without leukemia or cancer and humans with and without cancer have been done.   Dr. Gardner and his group did such a study in Los Angeles.   Although admittedly this study could not exclude rare events that might lead to cancer, there was no statistically significant evidence of significant associations or even any suggestion of such an association.   This was a rather extensive study and if cats were a significant factor in horizontal transmission of feline viruses and of the diseases they produce in cats, one would have expected to see it here.   There was no evidence of an association of human leukemia or other cancers in households that had cats or vice versa.   Similarly, Dr. Schneider of the California State Department of Public Health maintained a cat and dog cancer registry in the Bay area for a number of years.   He made a number of reports showing that there was no evidence of any excess of leukemia or cancer in those households that had cats or in those that had cats with leukemia or other cancers. This latter study, which was continued for 3 or 4 years, also would not have been able to exclude rare cases perhaps due to overwhelming infections.   I myself had an interesting experience.   Early during our studies with the University of Southern California, I encountered an apparent cluster of leukemia and sarcoma in South Pasadena.   The USPHS Surgeon General's wife told me that her children's school parent-teachers group knew of three cases of cancer, two osteosarcomas and a lymphoma in two adjacent houses, two of them fatal.   When we looked into the three cases, there was a cat in the household with the young boy who died of lymphoma.   The two children with sarcoma in the adjacent household played with the same cat.   The two boys who died were about eight years old.

The girl who had a giant cell sarcoma survived surgery.
The second house also had a parakeet which died of
sarcoma. On the next street was a dog who was running
around on three legs, one having been removed because
of a sarcoma. Down the street about five blocks was
another woman who had lost a leg some years earlier
because of sarcoma. These events occurred in a
population of fifty-people on two adjacent streets, so
it was a very unusual thing, but it is axiomatic that
everything that can happen by chance does sooner or
later. So we really don't believe that these events
really prove anything but it did greatly stimulate our
interest in cancer epidemiology. Unfortunately, we
obtained serum on only one of those cases. Subsequently,
Dr. Glass and Dr. Miller of NCI did a very extensive
study looking for clusters of leukemia in Los Angeles; no
really convincing clusters were found.

MILLER: Maybe epidemiological studies have failed
because you have been looking at the wrong age group.
From what we know, it seems as if the fetus and newborn
may be the most susceptible. Perhaps fetuses, when
pregnant women are exposed to cats, are more at risk
than people exposed as children or adults.

HULL: Yaba virus may be transmitted by mosquitoes.
The original outbreak in Yaba occurred in an outdoor
colony in which virus spread occurred. The group at
Buffalo observed no spread of infection in a laboratory
room with 200 to 300 animals, many of which were
experimentally infected with Yaba virus. In another
laboratory, however, where mosquitoes were present,
they encountered 11 spontaneous cases of Yaba tumors.
Elimination of the mosquitoes and other insects appeared
to solve the problem.

HUEBNER: I believe it possible that EB virus, which is
associated with infectious mononucleosis and is
universally found in Burkitt's tumors, should be considered as a possible trans-species viral agent. This
virus is found in a variety of simian hosts. It is
tempting to suggest that this virus was originally of
nonhuman origin, causing tumors in a heterologous host
like man but not in its own host. Thus herpes saimiri
and ateles do not produce lymphoma in their natural host,
but do so when inoculated in heterologous hosts.
Perhaps special concern should be directed against the
herpes viruses of nonhuman as well as human origin.

KALTER: It is important to recognize that EBV is
prevalent in only certain primates, primarily chimpanzees
and gorillas, which have a 100% incidence of EBV antibody.
Orangs and gibbons run about 50% positive, and as you go
on down the scale phylogenetically the prevalence
decreases. New World and lower monkeys appear to be
essentially negative.

OXMAN: The leukemic lymphoblasts which circulate in owl
monkeys inoculated with herpesvirus saimiri can
apparently induce leukemia in other owl monkeys. Is
there any data on transmission of disease by arthropod
vectors in primate centers?

GRIESEMER: No.

MARTIN: There are a couple of potential hazards that
Dr. Schneider did not make explicit. One is that if you
have an immunologically naive laboratory worker,
infectious mono can be a very serious disease. The
second thing is that you might have, admittedly very
rarely, a histocompatability between cell line and
technician. If we turn out to have a sort of $\underline{E}$. $\underline{coli}$-
type of human lymphocyte cell line which is disseminated
to hundreds and possibly thousands of laboratories, with
thousands of individuals involved, statistically there
is a chance for some histocompatability.

SCHNEIDER: We do have a rule that we don't let lab
workers work with their own cells. In other words, we
don't establish cell cultures from ourselves. Of course,
we are not certain how important this is.

MILLER: Do you mean that whatever rules or regulations
that have been made pertaining to virus laboratories and
the handling of their materials should be extended to
people who work with tissue cultures in general?

SCHNEIDER: Yes, we feel that any aneuploid cells should
be handled as if they contain viruses. The difficult
question comes when we consider diploid fibroblasts.
Do we have to be concerned with these cells? I don't
know the answer, but we certainly don't take risks that
can easily be avoided. For example, we never mouth
pipette and we don't work in horizontal laminar flow
hoods that blow cells right at us.

POLLACK: A common strategy in somatic cell genetics is
to select BrdU-resistant lines as parents for somatic
cell hybridization. BrdU-resistant cells die in· HAT
medium because they lack thymidine kinase. Based on
Todaro's remarks about BrdU induction of C-particles
from Balb-3T3, we now see that laboratories carrying
out this selection may also inadvertantly induce
endogenous C-particles from their cells.

SCHNEIDER: I think this is the common thought. Many
people are studying long-term lymphocyte lines, which
are valuable because of their biochemical markers, and
treating them with BrdU may enhance the appearance of
virus particles.

Session IV
Hazards Associated
with Modern Research Methodology

ARBOVIRUSES, ARENOVIRUSES AND HEPATITIS

Jordi Casals
Department of Epidemiology and Public Health
Yale University School of Medicine
New Haven, Connecticut

On first consideration it would seem unlikely that
non-tumor viruses constitute a hazard in cancer research.
Some of them may, however, be associated in nature with
hosts and vectors in circumstances such that the viruses
could turn up fortuitously; in addition, some, such as
Semliki, Sindbis and vesicular stomatitis are much used
for basic virological studies principally because they
have short replication cycles, reach high infective
titers and can be manipulated by inexperienced personnel
with little or no personal risk.

ARBOVIRUSES

Arthropod-borne viruses (i.e., arboviruses) are a
heterogeneous set defined on an ecological basis. They
are viruses which are maintained in nature principally
through biological transmission between susceptible
vertebrate hosts by blood-sucking insects. The viruses
multiply and produce viremia in the vertebrates, multiply
in the tissues of arthropods and are passed on to new
vertebrates by the bite of the arthropod after a period
of extrinsic incubation. This definition excludes
viruses that are disseminated by mechanical transmission
resulting from mere physical contamination of the
arthropod's mouth parts or other parts of its body. It
must not be assumed, however, that arboviruses cannot be
transmitted by means other than that which defines them:

223

under natural conditions arboviruses have been transmitted by the respiratory route--droplets and dust--and the alimentary route.

Currently there are well over 350 known arboviruses; a large proportion can be assembled on the basis of serological tests in antigenic groups. There are at this time about 45 recognized groups, varying in size from 2 to 40 or 45 viruses in each; a number of arboviruses are ungrouped.

The "typical arboviruses" are RNA-containing viruses with cubic or presumed cubic symmetry and an envelope. Two sets, A and B, have diameters between 50 and 65nm and 35 and 45nm, respectively.

Proposed attempts to classify arboviruses (1) are hindered to some extent by lack of information (2). There is enough knowledge, however, to realize that the arboviruses constitute a heterogeneous set which does not fit as an undivided whole in any of the groups of the general system. Conversely, other viruses that are not arthropod-borne have properties in common with some arboviruses. For example, Batai, Inkoo and Uukuniemi appear to be myxovirus-like or paramyxovirus-like; Nodamura, is a picornavirus; Colorado tick fever, Kemerovo and African horsesickness, are reovirus-like; vesicular stomatitis, Hart Park and Mokola, are rhabdoviruses; African swine fever, a DNA-containing virus, resembles iridoviruses.

ARBOVIRUS DISEASE IN MAN. It must be stressed that arboviruses in general, including those that are pathogenic for man, are maintained in nature by a cycle involving vertebrate hosts other than man. Infection of man and domestic animals is a tangential, dead-end transmission which is ineffective for virus preservation. Nevertheless of the approximately 350 currently identified arboviruses, 86 different ones are responsible for diseases of man in nature, the laboratory or both. The severity and prevalence of the natural disease or the severity of the laboratory-acquired one vary considerably from a mild, febrile episode with hardly any discomfort to fatal outcome, and from very few reported cases to hundreds of thousands. Between 26 and 28 arboviruses cause important disease of man, either in terms of numbers, severity or both; 14 are in group B, 5 in group A and the rest in other groups or ungrouped.

The disease may evolve in several forms. Following exposure and infection there is an incubation period of several days duration; the onset is usually abrupt and

may coincide with virus dissemination in the body and
viremia. The cardinal sign at this time is fever, with
generalized manifestations of variable intensity. Often
this is the entire disease. In other instances there is
a second phase of the illness with localization in
various organs and systems, resulting in different
clinical pictures. In some instances there are manifestations of capillary fragility, localization of aches and
pains in joints and muscles, generalized lymphadenopathy,
giving a dengue-like illness. In other cases the
manifestations of capillary fragility are pronounced and
result in conspicuous petechiae, echimoses and purpura,
alterations in the blood cell counts and hepatic
involvement; these are the hemorrhagic fevers, including
yellow fever. A number of the viruses have neurotropic
affinity in man giving rise to the encephalitides or
encephalomyelitides. A clinical picture that had lately
received much attention is the dengue shock syndrome;
its suggested pathogenesis is that two or more sequential
infections with different dengue serotypes result in an
antigen-antibody complex that combines with complement
at the capillary basement membrane, with damage to the
endothelium and increase in permeability.

NATURAL HOST RANGE. The vertebrate host range of arboviruses differs a great deal between serotypes; Eastern
equine encephalitis (EEE) and Western equine encephalitis
(WEE) viruses have probably the widest capacity to induce
illness in mammals and birds and silent infections in
birds and reptiles. Other arboviruses have a limited
host range; among these are the dengue viruses and
Neapolitan and Sicilian sandfly fever viruses. A large
proportion of arboviruses have been isolated only from
arthropods; hence their capacity as disease producing
agents in nature is unknown.

   Among invertebrates only mosquitoes, ticks,
phlebotomines and Culicoides have the capacity to transmit the virus biologically; some hematophagous mites have
been suspected to be vectors, but the evidence is largely
negative. Although there are instances in which several
species, even genera, of arthropods can propagate a
virus, one species is generally the main or exclusive
vector in a given ecological situation.

EXPERIMENTAL HOST RANGE. With rare exceptions, all
arboviruses are readily propagated in newborn white mice
by the intracerebral route of inoculation; older mice
are less generally susceptible. Cells in cultures,
either under fluid or an agar overlay, are extensively
used. Among the stable lines most generally used for

routine work are LLC-MK, BHK, Vero and primary chick
embryo fibroblasts.

OTHER MODES OF TRANSMISSION.   Arboviruses can be transmitted in nature by means other than the bite of an
arthropod.   Outbreaks of EEE among pheasants have been
observed under circumstances such that the likeliest
source of infection between birds was feather picking
and consequent ingestion of viremic blood (3).   Venezuelan
equine encephalitis (VEE) virus infection resulting in
disease has been transmitted from person to person, in
the absence of a vector, during the course of a severe
outbreak of this illness in Venezuela.   Overt infection
of man with tick-borne virus of group B (Central European,
Russian spring-summer virus) through ingestion of un-
pasteurized milk has been repeatedly reported by Soviet
and Czechoslovak scientists (4); these outbreaks have
been associated with goat's milk, or in the case at
Roznava, Czechoslovakia in 1951, with cow's milk to which
goat milk had been added.

     Under artificial laboratory conditions, accidental
infections have hardly ever occurred through an arthropod's bite; the most frequent routes have been inhalation
of dust or aerosols and through the skin by cuts or
injection.

OVERT LABORATORY-ACQUIRED INFECTIONS.

Etiology.   The magnitude of the overall problem of
laboratory-acquired infections among personnel handling
disease-producing agents has prompted the creation of
Committees to maintain a file on these infections.
One such group is a standing Committee on Laboratory
Infections and Accidents, American Public Health Association; a second is a Subcommittee on Laboratory Infections,
American Committee on Arthropod-borne Viruses (5).   One
of the latest published compilations (6) reports 788
overt infections by viruses with 38 deaths; arboviruses
were responsible for 438 infections with 18 deaths.   These
figures have been updated as of this writing (Sulkin,
pers. comm.) to 936 infections with 54 deaths for all
viruses, and 474 with 17 deaths for the arboviruses.   It
must be pointed out that owing to changes in classification of certain viruses there may exist discrepancies in
the figures assigned to various groups; nevertheless,
the last figures quoted for arboviruses represent a
minimum number of laboratory-acquired infections, not
including those acquired in the course of field studies.

     Detailed analysis of the responsible viruses is

available for 428 cases (5). The largest numbers of
infections were caused by the following viruses: VEE,
118; Kyasanur Forest Disease, 65; vesicular stomatitis,
38; Rift Valley Fever, 29; louping ill, 21; chikungunya,
19; group B tick-borne encephalitis, 18; and West Nile,
11. The remaining 25 listed arboviruses gave individual
lower figures. Yellow fever virus is considered apart;
it has been a high-risk virus in the past, having been
responsible for 38 infections with 8 deaths; it should be
noted, however, that these accidents occurred before
the advent of a vaccine. Other fatal infections have
been recorded, in one case with VEE, in 2 of 5 infections
with WEE, in 2 with group B tick-borne encephalitis and
in one with Rift Valley Fever. A recently available
report not included in any of the former tabulations
describes a fatal case of laboratory-acquired infection
with Crimean hemorrhagic fever virus (7).

Sources of infection. The main source from which arboviruses have been isolated from man in nature is blood;
other fluids or organs (CSF, mucous membranes and their
secretions, CNS tissues and viscera) can also be infective
but to much lesser degree.

     Among naturally infected lower vertebrates blood
is also the principal infective source; however, lung,
liver, spleen, kidney and CNS tissue have been found
infective with 25-50% of these viruses.

     Arboviruses, especially groups A and B, do not, on
the whole, cause latent persistent infections in
vertebrate hosts but rather a short-lived inapparent
or overt infection, with the virus disappearing from
the blood and probably also from other tissues within a
short period of time after infection; consequently, the
risk seems small that a virus will be accidentally
introduced in the laboratory through seemingly normal
animals or tissues derived from apparently normal
animals. There are, however, reports of persistent
infections, mainly after experimental inoculations but
also after natural infection. WEE virus has been
recovered from wild birds 10 months after the original
infection (8) and from inoculated garter snakes after
hibernation over the winter under simulated natural
conditions (9). Group B tick-borne virus (RSSE) has
been found for long periods during hibernation in the
blood of hedgehogs, dormice and bats (10-12); it has also
been claimed (13) that the virus was isolated from cases
of Kozhevnikov's epilepsy and other chronic neurological
syndromes in the Soviet Union in patients who had
developed these slowly progressive diseases years after
the acute infection.

Persistent infection of insect cells in cultures from Aedes aegypti and A. albopictus is caused by some arboviruses; the cultures, showing no cytopathic effect, appear to be latently infected (Buckley, pers. comm.).

In experimentally inoculated animals, particularly newborn mice, viremia generally develops. Sick or moribund mice are often eaten or partly eaten by the mother with spilling of blood or viscera and contamination of cage bedding. Some viruses are also excreted in the urine, feces and nasal-oral discharges resulting in additional contamination. Contaminated cage dust has been considered the cause of a number of laboratory-acquired infections (14); it may well be responsible for infection in a large number of cases when the probable source of infection is listed as "working with experimentally infected animals."

Infected bedding, just as much as close contact, may be responsible for cage-mate infection with viruses known to give high-titered viremia in newborn mice and to which mice are known to be highly susceptible on peripheral inoculation, such as VEE, RSSE and Germiston.

Work with cells in culture seems to be definitely less of a risk than where animal inoculations are involved; the two main causes of infection appear to be breakage of glassware and aerosols created by changing nutrient fluids and harvesting of materials.

Potentially dangerous procedures. The type of work done with arboviruses in the laboratory is varied; there are, however, a number of simple manipulations that are general, such as inoculation of virus suspensions to animals or to cells in cultures, dissection of animals, preparation of virus stock suspensions, lyophilization, preparation of antigens and immune sera, and in vitro tests of various types. In most, if not all, of these operations there is a risk of producing an infected aerosol, particularly when handling dried virus preparations or using mechanical blenders. Even such simple and recurring operations as pipetting, blowing out the last drop from a pipette or syringe, adjusting the amount of fluid in a syringe, or removing the plug or cap from a test tube are ways of producing infective aerosols. Nearly 20% of all laboratory-acquired infections with arboviruses have been attributed to aerosol inhalation (5). In fact, of all single mechanisms of acquiring an infection caused by an arbovirus, inhalation of contaminated dust and aerosols is undoubtedly the most frequent. Accidents involving inoculation through injured skin or spilling on skin abrasions are responsible for about 10% of infections (5,6).

ARENOVIRUSES

     Arenoviruses are a group of viruses (15) that have
the following properties in common: they contain RNA
and have an envelope with short, thin projections or
spikes; they mature by budding at the plasma membrane
or in cytoplasmic vacuoles; they are pleomorphic but
essentially round or ovoid particles with a diameter
between 80 and 110nm, although larger ones up to 200nm
are seen; the central part of the image, more opaque
than the periphery, contains a variable number (6 to 12)
of electron dense granules similar to ribosomes. No
capsid symmetry has been discerned. The viruses are
antigenically related, some rather distantly, by
complement fixation and fluorescent antibody tests;
included in the group are lymphocytic choriomeningitis
(LCM), Lassa, and the members of the antigenic group
Tacaribe (Amapari, Junin, Latino, Machupo, Parana,
Pichinde, Tacaribe and Tamiami). It was assumed at one
time on limited evidence--occasional isolation of some
of the serotypes from mites--that the Tacaribe group
viruses were arthropod-borne; currently the arenoviruses
are not considered to be arboviruses.

     Hemorrhagic fever with renal syndrome (HFRS),
occurring in the Soviet Union and adjacent areas, is known
to have viral etiology from transmission experiments in
human volunteers. The virus has not been propagated in
laboratory hosts. The disease is in many respects
similar to the South American hemorrhagic fevers.

HOST RANGE AND EPIDEMIOLOGY. Lymphocytic choriomeningitis virus is taken up in another chapter; of the
remaining viruses of this group, only those associated
with diseases of man will be considered here.

     An important, distinctive biological property of
arenoviruses is their capacity to induce latent, chronic
infections in some of their natural hosts and persistent
viruria in experimentally infected animals; these facts
have considerable epidemiological implications.

     Long lasting excretion of Machupo virus in the urine
has been observed in hamsters and colonized Calomys
callosus, even in the presence of circulating neutralizing antibodies; seemingly normal, wild-caught C.
callosus trapped in an endemic area have yielded virus
in the urine in 5 of 11 animals tested (16). Junin virus
has been isolated from the urine of patients on several

occasions and from the urine of inoculated guinea pigs
from the 7th post-inoculation day to the time of their
death 5 to 8 days later (17).   Lassa virus was present
in the urine of mice inoculated at birth on the 83rd
post-inoculation day, the mice having shown no signs of
illness in the meantime (18); the virus was present in
the urine of a patient 32 days after onset of illness
(19).   The urine from volunteers infected with HFRS has,
on inoculation to other volunteers, transmitted the
disease (20).

Junin, Lassa and Machupo viruses have been isolated
in nature from man; Junin has also been isolated from Mus
musculus, Hesperomys laucha laucha and Akodon aernicola
and Machupo from Calomys callosus; these rodents appear
to be reservoirs.   The reservoir of Lassa virus is
unknown, as it has only been isolated from man.   Rodents
of the genera Clethrionomys and Apodemus are considered
the reservoir of HFRS in the Soviet Union.   The other
members of the Tacaribe group have been isolated from
rodents, with the exception of Tacaribe virus which was
isolated from bats.

Junin and Machupo viruses cause overt infection in
newborn mice and hamsters, the former also in adult
guinea pigs; Lassa virus infects newborn mice asymptomat-
ically, but causes death in adult mice.   The three
viruses replicate in cells in culture, particularly well
in Vero cells, producing cytopathic effect in fluid
cultures and plaques in overlaid monolayers.

It is generally agreed that Argentinian hemorrhagic
fever (Junin virus) is not transmitted by personal
contact; there is evidence that Bolivian hemorrhagic
fever (Machupo virus) can be transmitted in that manner,
but this mechanism is considered of little importance
in nature.   In both diseases, excretion of virus in the
urine of a rodent, with contamination of human food,
drink and house dust, and penetration of the virus
by the oral-nasal route is the accepted mode of infection.
The method of transmission for Lassa fever is still undetermined; person to person contact certainly explains
some episodes; inhalation or inoculation through cuts
in the skin has been suggested as the route of infection
in the case of laboratory and autopsy accidents.
Exposure to rodents and contact with their excretions
has been suggested as the likeliest mechanism of infection
for HFRS.

OVERT LABORATORY-ACQUIRED INFECTIONS.   This group of
viruses has been particularly severe on laboratory
personnel and hospital staffs, even though work with at

least two of the agents, Machupo and Lassa, has been
restricted to very few laboratories and persons.

At least 5 persons have acquired infections in the
laboratory with Junin virus, with one death; one fatal
infection is recorded for Machupo; and two infections,
one fatal, for Lassa virus. There is in addition a
substantial number of reported infections acquired in
the field or in hospitals which, although strictly
speaking are not laboratory acquired, can be directly
ascribed to research work with the viruses. Four
persons, two of whom were technicians, were taken ill
with BHF while investigating the disease in Bolivia (21).
Contact between two of these patients and their respective
spouses resulted in two additional overt, hospital-
acquired infections. Overt infections of hospital staff
with Lassa virus have occurred in at least 10 documented
instances with 4 deaths, including a physician, nurses
and nurses aids (22).

One of the largest outbreaks of a laboratory-acquired
infection is that caused by the agent HFRS, reported by
Kulagin and associates in 1962 (23); the episode amply
illustrates the risks incurred when seemingly normal
wild-caught animals are introduced in a laboratory. A
large number of small rodents, including Clethrionomys,
Apodemus and Microtus, were brought to a laboratory in
Moscow in September, 1961, to study tick-borne
encephalitis; the animals were held in quarantine for
two weeks, after which they were assigned to holding
rooms. Ten to 12 days later, cases of HFRS began to
appear and continued for the next 4 weeks; a total of
113 persons of 186 employees or visitors were taken ill
and diagnosed as HFRS cases; none died. It is stated
that among a group of 24 persons who visited, on a given
day, the section where the animals were housed and
remained only 3 to 4 hours, 18 acquired the disease,
including a person who was in the section for only 5
minutes.

HEPATITIS

Numerous current sources of information give ample
evidence of the high prevalence of hepatitis, types A
and B, in the world population; and since its reporting,
even where it is a notifiable disease, is far from
complete, the actual incidence is undoubtedly much
greater than available morbidity figures indicate.
Personnel in laboratories that work on hepatitis are
aware of the risks involved, but persons in laboratories
that deal with other virus, or virus-suspected, diseases
may also be at risk owing to the accidental introduction

of materials contaminated with the hepatitis viruses.
Certain types of research, particularly those involving
seroepidemiological surveys, may represent substantial
exposure to danger. However, in view of the finding of
tumor-specific antigens and antibodies in experimental
animals, the day may not be far when seroepidemiology
may become a routine technique in cancer research.

PREVALENCE OF HEPATITIS.   The number of overt hepatitis
infections contracted in the laboratory currently stands
at 172 (Sulkin, pers. comm.).   For the year 1972 there
were 55,000-60,000 reported cases of viral hepatitis A
in this country; for viral hepatitis B, the number is
between 8000 and 9000 (24).   Most estimates, however,
place the actual numbers at a much higher level.
Transfused blood alone is estimated to cause 30,000 cases
of overt hepatitis each year in this country, and since
there are many cases of subclinical virus hepatitis, the
incidence has been estimated as high as 150,000 cases
annually (25).

    Hepatitis B antigen (HB Ag) and hepatitis B virus
(HBV) are found in the sera of individuals infected with
HBV.   Naturally, it is important that transfusion
services, blood banks and hospitals assay for this virus.
The seroepidemiological data obtained enables the mapping
of the extent of HBV infection.

    Available data are mainly based on surveys of blood
donors, professional, volunteer or both.   All surveys
show a much higher incidence of HB Ag among the
professional donors.   Most surveys for HB Ag have been
done by means of agar gel diffusion and precipitation
test or by immunoelectrophoresis technique; recently
developed tests, such as the radioimmunoassay, that have
a greater sensitivity than the others will, undoubtedly,
result in detecting higher prevalence rates.   Results
of a few surveys are given here in order to illustrate
the extent of the problem.

    In a survey conducted by the American National Red
Cross (26) of nearly 2,600,000 units of blood collected
in 1971 from volunteer blood donors, 1.17 per thousand
were positive for HB Ag; in two areas, the rates were
higher than the average, 1.70 in the Southeast U. S. and
3.44 in Puerto Rico.   In various surveys abroad the
rates have been generally higher than in this country:
Scotland, 1.3; Norway, 1.6; Denmark, 1.8; Germany, 2.0;
Spain, 4.0; France, 4.5 per thousand (27).   In other
countries, considerably higher rates have been reported:
in Japan, 12.3 of 4222 volunteer donors, 22.0 of 849
professional male donors (28); in Yugoslavia, 23.2 of

about 12,000 volunteer donors tested by the Ouchterlony method and 40.8 of a little over 2000 tested by immuno-electrophoresis were positive (29); and in Kenya, 12 of 200 apparently healthy blood donors--60.0 per thousand--were positive (30).

Infection by hepatitis A virus (HAV) cannot at present be assayed by laboratory tests. However, there are far many more hepatitis cases diagnosed as type A than there are type B; in this country the proportion is 6-8 to 1; it may even be greater in other countries where neither blood transfusions nor self-injections are as frequent as in the USA.

INFECTIVITY OF SERUM. In general, HAV is spread by the oral-intestinal route and HBV by parenteral inoculation; it is now well established, however, that HBV can also infect by ingestion and HAV by inoculation. Studies by Krugman and associates on the natural history of hepatitis (31) show that serum from patients with hepatitis type A generally contains virus from the 12th day of the incubation period until the 3rd day of jaundice, i.e., a length of 30-35 days, most of it in the incubation stage or asymptomatic. HB Ag is detected in the blood 70-80 days after ingestion of HBV, patients remain positive for variable periods of time, from one week to several months or even years. Some of the patients are assumed to become carriers.

The results of epidemiological surveys to detect HB Ag and the current knowledge on the natural history of hepatitis give ample evidence that there are at all times and in most places large numbers of individuals with hepatitis A and B viruses in their blood. Under the circumstances it would appear that individuals in research laboratories who handle large numbers of serum samples for virological or biochemical studies, not necessarily in conjunction with hepatitis work, are at risk of infection by hepatitis viruses.

OVERT LABORATORY-ACQUIRED INFECTIONS. One of the earliest warnings in this country of the increased risk to medical and laboratory personnel of acquiring viral hepatitis, compared with the risk of the general population, was given by Byrne (32). Nineteen cases of viral hepatitis occurred in employees of the Yale-New Haven Hospital in the period 1952 to 1965, including laboratory and X-ray technicians and a glass worker. It was calculated that for that period the average yearly employee attack rate was 69 cases per 100,000; the average annual attack rate (based on reported cases) for

the period for the general population of Connecticut was
15 per 100,000. Since easily half the hospital employees
had no contact with patients, the rate for those involved
in patient care or laboratory work, it was concluded,
could be easily doubled.

Two recent reports, although having a hospital
rather than laboratory background, indicate the frequency
with which hepatitis can be induced through contact
with infected blood. Fox et al. (33) describe that over
a period of 1-1/2 to 2 years, 8 of 50 persons in the
staff attached to the hemodialysis unit at a London
hospital contracted hepatitis, 4 of whom had HB Ag; in
addition 2 other persons had no symptoms but had HB Ag
in their serum. Allen (34) investigated the occurrence
of post-transfusion hepatitis in a 650-bed hospital
over a period of 10-1/2 years; considering only the
patients who had received one single transfusion, 66 of
4737 developed hepatitis, or 1.4%; in other words, one
unit of blood in 72 contained HBV.

Several recent reports stress laboratory-associated
infections. Feinglass et al. (35), while investigating
hospital-associated hepatitis in a large institution,
found that 74 cases had occurred over a 3-1/2 year period;
32 of these were in individuals working in the laboratories. Among the laboratory workers, those that had a
routine requiring that 10% or more of their determinations
be on plasmas from patients in the hemodialysis-transplant
unit had an attack rate of 5.9 per thousand employee-
month; in other laboratories the rate was only 0.4 per
thousand employee-month. Sutnick et al. (36) describe a
situation in which endemic hepatitis associated with HB Ag
developed in a hepatitis research laboratory. Following
the occurrence of an overt case of hepatitis in a staff
member, a surveillance program of monthly testing was
carried out over a 3-1/2 year period. Fifty-six persons
in the laboratory were tested (for HB Ag and SGPT) on
3-32 occasions; of these individuals 4 developed clinical
hepatitis with HB Ag in the blood, 15 developed high SGPT
but no antigen and were considered anicteric hepatitis
cases, and the remaining 37 had no significant rise in
SGPT. This study is of particular interest in that the
infections occurred in strictly research laboratory
personnel, most of whom had no contact with patients in
the hospital.

LoGrippo and Hayashi (37) summarized the data from
a survey in this country and abroad regarding the incidence of hepatitis and HB Ag antigenemia among laboratory
personnel; while several laboratories indicated more than
one function, the majority reported to be in research.
The incidence of icteric and anicteric hepatitis was 7.4%

in the USA, 5.2% abroad; the rate of HB Ag was 2.5% in
the USA and 2.7% in foreign countries. These percentages are far in excess of those seen in the normal
populations.

Contact with animals. Chimpanzee-associated hepatitis of
man was first described by Hillis (38) among handlers of
animals kept for experimental use; Ruddy et al. (39)
reported similarly associated cases observed during 1963.
Two hundred and thirty-four chimpanzees were imported
in the USA that year; contacts during the next 6 months
were identified for 202. Among 309 persons who had
contact with the animals, there were 9 cases of icteric
hepatitis. The reported incidence of viral hepatitis for
1963 was 23 per 100,000; even if only one case in 10
had been reported, the incidence among these handlers
was 10-15 times greater than in the general population.
The chimpanzees seem to have been infective only the
first 2-3 months after arrival. Retrospectively it
appears that from 1953 to the end of 1964, 85 cases of
hepatitis, all presumably type A, associated with
chimpanzees had occurred; a disease indistinguishable
from hepatitis has also occurred through association of
man with other primates, woolley monkeys, Celebes apes
and gorilla.

CROSS-INFECTIONS WITH OTHER BIOLOGICAL SYSTEMS

Cross contamination in laboratories occurs even
when every effort is made to minimize it. Whatever the
cause, the end result is usually the same: either one
serotype supplants another, or a latent, irrelevant virus
of etiological significance is activated. Constant
monitoring is required in a laboratory that handles many
different viruses in order to reduce the frequency of
these incidents, and, if they occur, to understand them.
Some situations that have resulted in relatively prolonged
mix-ups, rather than short-lived errors, involve both
arboviruses and non-arboviruses.

Kemerovo encephalitis and Newcastle disease viruses.
Strains of Kemerovo virus were first isolated in 1962 from
blood and CSF from patients with a febrile diphasic
meningo-encephalitis in Western Siberia and from Ixodes
ticks collected in the area; development of antibodies
was ascertained at the time in a number of recovered
patients.

Three strains of presumed Kemerovo virus were sent
to our laboratory for identification; the three strains

were alike and were identified as NDV. It was established that the three strains had at some time prior to
shipment been passaged either through cultures of
chick embryo fibroblasts, or through embryonated hen's
eggs from flocks vaccinated against NDV. Presumably NDV
supplanted the original Kemerovo virus in the strains
submitted; earlier passage material of the virus,
obtained from another laboratory where no chick embryos
had been used, established the validity of the isolation
of Kemerovo virus (Casals, unpubl.).

A similar, still unexplained episode involving NDV
occurred with four strains sent for identification from
Bangkok, Thailand; three of them had been isolated from
mosquitoes, the fourth from the spleen of a child who
had died in the course of an undiagnosed febrile illness
(Halstead, pers. comm.); the strains had been isolated
and propagated in newborn mice only. They were found
alike and indistinguishable from NDV in our laboratory;
it appeared most unlikely that the original materials,
mosquitoes and human spleen, had been the source of
these strains.

Vilyuisk encephalitis. This disease is a chronic,
slowly developing, progressive encephalitis occurring
only in persons of one ethnic group, Yakutsk, in Eastern
Siberia. Virus strains were isolated by inoculation into
mice of CSF and blood, perhaps other materials, taken
from patients at various times after onset of illness up
to 6 months. Arthropod transmission of the agent was
not truly considered, although a possible role of mites
had at some point been entertained. A strain of this
agent proved to be resistant to the action of sodium
deoxycholate; it was identified as mouse polioencephalitis virus (GD 7). The etiological role of the agent
in Vilyuisk encephalitis is now discounted; most likely
the virus was picked up during passages, including
blind passages, in mice from a colony infected with
polioencephalitis virus (40).

Ectromelia infections. The fact that the mouse is used
to a large extent in experimental work with arboviruses
may account for the relative frequency with which
ectromelia virus appears as a cross-contaminant in that
work. In mouse colonies contaminated with extromelia
virus, the virus, as a rule, maintains a low grade
enzootic situation with relatively few signs of the
disease or losses of animals; the infection may well
even go undetected for considerable periods of time.
Occasionally flare-ups occur and only then do typical
vesicles and foot lesions, accompanied by high mortality,

appear (Christensen, pers. comm.).

Over the years virus strains have been sent to our laboratory (YARU-WHO Reference Center) for identification that have been found either contaminated with ectromelia virus or were entirely this virus. A number of the strains originated in the Soviet Union, Rumania and France; only the French laboratory was aware of the possibility that ectromelia may have been present in their mouse supply. Six strains from two laboratories in the Soviet Union were obtained after inoculation of mice with human blood, a bird's spleen suspension and mosquitoes or ticks from widely separated areas in the country; three strains from Rumania and one from France were derived from mice inoculated with ground tick suspensions. The identification of the strains was done by complement fixation, fluorescent antibody technique and neutralization test.

The combined effect of ectromelia virus and a transplantable tumor of mice, sarcoma 180/TG, is note-worthy. The tumor has been used in our laboratory for production of immune ascitic fluids (41); at the time when one of the as yet unidentified ectromelia strains was being investigated, an increase in the early mortality rate was observed in the sarcoma-bearing mice. It was established that the cause of this increased mortality rate was ectromelia virus. It would appear as though the sarcomatous cells had offered the virus a more favorable medium for replication than the normal mouse tissues, with an increase in virus concentration in the tumor cells and the ascitic fluid; this view derives support from the observation that high-titered antigens can be obtained with some viruses using ascitic fluids as source material (42). Conceivably other neoplastic tissues may similarly offer a favorable substrate for replication of latent or adventitious viruses etiological-ly unrelated to the neoplasm.

Hepatitis. With the exception of one or two recent claims, as yet not universally accepted, hepatitis viruses A or B have not been maintained in a laboratory animal or cell culture. There is therefore no reason at this time to suspect that these viruses will con-taminate biological systems used for other work. On the other hand, there have been numerous reports in the past claiming isolation of new specific hepatitis viruses or association of already identified viruses with hepatitis. All these are the so-called "candidate" hepatitis viruses.

A brief examination of past claims suggests that

these candidate viruses may have been either cross
contaminations or agents that were in the original human
material or in the host systems but had no clear
etiologic association with the disease. It is evident
that the proposed etiologic agents are very dissimilar
in properties.

An agent was recovered from a plasma pool from
serum hepatitis cases by inoculation to rabbit kidney
cell cultures and maintained in these cultures; the agent
was subsequently identified as <u>Mycoplasma</u> <u>gallisepticum</u>
(43). Another transmissible isolate from the blood of
patients with infectious hepatitis by inoculation of
liver cells in culture, designated lipovirus (44), may
have been an ameba, <u>Hartmanella</u> genus (45).

Other agents were easily identifiable as specific
members of recognized virus groups. A relationship to
hepatitis has not been clear enough with any to support
etiologic significance; it is quite possible that the
conditions that provide for transmission of hepatitis
may also promote the transmission of other agents (45).
Among these viruses are:

a. Motol virus, first isolated in hen's eggs, then
transmitted to rhesus monkey kidney cell cultures and to
mice; the virus is considered to be a contaminant,
originally ectromelia and subsequently contaminated
again with a paramyxovirus that replaced ectromelia.

b. In at least two instances a paramyxovirus, SV_5,
was isolated from the blood of patients with infectious
hepatitis by inoculation of human and rhesus kidney
cells (46) or human diploid lung fibroblasts (WI38)
(47) in cultures.

c. Adenoviruses types 1 and 3 have been isolated
from stools of patients with hepatitis A, and adenovirus
type 5 from blood clots, by inoculation of human
embryonic lung cells; these are the viruses originally
called San Carlos viruses (48).

d. A number of strains designated hemoviruses were
isolated from stools or sera from patients with infectious
hepatitis by inoculation of a cloned line of neoplastic
human marrow cells, Detroit-6 (49). Determined efforts
to show an etiologic association with hepatitis have
met with failure. A recent attempt by Melnick and
Boggs (50), using Detroit-6 cells, led to the isolation
of an agent from the plasma of a volunteer after he had
ingested a human serum known to contain hepatitis A
virus; the agent has been subsequently identified as a
parvovirus and it was suggested that the strain was a

latent, reactivated tissue-culture contaminant, not
related to hepatitis.

e. Among other identified specific viruses isolated
during work with hepatitis and not etiologically associated with it are vaccinia, coxsackieviruses, reoviruses
and herpesviruses; how many more isolations have occurred,
but never published, is a matter for conjecture.

RECOMMENDATIONS

Recent guidelines to minimize the risk of laboratory-
acquired infections, as well as cross-contaminations, are
summarized by Wedum et al. (51). There is little doubt
that means are available such that, barring an occasional
accident, nearly all laboratory-acquired infections and
contaminations could be prevented. However, it must
be acknowledged that complete safety can only be
accomplished at a cost beyond the capacity of most private
research laboratories to finance and at a much diminished
productivity. The conflict between productivity and
possible errors or mix-ups is particularly acute in a
laboratory like YARU-WHO Reference Center; as part of a
service commitment, large numbers of serum specimens are
received for serological tests and many unidentified
suspect arboviruses or similar agents for identification,
as well as materials for virus isolations. These
materials are handled in addition to the on-going research
work with established viruses. Ideally only one identified or unidentified virus should be worked with at a
time; or at least, there should be as many suites of
animal-holding rooms and laboratories as there are
different specimens. In addition all unknown materials
submitted for identification should be considered highly
dangerous until reasonably proved otherwise; no one would
have considered the preparation of tissue cultures from
monkey kidneys to be a highly dangerous procedure before
the Marburg agent episode.

A number of precautions are taken at YARU which, at
the same time that they lessen the danger of personnel
infections, also diminish the risk of cross-contaminations:

1. When dealing with identified viruses, a number
of those that belong in the very high risk category
(Class IV) are totally excluded from the laboratory; no
work is done with the infective agent. The viruses
excluded are Lassa, Marburg, Machupo and low passage
Junin. In addition none of the viruses restricted by
the Department of Agriculture are knowingly handled.

2. High risk viruses (Class III) are handled as little as possible and then in restricted areas: animal rooms are set aside and inoculated mice or other small animals are kept in boxes, which in turn are placed in the Bauer-Horsfall type ferret cage. The airflow goes out of the cage through a filter and is not recirculated. Personnel are vaccinated against yellow fever, VEE, rabies and at one time when an experimental vaccine was available, against Russian spring-summer encephalitis.

3. Lesser risk viruses (Class I and II) are still handled with precaution, although inoculated animals are held in boxes on shelves.

4. All persons, while inoculating or checking animals, are required to use a mask, either a disposable type or the Dustfoe type; they also are required to wear gowns which are discarded on leaving the room. The use of gloves, though recommended for cleaning of cages and tending the animals, seems not practical for mouse inoculation or tissue culture work; the loss of dexterity seems to create more problems than the added safety solves.

5. Fiber-glass filters are placed in all exhaust ducts and changed once a week; the filters are also placed at special openings or portholes into sterile rooms. Ultraviolet light is kept on whenever a work room is not being used. Ultraviolet light is also kept on overnight in the animal holding rooms.

6. Serological work is done on an open bench with table top covered with anti-splash towels (Veratex); the same is done for tissue culture work. All pipettes have cotton plugs regardless of whether they are used in sterile or non-sterile operations.

7. The use of blenders is restricted to designated small cubicles and are operated only under a chemical hood and with a cloth towel dampened with a chlorox solution over the cup. Materials are withdrawn from the cup through a porthole with a rubber diaphragm, using a large syringe and needle when the material is considered particularly dangerous; otherwise the top of the blender is removed while the wet towel is held over it in the hood.

8. All new, unidentified specimens containing a virus, or presumed to have a virus, are handled one at a time by a given staff member and his technicians; inoculated animals are kept in an isolation room on a different floor from the rest of the animals until a stock seed is made. Similarly no more than one cell line

at a time is permitted in the sterile tissue culture room; and when inoculating cells in culture, only one virus is handled in the particular cubicle.

9. In order to reduce the risk of reintroducing ectromelia in the laboratory, all viruses or possible virus-containing specimens that come from other laboratories are put through an immune barrier on first inoculation. Dilutions of the problem material are mixed with a rabbit anti-vaccina serum, incubated as though for a neutralization test, and inoculated intracerebrally to mice. When mice appear sick, they are sacrificed and kept at -70°C. Material from the mice inoculated with the highest dilution of the original material are then used for the preparation of a stock suspension, after a complement fixation test has shown a negative result with a vaccinia immune serum.

10. The technical personnel are urged to report any possible error that they suspect may have been made, with the assurance given them that an error reported is never to be held against a person's record.

References

1. Wildy, P. 1971. Classification and nomenclature of viruses, p. VIII and 81. S. Karger, Basel.
2. Casals, J. 1971. Arboviruses: Incorporation in a general system of virus classification. Comparative virology (ed. K. Maramorosch and E. Kurstak) p. 307. Academic Press, New York.
3. Holden, P. 1955. Transmission of EEE in ring-necked pheasants. Proc. Soc. Exp. Biol. Med. 88:607.
4. Clarke, D. H. and J. Casals. 1965. Arboviruses: Group B. Viral and rickettsial infections of man (ed. F. L. Horsfall, Jr. and I. Tamm) p. 606. J. B. Lippincott, Philadelphia.
5. Hanson, R. P., S. E. Sulkin, E. L. Buescher, W. McD. Hammon, R. W. McKinney and T. H. Work. 1967. Arbovirus infections of laboratory workers. Science 158:1283.
6. Sulkin, S. E. and R. M. Pike. 1969. Prevention of laboratory infections. Diagnostic procedures for viral and rickettsial diseases (ed. E. H. Lennette and N. J. Schmidt) p. 66. Amer. Public Health Ass., New York.
7. Badalov, M. E., V. N. Lazarev, E. K. Koimchidi and G. A. Karinsakaya. 1970. Contribution to the problem of CHF infections in hospitals and laboratories. Crimean hemorrhagic fever (ed. M. P. Chumakov) p. 90. Oblast. Nauch. Prakt. Konf. Rostov-on-the-Don.

8. Reeves, W. C. 1959. Problems of overwintering and natural maintenance of mosquito-borne viruses. Proc. Sixth Int. Congr. Trop. Med. Malaria (Lisbon) 1958. 5:48.
9. Thomas, L. A. and C. M. Eklund. 1960. Overwintering of WEE virus in experimentally infected garter snakes and transmission to mosquitoes. Proc. Soc. Exp. Biol. Med. 105:52.
10. van Tongeren, H. A. W. 1959. Central European encephalitis: Epidemiology and vectors. Proc. Sixth Int. Congr. Trop. Med. Malaria (Lisbon) 1958. 5:174.
11. Nosek, J., M. Gresikova and J. Rhenacek. 1961. Persistence of tick-borne encephalitis virus in hibernating bats. Acta Virol. 5:112.
12. Kozuch, O., J. Nosek, E. Ernek, M. Lichard and P. Albrecht. 1963. Persistence of tick-borne encephalitis virus in hibernating hedgehogs and dormice. Acta Virol. 7:430.
13. Gajdusek, D. C. 1972. Slow virus infection and activation of latent infections in aging. Adv. Gerontol. Res. 4:201.
14. Lennette, E. H. and H. Koprowski. 1943. Human infection with VEE virus. A report on eight cases of infection acquired in the laboratory. J. Amer. Med. Ass. 123:1088.
15. Rowe, W. P., F. A. Murphy, G. H. Bergold, J. Casals, J. Hotchin, K. M. Johnson, F. Lehmann-Grube, C. A. Mims, E. Traub and P. A. Webb. 1970. Arenoviruses: Proposed name for a newly defined virus group. J. Virol. 5:651.
16. Johnson, K. M., R. B. Mackenzie, P. A. Webb and M. L. Kuns. 1965. Chronic infection of rodents by Machupo virus. Science 150:1618.
17. de Guerrero, L. B., M. C. Boxaca and A. S. Parodi. 1965. Fiebre hemorragica experimental en cobayos (Virus Junin). Rev. Asoc. Med. Argent. 79:271.
18. Buckley, S. M. and J. Casals. 1970. Lassa fever, a new virus disease of man from West Africa. III. Isolation and characterization of the virus. Amer. J. Trop. Med. Hyg. 19:680.
19. Leifer, E., D. J. Gocke and H. Bourne. 1970. Lassa fever, a new virus disease of man from West Africa. II. Report of a laboratory-acquired infection treated with plasma from a person recently recovered from the disease. Amer. J. Trop. Med. Hyg. 19:677.
20. Smorodintsev, A. A., L. I. Kazbintsev and V. G. Chudakov. 1964. Virus hemorrhagic fevers, p. VI and 245. Israel Program for Scientific Translations, S. Sivan Press, Jerusalem.
21. Stinebaugh, B. J., F. X. Schloeder, K. M. Johnson, R. B. Mackenzie, G. Entwisle and E. de Alba. 1966.

Bolivian hemorrhagic fever. A report of four cases.
Amer. J. Med. 40:217.
22. Casals, J. and S. M. Buckley. 1973. Lassa fever
virus. (in press).
23. Trencseni, T. and B. Keleti. 1971. Clinical aspects
and epidemiology of hemorrhagic fever with renal
syndrome, p. 1. Akademiai Kiado, Budapest.
24. Morbidity and Mortality. 1972. 21:373.
25. Marston, R. Q. 1972. Inagural remarks. Hepatitis
and blood transfusion (ed. G. N. Vyas et al.) p. 3.
Grune and Stratton, New York.
26. Dodd, R. Y., J. J. Levin, L. Ni, G. A. Jamieson and
T. J. Greenwalt. 1972. American National Red Cross
experience with HB Ag testing. Hepatitis and blood
transfusion (ed. G. N. Vyas et al.) p. 175. Grune
and Stratton, New York.
27. Taswell, H. F. 1972. Incidence of HB Ag in blood
donors: an overview. Hepatitis and blood transfusion (ed. G. N. Vyas et al.) p. 271. Grune and
Stratton, New York.
28. Okochi, K. and S. Murakami. 1968. Observations on
Australia antigen in Japanese. Vox Sang. 15:374.
29. Dejanov, I., B. Trajkovski, L. J. Sotirovska and P.
Ranic. 1971. Australia antigen and antibody in
14,379 volunteer blood-donors from S. R. Macedonia.
Lancet 2:164.
30. Bagshawe, A. F., A. M. Parker and A. Jindani. 1971.
Hepatitis-associated antigen in liver disease in
Kenya. Lancet 1:88.
31. Krugman, S. and J. P. Giles. 1972. Viral hepatitis:
Natural history of the disease. Hepatitis and blood
transfusion (ed. G. N. Vuas et al.) p. 9. Grune and
Stratton, New York.
32. Byrne, E. B. 1966. Viral hepatitis: An occupational
hazard of medical personnel. J. Amer. Med. Ass.
195:118.
33. Fox, R. A., S. P. Naizi, S. Sherlock, A. Knight and
F. J. Moorhead. 1970. Hepatitis-associated antigen
and antibody in haemodialysis patients and staff.
GUT 11:369.
34. Allen, J. G. 1972. The case for the single transfusion. New Eng. J. Med. 287:984.
35. Feinglass, E. J., S. V. Williams, J. C. Huff, J. M.
Matsen and M. B. Gregg. 1972. Preliminary report:
An investigation of hospital-associated hepatitis-B.
Hepatitis Scientific Memoranda, Memo H-359/1 (Oct.)
p. 2.
36. Sutnick, A. I., W. T. London, I. Millman, B. J. S.
Gerstley and B. S. Blumberg. 1971. Ergasteric
hepatitis: Endemic hepatitis associated with
Australia antigen in a research laboratory. Ann.
Int. Med. 75:35.
37. LoGrippo, G. A. and H. Hayasi. 1972. Incidence of

hepatitis and hepatitis-associated antigen among laboratory workers. Hepatitis Scientific Memoranda, Memo H-371 (Oct.) p. 13.

38. Hillis, W. D. 1961. An outbreak of infectious hepatitis among chimpanzee handlers at a U. S. Air Force base. Amer. J. Hyg. 73:316.

39. Ruddy, S. J., J. W. Mosley and J. R. Held. 1967. Chimpanzee-associated hepatitis in 1963. Amer. J. Epidemiol. 86:634.

40. Casals, J. 1963. Immunological characterization of Vilyuisk human encephalomyelitis virus. Nature 200:339.

41. Sartorelli, A. C., D. S. Fischer and W. G. Downs. 1966. Use of sarcoma 180/TG to prepare hyperimmune ascitic fluid in the mouse. J. Immunol. 96:676.

42. Mettler, N. E., J. Casals, D. H. Clarke, W. G. Downs and R. E. Shope. 1971. Use of sarcoma 180 to prepare hemagglutinating and complement fixing antigens for viruses in adult mice. Proc. Soc. Exp. Biol. Med. 136:1355.

43. O'Malley, J. P., H. M. Meyer, Jr. and J. E. Smadel. 1961. Antibody in hepatitis patients against a newly isolated virus. Proc. Soc. Exp. Biol. Med. 108:200.

44. Chang, R. S. 1961. Properties of a transmissible agent capable of inducing marked DNA degradation and thymine catabolism in a human cell. Proc. Soc. Exp. Biol. Med. 107:135.

45. McCollum, R. W. 1967. Present knowledge of the etiology of hepatitis. First Int. Conf. on Vaccines Against Viral and Rickettsial Diseases of Man. Scientific Publication No. 147. Pan American Health Organization, Washington, D. C. p. 500.

46. Hsiung, G. D., P. Isacson and R. W. McCollum. 1962. Studies of myxovirus isolated from human blood. I. Isolation and properties. J. Immunol. 88:284.

47. Liebhaber, H., S. Krugman, J. P. Giles and D. M. McGregor. 1964. Recovery of cytopathic agents from patients with infectious hepatitis: Isolation and propagation in cultures of human diploid lung fibroblasts (WI38). Virology 24:109.

48. Maynard, J. E., K. R. Berquist and D. H. Krushak. 1972. Studies of hepatitis candidate agents in tissue cultures and in animal models. Hepatitis and blood transfusion (ed. G. N. Vyas et al.) p. 393. Grune and Stratton, New York.

49. Rightsel, W. A., R. A. Keltsch, A. R. Taylor, J. D. Boggs and I. W. McLean, Jr. 1961. Status report on tissue-culture cultivated hepatitis virus. I. Virology laboratory studies. J. Amer. Med. Ass. 177:671.

50. Melnick, J. L. and J. E. Boggs. 1972. Cytopathogenic agent K-30 from patients with infectious

hepatitis.   International virology 2 (ed. J. L.
Melnick) p. 197.   S. Karger, Basel.
51. Wedum, A. G., W. E. Barkley and A. Hellman.   1972.
Handling of infectious agents.   J. Amer. Vet. Med.
Ass. 161:1557.

POTENTIAL FOR ACCIDENTAL MICROBIAL AEROSOL TRANSMISSION IN THE BIOLOGICAL LABORATORY

R. L. Dimmick, W. F. Vogl and M. A. Chatigny
University of California School of Public Health
The Naval Biomedical Research Laboratory
The Naval Supply Center, Oakland, California

Almost every activity of mankind produces airborne particles. Some are within the size range (0.5-5μm) where they penetrate the lungs and are retained (1). It is not unexpected, then, to find that aerosols are produced by procedures and methods employed in biological laboratories, and that some portion of the airborne particles undoubtedly find their way and are retained in the deeper recesses of the lungs of persons in the vicinity of these activities (2-5).

If some of the particles carry an infectious microbe or virus, there is a definite probability that exposed persons will become infected. The purpose of this paper is to provide data that might permit an estimation of the aerosol output and retained dosage under given circumstances.

Regardless of labels that define probabilities of infection, the person who actually becomes infected may well consider the probability to have been 100 percent. Consider the recent incident (6) where a scientist died, reportedly as a result of having breathed an aerosol of rabies virus created by blending infected animal tissue. The blender is said to have been closed, but opened shortly after blending stopped. It seems ironic that papers documenting the behavior of blenders had appeared prior to this incident (3,7,8).

Unfortunately, most incidents of laboratory-acquired infections are not so easily traced to the apparent cause, and other illnesses and minor complaints caused by breathing accidentally produced aerosols are probably not considered to have had such origins (9). An attempt to trace each incident to its origin would be a futile gesture unless the results were intended to influence changes in laboratory procedures. The human tendency is to ignore minor hazards.

One guiding principle of our research has been the idea that the cautious, concerned, skilled, attentive worker, as we all are, will create a minimal hazard whereas the hurried, sloppy, absent-minded, unconcerned worker, as the other fellow is, will present a maximal hazard; it is the other fellow we must be concerned about. So we deliberately attempted to reproduce worst-case conditions during testing procedures.

Our objectives in this study were:

1. To develop simple, quantitative methods to evaluate aerosol output from various laboratory operations;
2. To relate these values to data in the literature so that all tests would not have to be repeated;
3. To present simplified methods of calculating potential risks in given situations from values obtained by (1) or (2) above;
4. To include in the experimental parameters all important factors involved in a final dosage estimate.

EXPERIMENTAL DETAILS

Factors to be considered in dosage estimates. The important factors involved in the experimental evaluation of respiratory risk, some of which became apparent only after several experiments had been done, were considered to be:

1. The source strength, or the amount of infectious airborne material a given operation produces per unit of time. More precisely stated, this is the number of particles in the size range 0.5-5μm diameter produced per minute by a given operation.
2. The concentration of viable units (bacteria, phage, virus or other hazardous material) per unit volume of material being dispersed. It is obvious that the greater the number of viable units in a suspension, the greater the number that might become airborne.
These two parameters may be combined to form a single

number, which is dimensionless in terms of concentration, that we have labeled "spray factor." This value is determined by dividing the observed source strength by the concentration of microbes in the test fluid. The value has the dimensions of number of particles in the respirable range/min for continuous sources and number of those particles for momentary or "burst sources."

One can multiply the concentration of his suspension by the spray factor for this operation to derive a source strength (number of viable particles released per minute, or the total number released instantaneously).

3. The biological behavior of a given microbe. When assessing maximal risk, this factor is important only to the experimental phase. The exact humidity, temperature, and other environmental parameters influencing survival capabilities of a given microbe are usually unknown. For example, growth conditions (temperature, medium, pH, age of culture) can influence subsequent survival capacity of airborne bacteria and virus (10,11). Also, the greatest risk is during a short, initial interval; for longer periods, diffusion and ventilation may decrease the dosage faster than biological decay. To estimate field risk, maximal survival is assumed. To determine potential risk in a defined experiment, conditions must be such that maximal survival of the test organism is assured.

4. Type of operation, that is, whether the operation is a "burst source," such as a drop hitting the floor, or a continuous source, such as the output from a sonic homogenizer. This factor, again, is of major importance only in experimental situations. For field calculations of source strength, aerosols from a burst source can be assumed to have been present for 1 minute.

5. The influence of air volume. There are two volumes to consider. One is a spherical space about 3 feet in radius from the aerosol source. A person breathing in this space may be considered to have been exposed to the source strength diluted by 27 ft^3 if he stays in this volume at least one minute following start of aerosol generation, assuming no drafts. The other is the volume of the room. Volumes are more conveniently labeled in ft^3 rather than in metric units as a concession to architects and engineers.

The usual ventilated laboratory space can be considered to have a condition of air mixing such that within 1-3 minutes the aerosol concentration will be uniform throughout the room. Persons in the room will have been exposed initially to a concentration equal to the source strength divided by the room volume for the first 10 minutes after aerosol generation starts.

6. Ventilation rate. Few laboratories are without some forced movement of air in and out of the room. Ventilation serves to dilute and mix an aerosol more

rapidly throughout a room than if there were no
ventilation and hence to decrease dosage to some extent
over prolonged time periods.

7. Primary enclosures, such as hoods and cabinets.
Hoods and other enclosures are effective in varying
degrees in preventing aerosols from entering working
spaces (9). Class III systems (closed and sealed units
with filtered vents or vents to burners) are as near
100% effective as possible (12). Effectiveness of
other types of enclosures depends upon design and manner
of operation and use.

8. Miscellaneous factors, whose values are essentially constant. These include: (a) breathing rate of
man, 3×10^{-1} ft^3 per minute; (b) mean gravitational
settling rate of particles in the respirable range,
2×10^{-2} ft per minute; (c) the fraction of the breathed
dosage that remains for an appreciable time in the
alveoli, about 3×10^{-1} of the inspired particles (1).

METHODS AND MATERIALS

Tracer materials. Obviously a tracer material should
either directly stimulate the behavior of material used
in the laboratory or should be capable of being related,
both quantitatively and qualitatively, to this behavior.

A flavobacterium species was used as the bacterial
tracer in our laboratory. This organism is stable in
the airborne state over a wide range of humidities and
is non-pathogenic when tested in the mouse and guinea
pig. The bacterium was grown in yeast extract-glucose
medium and sampled on Casitone agar with 0.15µg/ml of
brilliant green.

T_1 bacteriophage was used as the virus tracer.
This phage is fairly stable in the airborne state.
To assure maximal recovery, tests were run at relative
humidities of 80% or higher. Airborne phage was
collected with an impinger (13) and assayed by the soft
agar overlay technique.

Polio virus (LSc-2ab) and EMC virus (37A) were
grown and assayed as described by Schaffer et al. (14)
and by Akers et al. (15). These viruses were used in
limited tests for corroborative purposes and always
at relative humidities of 80% or higher.

Chemical materials used were fluorescein, aqueous
solution, 10mg/ml, collected by impinger and assayed by
a Baird-Atomic Fluorispec (SF-1) with excitation at 475
nm and emission at 513 nm. A standard was run with each
set of samples. Another chemical, used only for tests
with a zonal centrifuge, was a fluorescent material

utilizing a fluorescent particle detector as described
by Goldberg (16).   The device, termed the microaero-
fluorometer (MAF), is capable of detecting one
fluorescent particle in the presence of thousands of
non-fluorescent particles.

Samplers.   Air samplers used were (a) the AGI-30
impinger (13) at 12.5 liters per minute, (b) the six-
stage Andersen sampler (17) at $1 ft^3$ per minute, (c) a
four-stage seive sampler at $1 ft^3$ per minute, (d) the
MAF (16), (e) a Climet particle detector (Climet
Instruments Company, Sunnyvale, Ca., 94086) and (f)
plastic petri dishes as settling plates.   In some
instances a Porton pre-impinger (18), which removes
particles larger than about 5µm in diameter, was used
upstream from the impinger.   When previous tests
indicated, or if we suspected that bacterial concentrations might overload the Andersen sampler, a 50/1
diluter was placed ahead of the sampler.   The diluter
consisted of an M 14 U. S. Army gas mask aerosol filter
canister attached to an adapter with an air sample
inlet 0.063 inches in diameter located between the
filter and the sampler.   Flow rate through the sampler
was not measurably restricted by this arrangement.   A
cascade impactor (19) was used for sizing purposes only.

Chambers.   We did not attempt to use a closed hood
system as a test container, as Kenny and Sabel (8)
had done, because we reasoned that the excess volume
would allow diffusion and eddy currents difficult to
predict or control.   Instead we constructed a $2 ft^3$ metal
box from which we could readily sample all of the
enclosed air and thereby measure the total output from
an operation.   The box had several inlet and outlet
ports and a plastic observation window.   It would be
purged with clean air or with air at a desired humidity
prior to and during an experiment.

Some experiments were conducted in a miniature
chamber that could accommodate all tests instruments
except the blender and the centrifuge; these will be
discussed separately.   Data from the two chambers were
equivalent, but the small chamber is more convenient
to use.

The miniature unit (Fig. 1) was a rectangular metal
box 6 x 13 x 15 cm, with holes at the two sides and top,
and a removable, gasketed door with a plastic window.
At the bottom, a plastic, threaded insert was sealed in
place.   A circular table with holding pins was fastened
to a threaded rod that extended through the insert.

FIGURE 1. Miniature chamber for testing aerosol output
            of various operations. See text for descrip-
tion and use.

This arrangement allowed a container to be raised or
lowered for testing operation of the sonic homogenizer
probe, which could be inserted through the top rubber
stopper. The top rubber stopper (#11) could also be
used to hold a glass tube of any length for testing
effects of dropping or splashing liquid, or various
size pipettes could be inserted through the stopper.

Figure 1 shows how pipetting operations were
tested, except the door (bottom left) is shown removed.
An M 14 canister, directly under the unit, filtered the
air replacing the volume removed by a sampler. A six-
stage Andersen sampler is shown (to the right of the
unit) connected to the box by a bend and tubing; the
sampling path was as short as practicable. Not visible
at the back is a rod connecting the unit to a ringstand
clamp.

This chamber is not unique, and units capable of
performing in the same manner could be constructed in a
variety of ways. The requirements are simply chambers
with a minimal enclosed volume to contain the aerosol-
generating point of an operation, and chambers from
which air to be replaced by filtered air can be withdrawn
through a short path into a sampler.

SPECIAL TEST CONDITIONS.

Blenders. Output from blenders was easily measured in the 2 ft^3 chamber. They were tested with lid off during operation, or with lid on during operation but removed at various times afterward, and different speeds. A different measure of source strength, however, was obtained by modifying a blender having a plastic bowl. Two holes were drilled on opposite sides of the bowl about 2 cm from the top. One hole was for air inlet through a filter, the other for sampling.

The centrifuge. To test the centrifuge (Beckman type S-2 ultracentrifuge) it was necessary to remove the top cover and provide a circular plastic plate which could be sealed to the "O" ring by means of weights placed around the plate. The plate was provided with holes for air and liquid interchange through tubing sealed with rubber stoppers. The arrangement was not ideal, for it required an upward sampling path of about 30 cm, but there was no practical way to provide holes through the thick steel shielding.

Fluid in an angle-head rotor is self-contained. The zonal rotor, however, must be loaded and unloaded while the rotor is spinning. With the cover open and the rotor spinning, the rotating seal was locked in place, the plate closed and sealed, and test fluid was pumped through the rotor while air samples were taken. The plate was then removed, the rotating seal disconnected, and the rotor was allowed to stop. Tests of effects of drops hitting the spinning rotor were conducted with the cover closed and the rotating seal removed. A bent pipette inserted through a rubber stopper was used to create the drops.

Dropping and pipetting. For testing effects of dropping liquid, 5 ml of fluid was ejected, drop by drop, onto a petri dish during a 2 minute period. For pipetting tests, the pipette was inserted beneath the liquid surface, and liquid and air was pumped vigorously in and out of the pipette during a one-minute period. In both instances, sampling was started 10 seconds prior to test and stopped 20 seconds after the test stopped.

RESULTS

Size estimates. Although the distributions varied, the mean-diameter of aerosols produced by the operations tested were surprisingly uniform, from 2.5-3.5μm.

Figures 2A-D illustrate size estimates from Andersen
sampler data and serve to illustrate the uniformity.
Of course, these data are only for particles containing
bacteria. In several experiments aerosols were sampled
by an optical particle analyzer; few particles larger
than 5μm were observed, but large numbers of particles
0.5μm and larger were present. The predominant number
of particles on cascade impactor slides were found, by
microscopic observation, to be 3μm or less; the smaller
the diameter, the greater the frequency.

For our experiments precise size estimates were
not essential. Observed source strengths were so
variable that adjusting for minor changes in particle
sizes would have made no statistical differences. It
was essential, however, to corroborate the fact that the

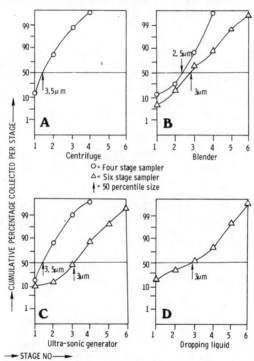

FIGURE 2. Size distribution of particles containing
           viable bacteria as collected by the Andersen
sampler (17) from 4 different operations. Values are
averages of at least 2 runs. Colonies per stage were
counted, adjusted for coincidence, and the total set
equal to 100%. The 50 percentile size was determined by
interpolation of graphical data from measured characteristic diameters of each stage.

means of the particle size distributions did cluster
near 3μm, that the predominate numbers were this size or
smaller, and that our sampling methods were collecting
that size.

Virus and phage can be contained in particles
smaller than 1μm. Unfortunately, our sampling apparatus
was not capable of collecting particles smaller than
about 0.8μm. Since most size distributions of aerosols
appear to be logarithmic, then great numbers of smaller
particles could have been present, and airborne phage
and virus could have existed in greater numbers than
shown. We believe that the similarity in output numbers
for bacteria and virus in a given experiment is probably
a result of sampling technique, and we intend to pursue
the matter further using techniques permitting recovery
of particles as small as 0.03μm.

The centrifuge. Under ordinary circumstances a closed,
angle-head rotor creates no aerosol output, even if
tubes break, because the liquid volume in the tubes is
less than that for the tube spaces. In an effort to
demonstrate leakage from an angle-head rotor, we finally
filled the rotor completely with liquid--a circumstance
impossible in ordinary usage. We did not test rotor
breakage, nor did we test what might happen if hanging-
bucket type centrifuges were improperly used since this
type of massive accident is not amenable to quantitative
testing.

The zonal centrifuge presents a special case.
Material used in zonal centrifugation may approach
titers of 10^{14} viable particles/cm^2 or higher. There
are three major steps involved in using the rotor:
(a) loading the rotor (or flushing the rotor) at low
speeds (1000 to 5000 rmp); (b) centrifugation at very
high speeds under vacuum; (c) unloading and analysis
at low speeds.

Initial tests with the zonal centrifuge indicated
that there was no leakage from either the rotating seal
or the centrifuge head when bacterial suspensions were
the test fluid. We suspected that shear forces in the
seal might have ruptured particles the size of bacteria,
so we tested the T_1 bacteriophage. Again, no leakage
was detected. We then sampled air from the centrifuge
with the Climet particle analyzer and found an appreciable
aerosol was being generated. We examined the seal
carefully and noted that the graphite lubricant from the
roller bearings was being deposited in the bearing
housing and appeared to be the source of the particles.
Finally, we pumped the fluorescent compound through the

seal and sampled the air with the MAF. During a 10-
minute test we found no fluorescent particles in the air.
After the 10 minute period we created an artificial
blockage in the fluid output tube. When pump pressure
reached an indicated value of 21 psi, the seal leaked
and a massive spill occurred. This situation might
happen if one were filling the rotor in a direction
opposed to the centrifugal force, and especially if
higher density fluids are used.

During these tests, we noted that a drop of fluid
would frequently fall from the rotating seal onto the
rotor during the removal step, regardless of attempts
to remove the seal carefully and even occasionally when
tubing was clamped off near the seal as recommended in
the instruction manual. With the system closed, we
tested the effects of this, using the approximate height
and position of the seal where drops emerged; output
data are included with other spray factors. It is
interesting that maximal output occurred at 3000 rpm,
the speed most workers prefer for the unloading operation
(personal communication, Beckman representative). At
1000 rpm the drops appeared to wet the rotor; at 3000 rpm
the drops disintegrated; at 5000 rpm the drops rolled
gently to the edge, apparently on a cushion of air and
were ejected.

Aerosol output. Rather than list all raw data, we will
present illustrative data and then report a list of spray
factors from our data and from the literature.

A sonic homogenizer was tested with maximal entrapped
air (probe tip raised to near surface of liquid), minimal
entrapped air (probe 1 cm from bottom of container) with
several different test materials and at approximately 40
watts output. In all tests except the miniature probe,
50 ml of fluid was placed in a 100 ml beaker. As a
precautionary measure, the beaker was chilled with ice
in tests with phage or virus. There was no loss of
bacteria during the 1-minute test period. In three
runs using the flavobacterium (Table 1) the spray factor
was 8, 10 and 13 x 10^{-5}, which is reasonable agreement.
Polio and EMC virus yielded spray factors of 10 and
7 x 10^{-5}, respectively. T_1 bacteriophage was lowest with
a spray factor of 4.2 x 10^{-5}. The mean was 8.7 x 10^{-5}
and the standard error was 3 x 10^{-5}. Minimal entrapped
air yielded spray factors about 2 logs less than that
for maximal entrapped air. Again, there was fair
agreement between the spray factors, whether phage or
bacteria were the test materials. As another example we
tested blenders having three types of bowls: a pint-
sized metal bowl, a half-pint-sized metal bowl and a

Table 1. Aerosol Generation by the Sonic Homogenizer

Test material	Organisms/ml*	Sampler Type	Output Organisms/min*	Spray Factor**
With entrapped air				
Flavobacterium sp.	1.5×10^8	AS(4)	2.0×10^4	1.3×10^{-4}
Flavobacterium sp.	1.4×10^8	AS(6)	1.4×10^4	1.0×10^{-4}
Flavobacterium sp.	1.0×10^8	IMP	8.0×10^3	8.0×10^{-5}
Phage (T_1)	7.0×10^7	IMP	3.0×10^3	4.2×10^{-5}
Polio virus	1.2×10^8	IMP	1.2×10^4	1.0×10^{-4}
EMC virus	4.0×10^7	IMP	2.8×10^3	7.0×10^{-5}
Flavobacterium sp.*	6.0×10^9	AS(4)	1.6×10^4	2.6×10^{-6}
With minimal entrapped air				
Flavobacterium sp.	6.0×10^8	AS(6)	7.0×10^1	1.2×10^{-7}
Phage (T_1)	2.0×10^9	IMP	1.8×10^3	9.0×10^{-7}

* A miniature probe with 10 ml of liquid in a 20 ml beaker.
**Spray factor is output divided by concentration.
AS is Andersen sampler; number of stages in parenthesis.
IMP is the AGI-30 impinger.

Table 2. Aerosol Generation by the Blender

Test material		Organisms/ml*	Sampler Type	Output Organisms/min*	Spray factor**
Laboratory blender, cover off					
Flavobacterium sp.	(H)	1.0×10^8	AS(6)	1.4×10^5	1.4×10^{-3}
Flavobacterium sp.	(M)	1.6×10^8	AS(6)	1.0×10^5	6.2×10^{-4}
Flavobacterium sp.	(L)	1.0×10^8	AS(6)	4.0×10^4	4.0×10^{-4}
Flavobacterium sp.	(L)	1.0×10^7	AS(6)	7.6×10^4	7.6×10^{-4}
EMC virus (half-pint bowl)	(L)	2.8×10^7	IMP	4.0×10^1	1.4×10^{-7}
Fluorescein (half-pint bowl)	(M)	1.0×10^{-2} g/ml	IMP	1.3×10^{-6} g	1.3×10^{-4}
Laboratory blender, cover on but removed immediately					
Fluorescein (half-pint bowl)	(M)	1.0×10^{-2} g/ml	IMP	1.4×10^{-7} g	1.4×10^{-5}
Laboratory blender, cover removed after 1 minute					
Fluorescein	(M)	1.0×10^{-2} g/ml	IMP	1.8×10^{-8} g	1.8×10^{-6}
Home blender (see text)					
Fluorescein	(M)	1.0×10^{-2} g/ml	IMP	1.0×10^{-4} g	1.0×10^{-2}
Flavobacterium sp.	(M)	3.0×10^6	AS(6)	2.0×10^4	6.6×10^{-3}
Flavobacterium sp.	(M)	8.0×10^6	AS(6)	1.3×10^4	1.6×10^{-3}

* Except for fluorescein.
**Spray factor is output divided by concentration.
(H) = highest speed.
(M) = medium speed.
(L) = lowest speed.

plastic pint-sized bowl. The cover was either off or
removed after intervals. Results were consistent within
a given operation (see Table 2). The higher the energy
input, the greater the aerosol output. In the test with
a plastic bowl, where aerosol could be sampled directly
from the bowl, the factor was 3×10^{-3}.

Using similar data, we have determined how height
affects the evolution of particles from single drops
(Fig. 3). These data indicate that drops falling from
about 10 cm or less produce no detectible particles,
but that there is an increase of particles as drops fall
from increasing heights to about 100 cm. Thereafter the
output is fairly constant as drops probably attain
terminal velocity.

One important finding is that output varied according
to the manner in which an operation was performed. For
example, the output from the sonic homogenizer varied over
3 logarithmic increments. Used properly with no entrapped
air, output from the homogenizer was minimal. Thus care
in proper operation of equipment can effectively reduce
potential risk.

Spray factors. The spray factor is a more useful indication of output intensity than is the measured output for
a given condition since the factor can be applied to
field conditions. The following example will demonstrate
this.

Assume a worker is sonicating, for 1/2 minute, a
suspension that titers 10^9 "viable" virus particles per

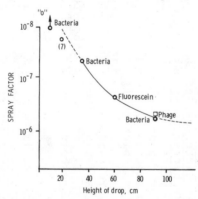

FIGURE 3. Variation of spray factor with height of drop.
           Material tested is noted in the figure, except
(7) is cited in legend, Table 4.

ml, and that the laboratory room volume is 2000 ft^3. Assume he has adjusted the instrument for proper operation (i.e. minimal entrapped air) and stands within 3 feet of the instrument.

The average spray factor is 5×10^{-7} (Table 1). Then: factor x concentration = source strength

or $5 \times 10^{-7} \times 1 \times 10^9 = 5 \times 10^2$

$$\frac{\text{source strength}}{\text{source volume}} \times \text{operation time} = \text{aerosol concentration}$$

or $\dfrac{5 \times 10^2}{27 \ ft^3} \times 5 \times 10^{-1} = 9$ virus particles per ft^3

Aerosol concentration x breathing rate x retention factor = dosage

or $9 \times 3 \times 10^{-1} \times 3 \times 10^{-1} \times 2 = 1.6$ virus particles.

This is a minimal value for dosage because, depending on air currents, the effective air volume could have been smaller than the theoretical value of 27 ft^3.

In a room the situation is different, for uniform diffusion occurs in about 5 minutes even with no ventilation. Assume a coworker, located more than 3 ft from that operation, remained in the room for 10 minutes. Then: $\dfrac{\text{source strength}}{\text{room volume}} \times \text{operation time} = \text{aerosol concentration}$

or $\dfrac{5 \times 10^2}{2 \times 10^3} \times 5 \times 10^{-1} = 1.3 \times 10^{-1}$ virus particles per ft^3

and as above dosage is

$$1.3 \times 10^{-1} \times 3 \times 10^{-1} \times 3 \times 10^{-1} \times 10 \ \text{min} = 1.2 \times 10^{-2},$$

or the probability of the coworker receiving a dosage of 1 virus particle is about 0.01.

Rationale for using the spray factor in this way is shown in Figs. 4 and 5. Figure 4 shows the calculated removal efficiency in ventilated and non-ventilated rooms. Even six changes per hour does not significantly influence the removal of aerosol for the first 3 minutes and is inconsequential for at least 10 minutes.

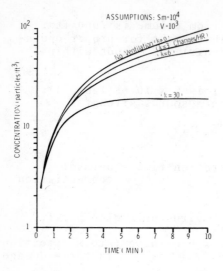

FIGURE 4.   Theoretical aerosol buildup in a ventilated space. Sm = source strength, v = volume.

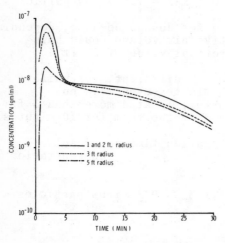

FIGURE 5.   Measured aerosol dispersion through a poorly ventilated room.

Figure 5 contains averaged data where samples were taken at the indicated times after either a blender or a sonic homogenizer were operated for 1 minute in a sealed room. Note that the concentration within a 3-ft radius is essentially the output in the immediate vicinity of the source. It is evident that ventilation does not reduce brief, intimate exposure levels, and even in an unventilated space the aerosol concentration becomes uniform through the room after about 5 minutes.

Spray factors can be estimated from values in the literature. The most comprehensive data are those of Kenny and Sabel (8). In 5 instances their experimental

procedures were similar enough to ours to allow valid
comparison (Table 3). Because some data were not
available, we had to assume certain values. We assumed
their hood volume was 25 ft^3. Since they used Serratia
marcescens as a test organism at relative humidities
about 50 percent, we doubled their reported outputs to
allow for biological decay. We multiplied our single
drop count by 10 to compensate for the continuous drop
method they used. We doubled the volume of the dropped
petri plate to allow for flask volume differences.
With these adjustments, the mean variation was± 0.4 logs.

One other datum in the literature which was
equivalent to ours is shown in Fig. 3; the values, after
similar adjustment, are almost identical.

It seemed reasonable, therefore, for us to use
reported values, and where possible, to adjust these
values with respect to sampled volume, sample rate and
time, expected biological loss, and to estimate spray
factors. The factors are probably within ± 0.5 logs of
a real value (Table 4). Massive spills, as from a
centrifuge, cannot be quantitated and one should assume
that a high dosage had been received by personnel near
such accidents.

DISCUSSION

It is evident that laboratory operations do produce
aerosols of a concentration that could be hazardous.
The authors are of the opinion that responsibility for
safety lies first with the individual and second with
supervisory personnel. We do not propose to recommend at
this time any specific or stringent rules governing
operations performed or materials handled. However, some
precautions seem self-evident. Blenders and homogenizers
are particularly dangerous. If material used in these
devices is even suspected of being pathogenic or
allergenic, then the device should be used in a hood or
other container that can be properly ventilated (9).
If a material is known to be highly pathogenic, then a
Class III system is required (see Barkley, this volume).

In the case of the zonal centrifuge, the rotating
seal has usually been washed by a gradient solution
prior to seal removal, so spilled drops would contain
fewer microbes than the original suspension. Washing in
this way should be standard practice in any case. One
should take precautions to assure that pump pressure is
not excessive. Regardless of our finding that a good
seal, properly assembled, failed to leak at low pump
pressures, the zonal rotor should not be used with

Table 3. Comparison of Equivalent Source Strengths Reported by Two Laboratories

	Kenny and Sabel (8)			This report	
	Init. Conc.	Output Per minute[a]	Spray Factor	Spray Factor	Approx. log Diff.
Mixing with pipette	1.2×10^9	2.0×10^3	1.7×10^{-6}	6.0×10^{-5}	1.5
Blender	3.5×10^9	7.6×10^4	2.0×10^{-5}	1.4×10^{-5}	0.0
Sonic homogenizer	3.6×10^9	2.5×10^2	6.9×10^{-8}	5.0×10^{-7}	1.0
Dropping liquid[b]	1.2×10^9	1.4×10^2	1.2×10^{-7}	1.0×10^{-7}[c]	0.0
Dropping flask[d]	7.0×10^9	2.0×10^5	3.0×10^{-5}	8.0×10^{-6}[e]	-0.5
				mean	+0.4

[a]The volume of the hood was estimated to be 25 ft^3. Reported output was doubled to allow
for biological decay at 50% relative humidity.
[b]Kenny allowed liquid to drop freely; we deliberately established single drops.
[c]A value of 1 x 10^{-8} for corresponding height was interpolated from Fig. 3 and multiplied
by 10 to adjust for single drop effect.
[d]Dropping a flask with liquid is compared to dropping a petri plate with liquid.
[e]Output was doubled to allow for volume differences.

Table 4. List of Adjusted Spray Factors

	Intermediate Values	Maximal Value
Continuous Sources		
Blender		
sampled internally	6.1×10^{-3} (A)	6×10^{-3}
lid open	2.3×10^{-5} (A)	
lid open after stop	2.0×10^{-5} (A)	
	1.2×10^{-4} (2,8)	1×10^{-4}
lid open after 1 min	1.8×10^{-6} (A)	2×10^{-6}
lid open after 5 min	2.0×10^{-5} (2)	
lid open after 30 min	1.2×10^{-7} (7)	1×10^{-7}
lid open after 1 hr	3.7×10^{-8} (21)	4×10^{-8}
loose lid		4×10^{-7} (2,7,21)
Sonic homogenizer		
maximal aeration		1×10^{-4} (A)
minimal aeration		5×10^{-7} (A,8)
Pipetting		
with bubbles		1×10^{-4} (A)
minimal bubbles		2×10^{-6}
Vortex mixer		
no overflow		0 (A,8)
overflow		8×10^{-8} (8)
intranasal inoculation of mice		2×10^{-7} (22)
transfer of lyophilized culture		3×10^{-11} (4)
streaking petri plate		4×10^{-8} (21)
Burst Sources		
Dropping flask culture		3×10^{-5} (2,21)
Splash on centrifuge rotor		1×10^{-5} (21)
Drop spilled on zonal centrifuge rotor		2×10^{-6} (A)
Drop spilled 100 cm or higher		2×10^{-6} (A)
Remove plug from test tube		3×10^{-8} (2,21)
Inoculating with loop		3×10^{-8} (2,21)
Hypodermic syringe with vaccine bottle		6×10^{-9} (2)
Opening lyophilized culture		2×10^{-9} (8,4)
Opening protected, lyophilized culture		4×10^{-12} (4)

Values are from averages reported in the literature (numbers) or from this paper (A).

highly virulent pathogens unless it is operated within the equivalent of a Class III system.

One real problem involved in safety, within bounds of practicality, is how safe is safe? Let me rationalize for a moment with an example. Suppose I were asked to use the zonal centrifuge with a suspension of a pathogen at a concentration of 10^{10} units per ml and with a known ID_{50} for humans of 10^4 units, without Class III equipment. I reason that if I wash at least 200 ml of fluid through the seal prior to seal removal, then the concentration remaining will be reduced by at least a thousand-fold. The spray factor is 8×10^{-7}, so if I spill a drop, the source strength is 8 units. I might breathe 3 or 4 of these organisms, so my chances of breathing 1 infectious particle is about 1 in 3000. I, personally, would be willing to accept that risk, whereas, another individual might not. What if the ID_{50} were known to be 10^2 units? Now that the odds are 1 in 30, I would decide to use a Class III system to contain the operation. The implication is that I have now decided the pathogen is highly virulent. Between these odds is a gray area where it is difficult, indeed, to establish an appropriate margin of safety or decide when a species should be considered to be highly virulent.

In this example an ID_{50}, needed to measure risk of infection, was assumed. But respiratory ID_{50} values for the human are unknown for most microbial species. This is especially true for oncogenic viruses, if they are infective. McKissick et al. (20) reported that mice can be infected with murine leukemia virus by the aerosol route. Therefore, the mechanism for potential infection of man by airborne oncogenic virus has been demonstrated.

The potential of contamination of cultures could pose a greater hazard than that of direct inhalation. If a supposedly non-pathogenic culture were to become contaminated by a virulent species and the investigator were unaware of the situation, the resulting suspension would be considered to be safe to manipulate in ways that could produce large numbers of air-borne pathogens.

References

1. Hatch, T. and P. Gross. 1964. Pulmonary deposition and retention of inhaled aerosols. Academic Press, New York.
2. Anderson, R. E., L. Stein, M. L. Moss and N. H. Gross. 1952. Potential infectious hazards of common bacteriological techniques. J. Bact. 64:473.

3. Sulkin, S. E. and R. M. Pike. 1951. Laboratory-acquired infections. J. Amer. Med. Ass. 147:1740.
4. Reitman, M., M. A. Moss, J. B. Harstad, R. L. Alg and N. H. Gross. 1954. Potential infectious hazards of laboratory techniques. J. Bact. 68:545.
5. Wedum, A. G. 1964. Laboratory safety in research with infectious aerosols. Public Health Rep. 79:619.
6. U. S. Dept. HEW. 1972. Morbidity and mortality report, vol. 21, no. 14.
7. Reitman, M., M. A. Frank, R. Alg and A. G. Wedum. 1953. Infectious hazards of the high speed blender and their elimination by a new design. Appl. Microbiol. 1:14.
8. Kenny, M. T. and F. L. Sabel. 1968. Particle size distribution of Serratia marcescens aerosols created during common laboratory procedures and simulated laboratory accidents. Appl. Microbiol. 16:1146.
9. Chatigny, M. A. and D. Clinger. 1969. Contamination control in aerobiology. Introduction to experimental aerobiology, ed. R. L. Dimmick and A. Akers, chap. 10. John Wiley and Sons, New York.
10. Hatch, M. T. and H. Wolochow. 1969. Bacterial survival: consequences of the airborne state. Introduction to experimental aerobiology, ed. R. L. Dimmick and A. Akers, chap. 11. John Wiley and Sons, New York.
11. Akers, T. G. 1969. Survival of airborne virus, phage and other minute organisms. Introduction to experimental aerobiology, ed. R. L. Dimmick and A. Akers, chap. 12. John Wiley and Sons, New York.
12. Chatigny, M. A. 1961. Protection against infection in the microbiological laboratory: Devices and procedures. Advances in applied microbiology, ed. W. W. Umbreit. Academic Press, New York.
13. Brachman, P. S., R. Erlich, H. F. Eichenwald, V. J. Cabelli, T. W. Kethley, S. H. Madin, J. R. Maltman, G. Middlebrook, J. D. Morton, H. I. Silver and E. K. Wolfe. 1964. Standard sampler for assay of airborne microorganisms. Science 144:1295.
14. Schaffer, F. L., M. R. Soergel and D. C. Straube. 1971. Electrophoretic analysis of ribosomal and viral ribonucleic acids with a simple technique for slicing low concentration polyacrylamide gels. Appl. Microbiol. 22:538.
15. Akers, T. G., S. B. Bond, C. Papke and W. R. Leif. 1966. Virulence and immunogenicity in mice of airborne encephalomyocarditis viruses and their infectious nucleic acids. J. Immunol. 97:379.
16. Goldberg, L. J. 1968. Application of the micro-aerofluorometer to the study of dispersion of a fluorescent aerosol into a selected atmosphere. J. Appl. Meteorol. 7:68.
17. Anderson, A. 1958. New sampler for the collection,

sizing and enumeration of viable airborne particles.
J. Bact. 76:471.
18. May, K. R. and H. A. Druett. 1953. The pre-impinger
    --A selective aerosol sampler. Brit. J. Indust. Med.
    10:142.
19. May, K. R. 1945. The cascade impactor: An
    instrument for sampling coarse aerosols. J. Sci.
    Inst. 22:187.
20. McKissick, G. E., R. A. Griesemer and R. L. Farrell.
    1970. Aerosol transmission of Rauscher murine
    leukemia virus. J. Nat. Cancer Inst. 45:625.
21. Reitman, M. and A. G. Wedum. 1956. Microbiological
    safety. Public Health Rep. 71:659.
22. Reitman, M., R. L. Alg, W. S. Miller and N. H. Gross.
    1954. Potential infectious hazards of laboratory
    techniques. J. Bact. 68:549.
23. Wedum, A. G. 1953. Bacteriological safety. Amer.
    J. Public Health 43:1428.

* * * * *

DISCUSSION

BALTIMORE: Are there measurements on various sorts of
viruses and their stability in dried aerosols?

DIMMICK: Each viral species behaves differently, just
as every bacterial species has different characteristics.

HELLMAN: Influenza virus can survive 10-15 minutes,
depending on relative humidity. Some tumor viruses like
RLV survive best at high humidity (40-50% is normal
room humidity). A Dutch group has infected animals with
nucleic acid of poliovirus and EMC. They claim that the
nucleic acid is more stable to room environment than the
whole virus. Of course infectious nucleic acids can
circumvent species barriers; that is, polio RNA will
infect chicken cells whereas the complete virus will not.

CURRENT METHODS FOR LARGE-SCALE PRODUCTION OF TUMOR VIRUSES

Howard E. Bond
Research and Development
Electro-Nucleonics Laboratories, Inc.
Bethesda, Maryland

There are two major objectives in large-scale production of tumor viruses, the propagation of viruses in large amounts and their concentration and purification.

VIRUS PROPAGATION

Suspension culture systems are much more amenable to scale-up than monolayer cultures. The approaches range from a simple increase in batch size to the use of fermentors. There are numerous references dealing with the general principles applicable to those tumor virus systems capable of being propagated in suspension culture. The review by Telling and Radlett (1) will serve as a useful introduction to the subject.

The consequences of a biological spill during large-scale production of tumor viruses can be of monumental proportions. For this reason we have constructed a laboratory building specifically to accommodate large-scale virus production and purification. Propagation of viruses requires the processing and transfer of fluids, procedures that produce aerosols. Consequently low and moderate risk viruses are processed in laminar flow biological safety cabinets (hoods) without glove ports. However these cabinets are notoriously sensitive to convective air disturbances from both outside and from

within the cabinet. There is considerable risk that
aerosols picked up on coat sleeves will be shaken off
into the room. To counter this problem we use sleeveless
paper coats and require that arms and hands be scrubbed
before and after each operation.

VIRUS PURIFICATION

Generally centrifugation has been the most effective
approach for recovering large quantities of viruses in
both a concentrated and purified form. Even when
precipitation procedures are used for primary recovery of
virus, centrifugal steps are involved. (See, e.g., a
typical application of this method using polyethylene
glycol [2].) Centrifuges are unique in their capacity
to produce aerosols. Instrument failure can result in
aerosols of catastrophic proportions. Moreover ultracentrifuge chambers are generally evacuated into the
laboratory through the vacuum pump, a device producing
additional aerosols. Even if steps are taken to contain
the vacuum pump exhaust, the sealed chamber cannot be
viewed as a secure barrier against sudden contamination
of the laboratory. Often in cases of rotor failure the
centrifuge drive is knocked out of the bottom of the
chamber, exposing the latter to the external environment.

The K-II zonal ultracentrifuge (3,4) has been
especially valuable as a primary means for initial
recovery and partial purification of viruses. The
smaller zonal ultracentrifuge rotors of Oak Ridge Nationa
Laboratory design, the B-XIV and higher numbered rotors
of this series (5,3), have removable seals which are
used only during loading and unloading. Since there is
no containment barrier surrounding the seal, there is a
possibility of releasing an aerosol. The larger
capacity centrifuges such as the K-II use seals of a
different, safer design which are not removable.
Consequently the seal system must be surrounded with an
envelope of cooling solution which may serve as a
barrier to the escape of materials released by seal
leakage. Decontamination of the cooling system can be
performed with chemical agents, and the ethylene glycol
used in the solution has mild disinfectant properties.

Although the K-II is usually unloaded at rest after
gradient reorientation (6,4), the B-series rotors are
unloaded through rotating seals. However both instrument
are typically unloaded in a continuous flow through
flexible tubing. This manner of collecting produces
both aerosols and spills as well as exposing the rotor
contents to possible air-borne contamination.

CENTRIFUGE CONTAINMENT. Reimer et al. (7) working with
live polio virus modified a centrifuge for remote
control operation and enclosed it in a stainless steel
dry-box with gloveports. The instrument panel was
mounted outside of the containment box and access to
the centrifuge was by means of rubber gloves sealed to
gloveports. During fractionation procedures, infectious
materials could be introduced to the system or withdrawn
through flexible tubing.

A more elaborate approach to centrifuge containment
evolved at the Oak Ridge National Laboratory (8),
where a complete environmentally controlled laboratory
was built with elaborate containment safeguards. A
negative pressure of one inch of water was maintained in
the glove box with a constant flow of filtered air
passing through it. The air, supplemented with an
additional flow of outside air, was exhausted through a
flame incinerator. Automatic controls valved the amount
of outside air admitted so that the air flow through the
incinerator remained constant. This assured that there
would be no reduction of temperature even if a glove
should come off of a port and create a surge of air from
the cabinet. All drainage from the glove box was passed
through an autoclave before entering a special septic
tank and drainage fleid. Germicidal ultraviolet lamps
were enclosed in the system, and the entire inside of
the glove box could be gas sterilized under specific
conditions of temperature and humidity. All equipment
entering and leaving the glove box could be sterilized,
and finished products could be packaged and surface
sterilized before removal.

We have taken a simpler approach in our own
laboratories, where we are working predominantly with
low to moderate risk agents. We have enclosed ultracentrifuges for preparative and small-scale zonal work
in specially designed safety cabinets fitted with
removable gloveports (Fig. 1). The inside of the cabinet
is kept at a pressure negative to the surrounding room.
Air is swept through the cabinet and exhausted through
HEPA filters. The large K-II ultracentrifuges are not
enclosed since seal design greatly reduces the aerosol
problem. These instruments are installed in a separate
room (Fig. 2), which is at a negative pressure with
respect to other laboratory space. The exhaust air
from the turbine is passed through an oil trap and a
HEPA filter prior to venting. The cooling system
(which also acts as a barrier) is chemically decontaminated when seal leakage occurs.

SAMPLE COLLECTION. This procedure is probably one of

FIGURE 1.   Containment cabinet for preparative and zonal
            ultracentrifugation.   The removable gloveports
are shown in place.

the most vulnerable in processing viruses relative to
potential contamination of the centrifuged product. We
deal with the problem using two procedures either
separately or together. For manual collection we simply
use laminar flow safety cabinets, with exit lines from
the centrifuge directed into the cabinets. A somewhat
more secure method entails the use of plastic bags with
gloves molded into them (Fig. 3). The bags are gas
sterilized and the sterile collecting vessels are placed
inside. The bags are then sealed and inflated with
nitrogen. The same port used to inflate the bag is
connected to the exit line from the zonal ultracentrifuge
and the rotor contents are collected in a sealed system.
At the completion of the collecting process, the lines
are chemically sterilized and disconnected. The bag with
its contents is then transported to a biological containment hood before being opened.

FIGURE 2.   K-II zonal ultracentrifuge and accessory
          equipment installed in a separate laboratory
room.   The instrument is shown being loaded with gradient
and sterile buffer solutions prior to acceleration to
operating speeds and introduction of virus-containing
sample.   Note the laminar flow hood to the left which
contains the sample feed and receiving vessels.

SUMMARY

     Many problems engaged in centrifuge containment
clearly point up the need for manufacturer involvement.
I have attempted to emphasize the almost inextricable
involvement of the centrifuge in several aspects of
virus production and the distinct hazards of their use.
Yet, there appears to have been few efforts on the part
of manufacturers to provide any solutions.   The typical
centrifuge is little short of a mechanical nightmare
when viewed from the standpoint of containment and
decontamination.   Besides the basic mechanism for spinning a rotor, it consists of a maze of parts and
accounterments, many of which are either inaccessible or
would be damaged by decontamination procedures.   The
centrifuge manufacturing industry should assume an
obligation to design the specialized equipment needed
for work with infectious and hazardous agents.

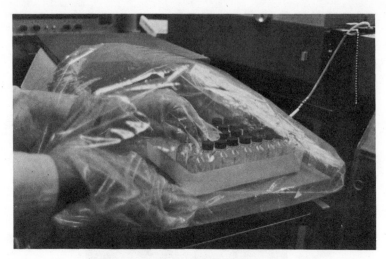

FIGURE 3.   A glove bag as used to collect infectious
            materials from a zonal ultracentrifuge.
These are available from Instruments for Research and
Industry, Cheltenham, Pa.

References

1. Telling, R. C. and P. J. Radlett. 1970. Large-scale
    cultivation of mammalian cells. Adv. Appl. Microbiol.
    13:91.
2. Syrewicz, J. J., R. B. Naso, C. S. Wang and R. B.
    Arlinghaus. 1972. Purification of large amounts of
    murine RNA tumor viruses produced in roller bottle
    cultures. Appl. Microbiol. 24:488.
3. Anderson, N. G., C. E. Nunley and C. T. Rankin, Jr.
    1969. Analytical techniques for cell fractions. XV.
    Rotor B-XXIX--A new high resolution zonal centrifuge
    rotor for virus isolation and cell fractionation.
    Anal. Biochem. 31:255.
4. Anderson, N. G., D. A. Waters, C. E. Nunley, R. F.
    Gibson, R. M. Schilling, E. C. Denny, G. B. Cline,
    E. F. Babelay and T. E. Perardi. 1969. K-series
    centrifuges. I. Development of the K-II continuous-
    sample-flow-with-banding centrifuge system for vaccine
    purification. Anal. Biochem. 32:460.
5. Anderson, N. G., D. A. Waters, W. D. Fisher, G. B.
    Cline, C. E. Nunley, L. H. Elrod and C. T. Rankin, Jr.
    1967. Analytical techniques for cell fractions. V.
    Characteristics of the B-XIV and B-XV zonal centrifuge
    rotors. Anal. Biochem. 21:235.
6. Anderson, N. G., C. A. Price, W. D. Fisher, R. E.
    Canning and C. L. Burger. 1964. Analytical

techniques for cell fractions.   IV.   Reorienting
gradient rotors for zonal centrifugation.   Anal.
Biochem. 7:1.

7.   Reimer. C. B., T. E. Newlin, M. L. Havens, R. S.
     Baker, N. G. Anderson, G. B. Cline, H. P. Barringer
     and C. E. Nunley. 1966. An evaluation of the B-V
     (continuous-flow) and B-IV (density gradient) rotors
     by the use of live polio virus. Nat. Cancer Inst.
     Monogr. 21:375.

8.   Cho, N., H. P. Barringer, J. W. Amburgey, G. B.
     Cline, N. G. Anderson, L. L. McCauley, R. H. Stevens
     and W. M. Swartout. 1966. Problems in biocontain-
     ment. Nat. Cancer Inst. Monogr. 21:485.

IMMUNOSUPPRESSION AND CANCER

Anthony C. Allison
Clinical Research Center
Harrow, Middlesex, England

During the past decade information has accumulated
showing that immunosuppression is associated with an
increased incidence of certain types of cancer. These
observations are of theoretical interest and clinical
importance. From the theoretical point of view it has
been postulated by Thomas (1), Burnet (2) and others
that in intact animals an immunological surveillance
mechanism prevents many neoplastic cells from growing
into tumors, and the results of experiments using immunosuppression provide a test of this hypothesis. From
the clinical point of view it is becoming apparent that
the development of neoplasia is a major risk associated
with long-term immunosuppressive therapy in kidney
transplant recipients. The literature on this subject
is already extensive. In this article it will be possible
to describe only some of the more striking results which
have been obtained with experimental animals, which
reflect my own interests and illustrate some general
principles. The published information on human patients
will then be discussed.

POLYOMA VIRUS IN MICE

Polyoma virus is a small DNA virus (about 50nm in
diameter) which, when injected into newborn mice of many
strains, induces salivary and mammary adenocarcinomas,
osteosarcomas and other tumors in a high proportion of
recipients. Normal adult mice of all strains are

resistant to polyoma oncogenesis, even though the virus
multiplies extensively in host organs. Many colonies
of mice, in both the laboratory and the wild state, are
infected with polyoma virus as shown by the presence of
serum antibodies. Unlike the indigenous leukemogenic
viruses and the mammary tumor agent, polyoma virus is
not vertically transmitted from mothers to offspring.
Indeed a mother who has been infected passes antibodies
to her offspring, which are then protected for some weeks
against infection by the virus. Only when passive
protection has waned do they become horizontally infected
by polyoma virus from other mice in the colony. The
lateness of the infection probably explains why tumors
hardly ever arise.

A basic question which must be asked is, therefore,
why newborn animals are susceptible to polyoma virus
oncogenesis (and also to oncogenesis by many other
viruses) whereas adult animals are resistant. Two main
explanations can be considered. There might be some
feature of differentiation by which adult cells become
relatively resistant to polyoma virus transformation.
Alternatively there might be an efficient immune response
in the adult against the virus-specific transplantation
antigens on the surface of transformed cells which prevent
them from growing into a tumor, whereas this immune
response is absent, weak or delayed in newborn animals.

These alternatives can be tested by the use of
immunosuppression in adult mice infected by contact or
inoculated with polyoma virus. The most efficient
immunosuppressive procedure in our hands has been
thymectomy (3 days or 6 weeks after birth) followed by
repeated injections of antilymphocytic globulin (ALG).
All mice so treated develop polyoma tumors after
infection as adults (see Table 1). The tumors have the
virus-specific transplantation antigen, and the importance
of cell-mediated immunity in the prevention of tumor
growth is emphasized by the results of restoration
experiments. Hyperimmune serum from infected mice confers
some protection when given 24 hours after the virus--
apparently because it limits the dissemination of the
infection--but when given 7 days after the virus, it has
no protective effect. Likewise normal syngeneic lymphoid
cells given after the cessation of the ALG treatment
provide no protection. In contrast, transfer of lymphoid
cells from syngeneic donors immunized with a polyoma
tumor protects the animals very efficiently. Immunization
with a tumor induced in the same strain by an unrelated
virus (adenovirus type 12) is ineffective. Treatment of
the specifically sensitized lymphoid cells by anti-θ serum
and complement, to destroy thymus-dependent (T)
lymphocytes, abolishes the protective effect (3).

Table 1.   Development of Tumors in CBA Mice Infected
as Adults with Polyoma Virus

Preliminary Treatment	Restoration at 7 Weeks	No. of Animals	% with Tumors
Normal rabbit globulin	None	24	0
Thymectomy & antilymphocyte globulin	None	14	100
	Normal lymphoid cells	10	90
	Sensitized lymphoid cells	11	0
	Sensitized lymphoid cells	10	0
Thymectomy & antilymphocyte globulin	Sensitized lymphoid cells treated with anti-θ serum & complement	6	83
Thymectomy & antilymphocyte globulin	Antibody at 24 hours	12	17
	Antibody at 7 days	10	90

Thus the main reason why adult mice contracting
infections with polyoma virus do not develop tumors is
that they mount an effective cell-mediated immune
response against the tumor cells. This response is
mediated by T lymphocytes, but whether these lymphocytes
are themselves responsible for killing polyoma tumor
cells is not known. However, there is substantial
evidence that T cells transformed in the presence of
antigen can kill target cells (see [4]), so it is
reasonable to suppose that this can happen with polyoma
tumor cells.

There are at least two explanations for the
susceptibility of newborn animals to viral oncogenesis.
The traditional explanation is that early exposure to
the tumor antigens induces immunological tolerance.
An alternative interpretation is that in the adult animal
a cell-mediated immune response is mounted rapidly,
whereas in the newborn animal the response develops only
slowly. These possibilities can be distinguished by
transferring spleen cells from mice inoculated as
newborns with polyoma virus into immunosuppressed
recipients. The results show that in both mice (5)
and rats (6) inoculation of newborns results in efficient
cell-mediated immunity, but only after several weeks.
Hence there is no tolerance to the polyoma-specific
transplantation antigen, but there is a delay in mounting
cell-mediated immunity which allows the virus-transformed
cells to grow into masses too large for control by the

developing immune response.

Similar results have been obtained in other situations. For example, newborn hamsters inoculated with SV40 all develop tumors in about 100 days, whereas only about one-fifth of animals inoculated as adults develop tumors--and only after a very long interval representing the greater part of the lifespan of the animals. The incidence of tumors is increased and the latent period reduced if the adults are irradiated (7).

LEUKEMOGENIC VIRUSES IN MICE

Effects of immunosuppression on leukemias and lymphomas of mice are of special interest in view of the raised incidence of lymphoreticular malignancies in immunosuppressed human patients. There are several groups of leukemogenic viruses in mice. One consists of variants of Gross virus that are vertically transmitted in most, if not all, strains of mice. These variants of Gross virus share major antigens but differ in certain properties, e.g. host range. In addition there are several varieties of laboratory-passaged murine leukemogenic viruses (Friend, Moloney and Rauscher) that share major transplantation and other antigens.

It has long been known that early thymectomy greatly decreases the incidence of leukemia in mice inoculated with Gross or Moloney viruses. The most plausible explanation is that the target cells transformed by the virus are T lymphocytes, which are depleted after thymectomy, and the presence of thymus-specific antigens on tumor cells supports this interpretation. Hence thymectomy cannot be used as an immunosuppressive procedure with these viruses. However, ALG was shown by Allison and Law (8) to greatly increase the leukemogenic effect of Moloney virus in Balb/c mice. A high incidence of reticulum-cell sarcomas was seen in these animals, many at the subcutaneous site of injection of the ALG. Leukemia can be delayed or prevented by transferring normal syngeneic lymphoid cells to the mice after the cessation of ALG treatment (9). These results support the view that the immunosuppressive effect of ALG was responsible for the increased susceptibility observed, rather than other effects such as ALG-induced proliferation of lymphoid cells.

Again, the question of tolerance arises. If mice were fully tolerant to antigens of vertically transmitted leukemogenic viruses, they should show no antibody and no cell-mediated immunity against these antigens. Mice of the Balb/c strain with vertically-transmitted Moloney

virus were found to have complexes of viral antigen and
antibody in the circulation and especially in the renal
glomeruli (10).   Similar results have been obtained with
Gross virus antigens in the high-leukemic AKR strain of
mice (11).   Thus these mice are certainly capable of
producing some antibody against leukemogenic virus
antigens.   Dore et al. (12) reported that AKR mice, after
immunization with a syngeneic Gross tumor (K36), produce
antibodies cytotoxic for Gross leukemia cells.   Two
investigations suggest that, contrary to earlier reports,
these mice are also able to mount a cell-mediated
response against the virus-specific antigens of leukemic
cells.   Wahren and Metcalfe (13) found that lymphoid
cells from preleukemic AKR mice can exert a cytotoxic
effect on target cells bearing Gross virus-specific
antigens.   Haran-Ghera (14) has found that C57Bl mice
can, by appropriate manipulations, be immunized with the
strain of Gross virus which they themselves carry
(radiation leukemia virus) and are then resistant to
transplantation of syngeneic Gross tumor cells.   Thus
they are not tolerant, an important point when the
effects of immunosuppression are considered.   In fully
tolerant animals immunosuppression would not be expected
to increase the incidence of malignancies.   In several
of these mouse strains, ageing animals spontaneously
develop leukemias, sometimes (as in the AKR strain)
when they are still relatively young.   Presumably immune
responses help to delay the onset of leukemia until the
animals have had an opportunity to reproduce.   Under
these circumstances it would be expected that the incidence of leukemia would be raised by immunosuppression.
Casey (15) showed that in NZB mice treatment with
azathioprine increases the incidence of lymphoreticular
malignancies, and C57Bl mice treated with 6-mercapto-
purine show a higher proportion with leukemia than
untreated mice (16).   Under appropriate conditions
repeated administration of ALG to AKR mice accelerates
the onset of leukemia (5).   However, this result is not
always observed, apparently because some batches of ALG
are cytotoxic for Gross leukemia cells.   Administration
of ALG also accelerates the onset of lymphoreticular
malignancy, which is not known to be related to a virus,
in the SJL/J strain of mice (17).   One effect of
radiation under conditions where it elicits leukemia in
strains such as the C57Bl is immunosuppression (14).

MURINE SARCOMA VIRUS

The different strains of murine sarcoma virus (MSV)
are antigenically related to Moloney leukemogenic virus.
All appear to be defective in that, for the sarcoma
virus to be liberated from cells, they must be concom-

itantly infected with MLV (18). When MSV is inoculated
intramuscularly or subcutaneously into young mice or
rats, progressively growing tumors soon develop at the
sites of inoculation; there is usually also splenomegaly.
In some mice the local tumors continue to grow until they
kill the animal (progressors), while in other animals
they regress (regressors). The regression of palpable
tumors is a striking phenomenon and suggests that immune
reactions against tumor cells are strong. This is
confirmed by the immunity of such animals to the
subsequent transplantation of syngeneic tumor cells
bearing virus-specific antigens (MSV or MLV).

Adult mice are relatively resistant to tumor
induction by MSV, but they become susceptible after
thymectomy, ALG administration or both (19), or after
treatment with cortisone or X-radiation (20). In contrast
to the situation with polyoma virus, injections of serum
from regressor or hyperimmune animals even several days
after virus inoculation, protect the animals against
sarcoma development, but not against the later onset of
leukemia (9, Allison, unpubl.). The protective effect
of sera is not correlated with the levels of antiviral
antibody; such serum also confers protection against
transplants of tumor cells bearing Moloney-specific
antigens. Hence the effect is exerted against the tumor
cells rather than the virus itself. Our preliminary
results suggest that antibody in the regressor serum
becomes attached to the target cells and collaborates
with non-specific effector cells in the host to destroy
the target cells, in a manner analogous to that described
by Perlmann and Holm (21). The identity of the effector
cells is still uncertain: they do not appear to be T
lymphocytes (they are present in normal numbers in nude
mice and lack the θ antigen) or classical B lymphocytes
(they lack surface immunoglobulins) or macrophages (they
are not phagocytic). It has been suggested (22) that the
effector cells may represent an independent lineage of
bone-marrow-derived cells, for which the name A cells
was suggested, because the effector cells in the mouse
are adherent to glass and other substrates and are
involved in antibody-dependent reactions.

Antibody-dependent effector cell activity in human
peripheral blood can be quantitated (23), and it is of
interest that this activity is reduced after treatment of
human patients with immunosuppressive drugs (azathioprine,
cyclophosphamide) or X-radiation. Thus there appear to
be at least two major immunological defense mechanisms
against tumors: T-lymphocytes and non-specific effector
cells collaborating with antibody. Possibly other
mechanisms, including macrophages (3) and polymorphonuclear leukocytes (24) are also operating. The relative

importance of these defense mechanisms may vary from
tumor to tumor, and immunosuppressive procedures may
affect them differently. Empirically a combination of
azathioprine and corticosteroid therapy is usually
given to all human transplant patients, and in some
cases this is supplemented, for example, by the use of
ALG. The effects of these treatments on the different
mechanisms of graft and tumor rejection should be
further clarified.

EFFECTS OF IMMUNOSUPPRESSION ON OTHER TUMORS AND METASTASIS

The most striking effects of immunosuppression in
experimental animals are seen on virus-induced tumors
bearing relatively strong antigens of the transplantation
type. As a rule, effects on spontaneous and chemically
induced tumors are less marked, although several examples
have been reported. These include a striking increase
following neonatal thymectomy of late-appearing,
apparently nonviral mammary tumors in mice (25) and
facilitation by ALG of the induction by chemical
carcinogens of skin tumors in mice. Facilitation by ALG
of the spread of a hamster lymphoma (26), rat tumors
(27) and a mouse tumor (28) have also been described.

IMMUNOSUPPRESSION IN HUMAN TRANSPLANT PATIENTS

The increasing use over the past decade of powerful
immunosuppressive drugs has had obvious clinical benefits,
the most notable being the possibility of kidney
transplantation. However, the use of long-term immunosuppression in transplantation has had complications,
including severe infections with viruses that are normally
controlled by cell-mediated immune responses, such as
cytomegalovirus (29) and herpes simplex virus (30).

Five years ago reports from widely separated centers
in the U.S.A., New Zealand and Scotland of reticulum-cell
sarcomas in patients on immunosuppressive therapy after
renal transplantation drew attention to an even more
serious problem. The occurrence of several cases of so
rare a disease in a relatively small group of transplant
recipients was very unlikely to be due to chance.
Informal registries of malignant neoplasms in organ
transplant recipients have been kept by I. Penn and his
colleagues in Denver, Colorado, and L. Kinlen and W. R.
S. Doll, Oxford, England. Penn and Starzl (31) have
recently reported that in a group of 7581 renal and 179
cardiac homograft recipients on immunosuppressive therapy
until the end of 1971, 31 had mesenchymal tumors. The

striking fact is that 28 of these had lymphoreticular
malignancies, 20 being reticulum cell sarcomas, while
4 were other lymphomas (3 unclassified and 1 lymphosarcoma). Since lymphoreticular tumors within the
central nervous system are uncommon (the incidence
being 0.04-1.5% in the large series of Richmond et al.
(32) and Rosenberg et al. (33), it is also of interest
that the brain and spinal cord were involved in 11 of
the lymphoreticular tumors in transplant patients. The
precise risk cannot yet be accurately assessed because
too few cases have been followed sufficiently long, but
the increased incidence of reticulum-cell sarcomas in
immunosuppressed transplant recipients is highly
significant statistically and is conservatively estimated
to be at least 100 times the incidence expected in an
untreated population of comparable age (34).

    Several possible explanations for the lympho-
reticular malignancies can be considered. The first is
that a surveillance mechanism is normally operating in
man against leukemogenic viruses or lymphoma cells and
that this mechanism is inhibited by immunosuppression.
That immunosuppression can facilitate the growth of
malignant cells in man is shown by cases of inadvertent
transplantation of cancer in grafted organs. One such
tumor, secondary to a primary in the donor's bronchus,
presented as an abdominal mass (35). Only limited
surgical excision was possible, but termination of
immunosuppression was followed by disappearance of
malignant tissue as shown by later laparotomy. In another
case (36) the primary was in the donor's liver and
pulmonary metastases developed but resolved after with-
drawal of immunosuppressive therapy. However, an
alternative explanation is that lymphoreticular
malignancies in transplant recipients may be due to
carcinogenic effects on lymphoreticular cells of the
azathioprine and steroids used in all cases.

    The possible role of the grafted tissue itself has
to be borne in mind. In mice, graft-versus-host reactions
(37) and prolonged antigenic stimulation (38) have been
associated with lymphoreticular malignancy. For this
and other reasons it would be important to know whether
patients receiving immunosuppressive therapy for
conditions other than transplantation also develop
malignancies.

    Reports have been published of a cerebral lymphoma
in a patient with systemic lupus erythematosus treated
with azathioprine (39), of a lymphosarcoma in a 66-year-
old man with glomerulonephritis a few months after
completing a year's treatment with azathioprine and
prednisone (40), and of a reticulum-cell sarcoma in an

unusual site (the vulva) in a 34-year-old woman with
dermatomyositis treated with prednisolone and azathioprine
(41). Other cases have been recorded in the Oxford
registry, which includes medical as well as transplant
patients treated with immunosuppressive drugs, so that
it will eventually be possible to assess whether there
is an increased risk of malignancies in immunosuppressed
medical patients.

The observations on effects of immunosuppression
on the incidence of tumors in experimental animals and
man have already provided important information about
the operation of the surveillance mechanism and the
risk to man of prolonged immunosuppression. However,
further information is needed to assess these risks
more precisely so that the benefits and drawbacks can
be balanced. This is especially important when the
immunosuppressive treatment is given for diseases that
do not endanger life, such as psoriasis.

Spencer and Andersen (42) have reported that in a
group of Danish renal transplant recipients under
immunosuppression for more than one year, the incidence
of common warts was 42%, significantly higher than in
control populations. Whether cell-mediated immunity
helps in recovery from human papilloma infections has
not been established and further investigations would
be of interest.

VIRUSES OF THE SV40-POLYOMA GROUP

Another group of viruses that produce well-controlled
infections in normal persons but overt infections in
patients with immunodeficiency or undergoing immunosuppressive therapy has recently been defined. Progressive multifocal leucoencephalopathy (PML) is a rare
demyelinating disease occurring in patients with
Hodgkin's disease, sarcoidosis and other conditions in
which cell-mediated immunity is often defective. The
suggestion that this disease is the result of a viral
infection was strengthened when electron microscopic
studies revealed papovavirus-like particles in nuclei of
oligodendrocytes in demyelinated areas of brain. Padgett
et al. (43) isolated a papovavirus from the brain of a
case of PML complicating Hodgkin's disease. Primary
cultures of fetal glial cells were inoculated with
extracts of brain made at necropsy. The virus (JC) was
similar in size and shape to the polyoma-SV40 group of
viruses. It did not produce a cytopathic effect in
African geeen monkey kidney cells and infected cells did
not show immunofluorescence with antibody against SV40.
Other viruses isolated from cases of PML (44) are anti-

genically closely related to SV40.   (See Takemoto and Mullarkey, this volume.)

Another virus of the same group was isolated from the urine of an immunosuppressed renal transplant recipient by Gardner et al. (45). The patient had a fibrotic obstruction of the donor ureter, which was excised, and many virus particles were seen in nuclei of epithelial cells bordering the lumen. This virus (BK) also resembled polyoma and SV40 in size and shape. Although antigenically related to SV40, BK virus was clearly distinct from it on the basis of antigenic differences and its ability to agglutinate human erythrocytes. (See Takemoto and Mullarkey, this volume.) Four similar isolates have been made from immunosuppressed transplant recipients, and BK virus has been found to induce tumors after inoculation into newborn hamsters (46).

The BK and JC viruses are smaller than human papilloma virus and serologically unrelated to it. The morphological similarity and partial serological relationship to SV40 suggests that these viruses are representatives of a group of human papovaviruses. The frequent demonstration of antibodies reacting with SV40 in sera from normal humans (47,48) suggests that members of this group may produce latent infections in man, just as SV40 does in rhesus monkeys. It seems reasonable to conclude that normally the virus infection is controlled by an immune response, although prolonged latent infections can occur, and that only in subjects with secondary immunodeficiency (as in Hodgkin's disease) or immunosuppression (renal transplant recipients) do infections become overt, with virus demonstrable by electron microscopy in the tissues and available for isolation. Whether these viruses are oncogenic in humans is another problem of interest that requires further investigation. There is no evidence that SV40 induces tumors in its natural host, the rhesus monkey, but the BK/JC group of viruses might produce tumors in some humans. Studies on immunosuppressed monkeys would themselves be of interest, but would have to be undertaken with strict precautions because of the danger of activating other infections, e.g., herpesvirus B.

References

1.   Thomas, L. 1959. Cellular and humoral aspects of the hypersensitive states, ed. H. S. Lawrence, p. 529. Cassell, London.
2.   Burnet, F. M. 1970. The concept of immunological surveillance. Progr. Exp. Tumor Res. 13:1.

3.   Allison, A. C. 1972. Interactions of antibodies and effector cells in immunity against tumors. Ann. Inst. Pasteur 122:619.

4.   Cerottini, J. C. and K. T. Brunner. 1973. Identification of cytotoxic lymphocytes. Adv. Immunol. (in press).

5.   Allison, A. C. 1970. On the absence of tolerance in virus oncogenesis. Proc. IV Quad. Int. Conf. on Cancer, Perugia, p. 563.

6.   Vandeputte, M. and S. K. Datta. 1972. Cell-mediated immunity in polyoma oncogenesis. Europ. J. Cancer 8:1.

7.   Allison, A. C., F. C. Chesterman and S. Baron. 1967. Induction of tumors in adult hamsters with SV40. J. Nat. Cancer Inst. 38:567.

8.   Allison, A. C. and L. W. Law. 1968. Effects of antilymphocyte serum on virus oncogenesis. Proc. Soc. Exp. Biol. Med. 127:207.

9.   Law, L. W. 1972. Influence of immune suppression on the induction of neoplasms by the leukaemogenic virus MLV and the variant MSV. The nature of leukaemia, p. 23. Proc. Int. Conf., Sydney, Australia (ed. P. C. Vincent). Publ. U. C. N. Blight, Sydney.

10. Hirsch, M. S., A. C. Allison and J. J. Harvey. Immune complexes in mice infected with Moloney leukemogenic and murine sarcoma viruses. Nature 223:739.

11. Oldstone, M. B. A., T. Aoki and F. J. Dixon. 1972. The antibody response of mice to murine leukemia virus in spontaneous infection: Absence of classical immunologic tolerance. Proc. Nat. Acad. Sci. 69:134.

12. Dore, J. F., M. Schneider and G. Mathe. 1969. Reactions immunitaires chez les souris AKR leucemiques ou preleucemiques. Rev. Franc. Etud. Clin. Biol. 14:1003.

13. Wahren, B. and D. L. Metcalf. 1970. Cytotoxicity in vitro of preleukaemic lymphoid cells on syngeneic monolayers of embryo or thymus cells. Clin. Exp. Immunol. 7:373.

14. Haran-Ghera, N. 1971. Influence of host factors on leukemogenesis by the radiation leukemia virus. Israel J. Med. Sci. 7:17.

15. Casey, T. P. 1968. Azathioprine (Imuran) administration and the development of malignant lymphomas in NZB mice. Clin. Exp. Immunol. 3:305.

16. Doell, R. G., C. de Vaux St Cyr and P. Grabar. 1967. Immune reactivity prior to development of thymic lymphoma in C57Bl mice. Int. J. Cancer 2:103.

17. Burstein, N. A. and A. C. Allison. 1970. Effect of antilymphocytic serum on the appearance of reticular neoplasms in SJL/J mice. Nature 225:1139.

18. Harvey, J. J. and J. East. 1971. The murine sarcoma virus (MSV). International review of experimental

pathology, vol. 10 (ed. G. W. Richter and M. A. Epstein), p. 266. Academic Press, New York.

19. Law, L. W., R. C. Ting and A. C. Allison. 1968. Effects of antilymphocyte serum on tumor and leukemia induction by murine sarcoma virus (MSV). Nature 220: 611.

20. Fefer, A. 1970. Immunotherapy of primary Moloney sarcoma virus-induced tumors. Int. J. Cancer 5:327.

21. Perlmann, P. and G. Holm. 1969. Cytotoxic effects of lymphoid cells in vitro. Adv. Immunol. 11:117.

22. Allison, A. C. 1972. Immunity and immunopathology of virus infections. Ann. Inst. Pasteur 123:585.

23. Campbell, A. C., I. C. M. MacLennan, M. L. Snaith and I. G. Barnett. 1972. Selective deficiency of cytotoxic B lymphocytes. Clin. Exp. Immunol. 12:1.

24. Pickaver, A. H., N. A. Ratcliffe, A. E. Williams and H. Smith. 1972. Cytotoxic effects of peritoneal neutrophils on a syngeneic rat tumor. Nature New Biol. 235:186.

25. Burstein, N. A. and L. W. Law. 1971. Neonatal thymectomy and non-viral mammary tumors in mice. Nature 231:450.

26. Gershon, R. K. and R. L. Carter. 1970. Facilitation of metastatic growth by antilymphocyte serum. Nature 226:328.

27. Fisher, B., O. Soliman and E. R. Fisher. 1970. Further observations concerning effects of anti-lymphocyte serum on tumor growth: With special reference to allogeneic inhibition. Cancer Res. 20:2035.

28. James, S. E. and A. J. Salsbury. 1973. Immuno-suppression and metastases. Brit. J. Cancer (in press).

29. Craighead, J. E. 1969. Immunologic response to cytomegalovirus infection in renal allograft recipients. Amer. J. Epidemiol. 90:506.

30. Montgomerie, J. Z., D. M. O. Becroft, M. C. Croxson, P. B. Doak and J. D. K. North. 1969. Herpes-simplex-virus infection after renal transplantation. Lancet 2:867.

31. Penn, I. and T. Starzl. 1972. Malignant tumors arising de novo in immunosuppressed organ transplant recipients. Transplantation 14:407.

32. Richmond, J., R. S. Sherman, H. D. Diamond and L. F. Craver. 1962. Renal lesions associated with malignant lymphomas. Amer. J. Med. 32:184.

33. Rosenberg, S. A., H. D. Diamond, B. Jaslowitz and L. F. Craver. 1961. Lymphosarcoma: 1269 cases. Medicine (Baltimore) 40:31.

34. Immunosuppression and malignancy. Editorial, Brit. Med. J. 3:713 (1972).

35. Wilson, R. E., E. B. Hager, C. L. Hampers, J. M. Carson, J. P. Merrill and J. E. Murray. 1968.

Immunologic rejection of human cancer transplanted with a renal allograft. New Eng. J. Med. 278:479.

36. Zukoski, C. F., D. A. Killen, E. Ginn, B. Matter, D. O. Lucas and H. F. Seigler. 1970. Transplanted carcinoma in an immunosuppressed patient. Transplantation 9:71.

37. Schwartz, R. and J. Andre-Schwartz. 1968. Immunoproliferative diseases. Interactions between immunologic abnormalities and oncogenic viruses. Ann. Rev. Med. 19:269.

38. Metcalfe, D. 1963. Induction of reticular tumors in mice by repeated antigenic stimulation. Acta Unio Int. Cancer 19:657.

39. Lipsmeyer, E. A. 1972. Development of malignant lymphoma in a patient with systemic lupus erythematosus treated with immunosuppression. Arth. Rheum. 15:183.

40. Sharpstone, P., C. S. Ogg and J. S. Cameron. 1969. Nephrotic syndrome due to primary renal disease in adults. II. A controlled trial of prednisolone and azathioprine. Brit. Med. J. 2:535.

41. Sneddon, I. and J. M. Wishart. 1972. Immunosuppression and malignancy. Brit. Med. J. 4:235.

42. Spencer, E. S. and H. K. Andersen. 1970. Clinically evident, nonterminal infections with herpesviruses and the wart virus in immunosuppressed renal allograft recipients. Brit. Med. J. 3:251.

43. Padgett, B. L., G. M. zu Rhein, D. L. Walker, R. J. Eckrode and B. H. Dessel. 1971. Cultivation of papova-like virus from human brain with progressive multifocal leucoencephalopathy. Lancet 1:1257.

44. Weiner, L. P., R. M. Herndon, O. Narayan, R. T. Johnson, K. Shah, L. J. Rubinstein, T. J. Preziosi, and F. K. Conley. 1972. Isolation of virus related to SV40 from patients with progressive multifocal leukoencephalopathy. New Eng. J. Med. 286:385.

45. Gardner, S. D., A. M. Field, D. V. Coleman and B. Hulme. 1971. New human papovavirus (BK) isolated from urine after renal transplantation. Lancet 1:1253.

46. Gardner, S. D. 1973. Private communication.

47. Shah, K. V., H. L. Ozer, H. S. Pond, L. D. Parma and G. P. Murphy. 1971. SV40 neutralizing antibodies in sera of U.S. residents without a history of polio immunization. Nature 231:448.

48. Shah, K. V., F. R. McCrumb, Jr., R. W. Daniel and H. L. Ozer. 1972. Serologic evidence for a SV40-like infection of man. J. Nat. Cancer Inst. 48:557.

DISCUSSION

DIXON: Tony, in your experiments with immunosuppression, do you find an actual increase in the amount of virus in the immunosuppressed animals as compared with controls? I ask this because in our experience, using a variety of immunosuppressive regimes in animals infected with LCM or Gross virus, we do not see a change in the level of the viruses in spite of the fact that we have applied maximal immunosuppression.

ALLISON: In the case of the nononcogenic viruses, such as Coxsacki or herpes, we certainly see an increase in the amount of virus, particularly in the amount of virus associated with a susceptible target organ such as heart in the case of the Coxsacki virus.   In the case of polyoma virus, there is no obvious increase that we can see and we haven't quantitated accurately the leukemogenic viruses.   Actually we have some information on this and the amount of virus in the plasma is increased in the immunosuppressed animals.

MILLER: On the basis of the information you have presented, would you recommend that people who are immunosuppressed for any reason--on drugs, for example-- be excluded from laboratories of viral oncology?

ALLISON: Well, I certainly think that this is a fair presumption.   I wouldn't like to have one in my laboratory and not only in viral oncology; I think that it would be undesirable for them to work with infectious agents at all.

KURU AND VIRUS DEMENTIAS

D. C. Gajdusek and C. J. Gibbs, Jr.
National Institute of Neurological Diseases and Stroke
Bethesda, Maryland

Kuru and Creutzfeldt-Jakob (C-J) disease are the two human representatives of a group of aberrant infections which we have called the "subacute spongiform virus encephalopathies" (1,2). We have recently preferred to call C-J disease a "transmissible virus dementia" because there is considerable synonymy in neurology for the spectrum of clinical and pathological entities which we are able to transmit after long incubation periods to chimpanzees and new world monkeys (3). Both diseases may be transmitted by intracerebral or peripheral inoculation of highly diluted suspensions of brain from patients, even after filtration through 220 nm Millipore filters (4,5). Both viruses have been maintained for long periods in vitro in explant and trypsinized cultures of affected human or chimpanzee brain tissue (6).

Two animal diseases are caused by closely related agents: scrapie of sheep and goats and transmissible mink encephalopathy. All four diseases are unusual among virus infections in that they have incubation periods of many months to many years before the slowly progressive, invariably fatal symptoms develop; the pathology is noninflammatory, without any perivascular cuffing with mononuclear cells; there is usually no pleocytosis or significant elevation of protein in the cerebrospinal fluid during the course of the disease; and there is no evidence of an immune response or hypersensitivity to these viruses. The viral agents causing the diseases are atypical in that they remain

288

stable at temperatures up to about 80°C. At 85°C or
higher inactivation curves approach the time axis
asymptotically and it is difficult, even at 100°C, to
totally inactivate the agent, although suspensions
titering 10^{-6}-10^{-8} are quickly reduced to titers under
10^{-1}. The viruses have shown remarkable resistance to
a number of agents that quickly inactivate most other
viruses. In most cases this resistance is reduced by
partial purification with fluorocarbons or zonal density
gradient banding. Proteases (7), acetylethyleneamine
(8), β-propriolactone (9), 10% formaldehyde (10),
and nucleases (11) fail to inactivate crude suspensions
of these viruses, whereas inactivation is quickly
accomplished with chlorobutanol (12), ether (13),
6-8M urea, periodate, and 95% phenol (14). The viruses
show remarkable resistance to UV inactivation (15,16)
and with ionizing radiation (neutron beam) they exhibit
an unusually small inactivation target size (17).

A vast speculative literature has rapidly appeared
in major journals, suggesting that the cause of these
unusual findings may be that the agents do not contain
nucleic acid (18,16) and that they might represent
replicating basic proteins (19), polysaccharides (20,18),
or plasma membrane fragments (21). Much of this speculation has been based on earlier incorrect reports of
very small sizes, as measured by filtration through
semipermeable membranes, and a premature, over-rigorous,
interpretation of the UV inactivation data, i.e., that
it indicates the total absence of nucleic acid. The
significant failure of many electron microscopists
to demonstrate recognizable virions in high-titer
suspensions of zonal density gradient bands or in
tissue suspensions of high infectivity, and in optimally
fixed pathological specimens, has further stimulated
such speculation.

We believe that the scrapie, kuru, C-J, and mink
encephalopathy viruses probably consist of small pieces
of genetically active nucleic acid tightly bound to
fragments of plasma membrane. In the disease processes
they evoke in vivo, in the cell cultures in which they
are maintained in vitro, and in semi-purified suspensions
they are not present as recognizable encapsidated
virions. However, fragmented plasma membranes are
clearly visible in the EM studies of the processes of
neurons (22). These vacuoles are not empty, but contain
fragmented sheets of the same plasma membrane which
forms the wall of the vacuole and the envelope of
many more conventional viruses.

The long incubation periods, the high cost of the
laboratory animals required, and the absence of any in

vitro assay methods for the viruses has made the
accumulation of necessary data difficult, slow and
expensive. However Diener (23) has recently demonstrated
that the viruses causing three plant virus diseases
(potato spindle tuber disease, chrysanthemum stunt
disease, and citrus exocortis disease) are very small
uncoated fragments of RNA of about 50,000 daltons
molecular weight, which when purified are very sensitive
to RNase and to inactivation by UV radiation, as are
other conventional viruses. He finds, however, that in
suspensions of crude plant sap the infectivity is even
more resistant to UV inactivation than is that of the
scrapie agent in crude brain suspensions. Further, the
purified RNA "viroids" of Diener, with conventional UV
inactivation properties, show enormously increased
resistance to UV inactivation when mixed with suspensions
of normal plant sap (24). These findings greatly
strengthen our long-held contention that small,
genetically active nucleic acid moieties strongly bound
to plasma membrane fragments might be protected from
UV inactivation.

It is the mystery which has been engendered by the
erroneous laboratory data and by the premature interpretation of UV inactivation results that has led to
much of the fear of these viruses--that they cannot be
inactivated by conventional means--and to legends of
their possible aberrant patterns of communicability.

THE NATURAL DISEASES IN MAN

There has been no evidence in clinical or
epidemiological studies that either kuru or C-J disease
has ever been communicated by contact with ill patients
or with patients during their asymptomatic incubation
period. Fifteen years of careful surveillance of over
2500 kuru patients in New Guinea has failed to reveal a
single case of contact infection in an outsider. The
world experience up to 1968 with C-J disease, consisting
of 150 recorded cases (25) and an almost equivalent
number in the series we have accumulated since then,
has failed to reveal any pattern of association between
cases in this usually sporadic disease.

The mode of spread of kuru seems to be totally
explained by contamination of women and children,
occasionally older boys and men, with highly infectious
brain tissue (10^7 ID per gram) during the butchery of
the patient's body in the mortuary rite of ritual
cannibalism. The progressively complete disappearance
of the disease in children, with the youngest victim
becoming progressively older since the cessation of

cannibalism in the kuru region between 1957 and 1962, is in total agreement with this hypothetical mode of spread.

The mode of transmission of C-J disease, on the other hand, is totally unknown. Most cases are sporadic, with no memory of similar disease in relatives or in close acquaintances. However there are three well-documented families in which the disease has a high familial incidence, and we are now reporting four other families in which the disease occurs in a heredofamilial pattern (26). We have transmitted the disease to chimpanzees and squirrel monkeys with brain tissue from patients in two of these families (Fig. 1). This raises the possibility of genetic control of susceptibility to infection itself, or to the pathological expression of these viruses.

THE NATURAL DISEASES IN ANIMALS

SCRAPIE. The mode of transmission of scrapie is still in doubt, in spite of four decades of investigation and over two centuries of farm experience. Folklore has maintained that the disease is a breeding problem and this has been supported in the last decade by the genetic studies of Parry, Darlington, and others, which indicate that it behaves, on Scottish farms at least, as a Mendelian recessive trait. Contagion has been very difficult to demonstrate experimentally. Rarely, lambs or kids placed in extensive and long-term contact with sheep with natural scrapie have later developed the disease. Recently demonstration of the agent in fetal membranes from scrapie-affected animals and transmission of the disease to sheep and goats by the oral administration of such membranes suggest that contamination of the young animals with these tissues may be the source of some infection in nature (19).

MINK ENCEPHALOPATHY. The epidemiology of this disease, which has appeared almost explosively on several mink farms, is that of massive contamination of ingested meat, rather than animal to animal infection (27).

EXPERIMENTAL VIRUS ENCEPHALOPATHIES

To date there has been no recognized case in man of acute or chronic disease associated with long-term exposure of either laboratory workers or animal caretakers to any of these diseases.

C FAMILY (France)

A

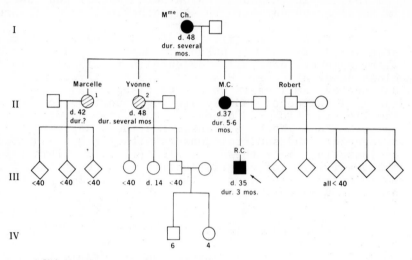

1 Died with depression and cachexia
2 Died with mutism and apraxia

G FAMILY

B

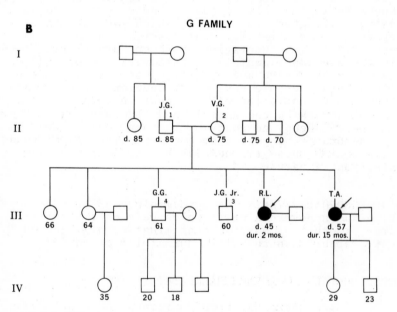

1 "Senile tremor," blind post corneal transplant, death following pneumonia
2 A.S.C.V.D.
3 "Mental disorder," depression since 1930's
4 Infantile paralysis

Animal to animal transmission of experimentally induced kuru or C-J disease in monkeys has not been observed.

In the case of scrapie, spontaneous transmission of the disease from experimentally infected animals to uninoculated animals has been reported (28-31). It has occurred rarely, as indicated above, and the exact mode of transmission has not been established. Since oral and other peripheral routes of inoculation have been demonstrated as adequate for establishing infection, many possibilities, including fighting and cannibalism, are suggested. In view of the rarity of such transmissions, the possibility that skin or bucal mucosal laceration and close direct contact are required seems likely. Field transmission of scrapie to lambs reared on fields where scrapied animals have previously been quartered has occurred, again, however, as a rare event, in spite of the many animals which have been exposed for many years.

VECTORS AND INTERMEDIARY HOSTS. No arthropod vectors have been demonstrated for any of these diseases. The epidemiology of the human and animal diseases does not suggest that ectoparasites or intermediary hosts are involved. Experimental attempts to infect arthropods with scrapie have been unsuccessful.

PRESENCE OF VIRUS IN TISSUES OTHER THAN THE BRAIN. Both kuru and C-J virus have been demonstrated in the lymph nodes, spleen and other abdominal viscera of affected animals, but not in these organs in man. Moreover the

------

FIGURE 1A. Brain biopsy suspension (5%) from patient R. C. inoculated intracerebrally and intravenously into a chimpanzee has produced the clinical picture and pathology of Creutzfeldt-Jakob disease in the chimpanzee. The incubation period was 13 months; the animal was killed in the terminal disease after 4 months of illness. Squares, circles and diamonds represent male, female and sex unknown, respectively. Black figures represent patients with C-J disease, and the diagonally-lined figures represent patients with probable C-J disease. The symbol d = died.

FIGURE 1B. Autopsy brain suspension (10%) from patient T. A. has produced Creutzfeldt-Jakob disease syndrome in a squirrel monkey after 25 months incubation. The animal was killed after 1-1/2 months of disease and demonstrated typical pathology of subacute spongiform encephalopathy. Both the patient and her sister had classical C-J disease.

brain of man and of experimentally infected animals
regularly contains over 10^6 infectious particles per gram.
Urine, serum, whole blood, leucocytes, milk, placenta
and amnion from kuru patients, and these tissues from
experimentally infected animals, have not been demonstrated to contain the agent. The same applies to C-J
disease as far as this has been studied. However, in
sheep affected with scrapie, the placenta and fetal
membranes from affected animals have been shown to
contain the agent (19).

THE ABSENCE OF AN IN VITRO MARKER FOR THESE VIRUSES.   No
in vitro marker system has been found to identify the
viruses of kuru and C-J disease. It has not been possible
to develope a fluorescent antibody, a complement fixation,
neutralization, or precipitin test using sera from
animals that received repeated large doses of infectious
virus (even density gradient banded and concentrated
virus) with or without adjuvants. Many different protocols of immunization have been attempted, and many
species, such as mice, guinea pigs, rabbits, and chickens
have been used.

The major problem limiting work with these viruses
is the lack of serological identification or any other
method for their in vitro detection. Since the plasma
membrane fragments that appear to comprise an important
part of the virus are host derived, they may well lack
any antigenic distinction from normal host cell membranes.
This could account for the lack of any immune response
to these viruses. A further implication of this membrane-
association would be the alteration of the properties of
the virus on passage through each new host.

There have recently been reports of rapid methods
of detecting the scrapie agent, using complex indirect
indicators based on an intact animal. One study reports
the alteration of macrophage electrophoretic migration
by a factor released from specifically sensitized
lymphocytes early after inoculation of guinea pigs with
the scrapie virus (32). In the second study a depression
in the polymorphonuclear leucocyte count only a few days
after inoculation of mice with scrapie is reported (33).
Both procedures have been difficult to reproduce and
confirm, and neither seems capable of rapidly indicating
the presence of the virus with certainty.

IN VITRO CULTIVATION OF THE AGENTS. All four viruses
have been maintained for over a year in explant cultures
of brain cells from the affected patients or from animals
with natural or experimentally induced disease (6,7,34,

35). The infective cells seem to grow better in tissue culture than do cells from normal brain and thus a cell stimulation effect has been attributed to all the agents (36). In vivo, the enormous hyperplasia and hypertrophy of glial cells without obvious cause has long been taken as evidence for a primary effect of the agent on the cells.

EVALUATION OF THE BIOHAZARD

The epidemiology of the human diseases gives no cause for suspicion that the diseases are naturally contagious. However, once the infected tissues are handled, as was the case with kuru-infected tissues in New Guinea, infection through skin abrasions may occur. There is one example of C-J disease in a man and wife; one neurosurgeon has suffered the disease. With the exception of those cases in families in which the disease appears to be inherited, all other cases are sporadic, without a known association.

There has been no case of contact infection in uninoculated animals quartered for over a decade in close contact with successive experimentally infected primates during the incubation period and the entire course of the clinical disease. No virus shedding or excretion has been detected; urine, blood and cerebro-spinal fluid are without infectivity.

Both scrapie, which is found on farms from coast to coast in the United States, and mink encephalopathy are transmissible by the oral route. Infection from contact with naturally infected mice or experimentally infected goats or mice, although rare, may occur. These animal viruses have also caused disease in inoculated laboratory monkeys. Whether or not man is susceptible to these viruses is presently unknown. High heat stability and resistance to many agents which usually inactivate viruses, such as UV radiation and 4% formaldehyde, demand that special attention be given to decontamination of laboratory work surfaces and equipment.

The phenomenon of silent incubation periods that may be as long as two or more decades greatly complicates the formulation of sensible biohazards control. Chimera of hazards, the reality of which cannot be assessed, are too easily suggested, while accidental infection of laboratory workers may be years in declaring themselves.

References

1. Gibbs, C. J., Jr. and D. C. Gajdusek. 1970.
    Characterization and nature of viruses causing
    subacute spongiform encephalopathies. 6th Int. Congr.
    Neuropath., Paris, p. 779.
2. Gajdusek, D. C. and C. J. Gibbs, Jr. 1972. Subacute
    and chronic diseases caused by atypical infections
    with unconventional viruses in aberrant hosts.
    Perspectives in virology, vol. 8, p. 279. Academic
    Press, New York.
3. Gajdusek, D. C. and C. J. Gibbs, Jr. 1973. Trans-
    missible virus dementias. Proceedings, 5th world
    congress of psychiatry, Mexico, D. F., Excerpta
    Medica (in press).
4. Gajdusek, D. C. 1972. Spongiform virus encepha-
    lopathies. J. Clin. Path. 25:(suppl.)72.
5. Gibbs, C. J., Jr. and D. C. Gajdusek. 1972. Isola-
    tion and characterization of the subacute spongiform
    virus encephalopathies of man: Kuru and Creutz-
    feldt-Jakob disease. J. Clin. Path. 25:(suppl.)84.
6. Gajdusek, D. C., C. J. Gibbs, Jr., N. G. Rogers, M.
    Basnight and J. Hooks. 1972. Persistence of
    viruses of kuru and Creutzfeldt-Jakob disease in
    tissue cultures of brain cells. Nature 235:104.
7. Haig, D. A. and M. C. Clarke. 1965. Observations
    on the agent of scrapie. Slow, latent and temperate
    virus infections. NINDB Monogr. No. 2, PHS Publ. No.
    1378, p. 215. Gov. Print. Office, Wash., D. C.
8. Stamp, J. T., J. G. Brotherston, I. Zlotnik, J. M. K.
    Mackay and W. Smith. 1959. Further studies on
    scrapie. J. Comp. Path. Therap. 69:268.
9. Haig, D. A. and M. C. Clarke. 1968. The effect of
    β-propiolactone on the scrapie agent. J. Virol.
    3:281.
10. Hunter, G. D. 1970. The biochemical properties and
    nature of the scrapie agent. 6th Int. Congr.
    Neuropath., Paris, p. 802.
11. Gibbs, C. J., Jr. 1967. Search for infectious
    etiology in chronic and subacute degenerative
    diseases of the central nervous system. Current
    topics in microbiology and immunology 40:44.
    Springer-Verlag, New York.
12. Hunter, G. D. 1971. Scrapie: A prototype slow
    infection. J. Infect. Dis. 125:427.
13. Gibbs, C. J., Jr., D. C. Gajdusek and J. A. Morris.
    1965. Viral characteristics of the scrapie agent
    in mice. Slow, latent and temperate virus in-
    fections. NINDB Monogr. No. 2, PHS Publ. No. 1378,
    p. 195. Gov. Print. Office, Wash., D. C.
14. Hunter, G. D. 1969. The size and intracellular
    location of the scrapie agent. Biochem. J. 114:22.
15. Haig, D. A., M. C. Clarke, E. Blum and T. Alper.

1969. Further studies on the inactivation of the scrapie agent by ultraviolet light. J. Gen. Virol. 5:455.

16. Latarjet, R., B. Muel, D. A. Haig, M. C. Clarke and T. Alper. 1970. Inactivation of the scrapie agent by near monochromatic ultraviolet light. Nature 227:1341.

17. Alper, T., D. A. Haig and M. C. Clarke. 1966. The exceptionally small size of the scrapie agent. Biochem. Biophys. Res. Commun. 22:278.

18. Adams, D. H. and E. A. Caspary. 1968. The incorporation of nucleic acid and polysaccharide precursors into a post-ribosomal fraction of scrapie-affected mouse brain. Biochem. J. 108:38.

19. Pattison, E. H. and K. M. Jones. 1967. The possible nature of the transmissible agent of scrapie. Vet. Rec. (London) 80:2.

20. Adams, D. H. and E. A. Caspary. 1967. Nature of the scrapie virus. Brit. Med. J. 3:173.

21. Gibbons, R. A. and G. D. Hunter. 1967. A consideration of the nature of the scrapie agent--a membrane hypothesis. Biochem. J. 105:7.

22. Lampert, P. W., D. C. Gajdusek and C. J. Gibbs, Jr. 1972. Subacute spongiform virus encephalopathies. Scrapie, kuru and Creutzfeldt-Jakob disease: A review. Amer. J. Path. 68:626.

23. Diener, T. O. 1972. Viroids. Adv. Virus Res. 17:295.

24. Diener, T. O. 1973. Similarities between scrapie agent and the agent of the spindle tuber disease. Ann. Clin. Res. (in press).

25. Kirschbaum, W. R. (ed.) 1968. Jakob-Creutzfeldt disease. Elsevier Press, New York.

26. Ferber, R. A., S. L. Wiesenfeld, R. Roos, A. R. Bobowick, C. J. Gibbs, Jr. and D. C. Gajdusek. 1973. Familial Creutzfeldt-Jakob disease. Transmission of the familial disease to the chimpanzee. Proceedings of the international congress of neurology, Barcelona. Excerpta Medica (in press).

27. Hartsough, G. R. and D. Burger. 1965. Encephalopathy of mink. I. Epizootiologic and clinical observations. J. Infect. Dis. 115:387.

28. Brotherston, J. G., C. C. Renwick, J. T. Stamp and I. Zlotnik. 1968. Spread of scrapie by contact to goats and sheep. J. Comp. Path. 78:9.

29. Dickenson, A. G., J. M. K. Mackay and I. Zlotnik. 1964. Transmission by contact of scrapie in mice. J. Comp. Path. 74:250.

30. Morris, J. A., D. C. Gajdusek and C. J. Gibbs, Jr. 1965. Spread of scrapie from inoculated to uninoculated mice. Proc. Soc. Exp. Biol. Med. 120:108.

31. Zlotnik, I. 1968. Spread of scrapie by contact in

mice. J. Comp. Path. 78:19.
32. Field, E. J. and B. K. Shenton. 1972. Rapid
    diagnosis of scrapie in the mouse. Nature 240:104.
33. Licursi, P. C., P. A. Merz, G. S. Merz and R. I.
    Carp. 1972. Scrapie-induced changes in the
    percentage of polymorphonuclear neutrophilis in
    mouse peripheral blood. Infec. Immun. 6:340.
34. Gustafson, D. P. and C. L. Kanitz. 1965. Evidence
    of the presence of scrapie in cell cultures of the
    brain. Slow, latent and temperate virus infections.
    NINDB Monogr. No. 2, PHS Publ. No. 1378, p. 221.
    Gov. Print. Office, Wash., D. C.
35. Marsh, R. F. K. and R. P. Hanson. 1969. Trans-
    missible mink encephalopathy: Neurological
    response. Amer. J. Vet. Res. 30:1643.
36. Caspary, E. A. and T. M. Bell. 1971. Growth
    potential of scrapie mouse brain in vitro. Nature
    229:269.

* * * * *

DISCUSSION

DIMMICK: Are the slow viruses extraordinarily stable
and how can they be inactivated?

GAJDUSEK: We have demonstrated that the scrapie virus,
even in crude tissue suspensions or fragments of whole
tissue, is completely inactivated by autoclaving at
115°C for 30 minutes or by dry heat of 190°C for 20
minutes. It is not inactivated by usual 10% formal
saline (4% formaldehyde). It is moderately sensitive to
ether, phenol, iodine and detergents.

DIMMICK: Have you tried hypochlorite as a disinfectant?

GAJDUSEK: Yes. We are using it as Chlorox diluted from
1:3 to 1:10.

HELLMAN: Are transformed cells capable of transmitting
kuru?

GAJDUSEK: We have only recently inoculated chimpanzees
and monkeys with late passages of these transformed
cell lines and it is too early to know whether they will
transmit C-J disease. Even if they do, it will be
impossible to conclude that the MP-MV-like virus in them

is the cause, since long-maintained brain cell cultures which do not contain such virions also transmit the disease.

SLY: Are the long-lived glial cells all transformed? Are there any controls to indicate the lifespan in culture of normal glial cells?

GAJDUSEK: That is a very dicey question.   Glial cells in culture from kuru or C-J disease brains appear to grow better and faster than those from "normal" brains. They appear to be growing indefinitely, even after over 2 years of in vitro cultivation, although cells do not pile up and divide as rapidly as do the MP-MV-like virus-containing transformed cells.

YATVIN: When kuru is put into a primate, can it be passed before clinical manifestations of disease appear?

GAJDUSEK: Yes.   We have passed the disease to other primates using brain from animals which have been killed during the incubation period, before they have developed clinical manifestations of disease.

YATVIN: What information is available pertaining to mink encephalopathy?

GAJDUSEK: We now assume that mink encephalopathy is the scrapie virus modified by serial passage through mink.   It may have originated on mink farms from the feeding of scrapie-contaminated meat.   Changes in host range and other biological properties of scrapie after passage through mink are to be expected.   Mink encephalopathy affects rhesus, stump tail and squirrel monkeys.

Scrapie in a mouse-adapted line has been transmitted to the cynomolgus monkey after 5 1/2 years incubation, and it has also taken in the squirrel monkey after 14 months incubation.

PANEL IV
MODIFICATIONS OF VIRUSES
AND THE IMPORTANCE
OF THE SPECIES BARRIER

Anthony C. Allison
Clinical Research Center

Jordi Casals
Yale University School of Medicine

D. Carleton Gajdusek
National Institute of Neurological Diseases and Stroke

Anthony J. Girardi
Wistar Institute

Wallace P. Rowe
National Institute of Allergy and Infectious Diseases

COMMENTS BY

Anthony J. Girardi

Virus effects may be modified by mutation of the
viral genome, and also by infection of a new host or
unfamiliar tissue, by host immunosuppression, or by
infection with a second virus interacting at a molecular
or mechanical level. Examples are noted in slow virus
infections of the central nervous system (CNS) such as
subacute sclerosing panencephalitis (SSPE), progressive
multifocal leukoencephalopathy (PML), and possibly in
Jakob-Creutzfeldt (JC) disease. In these diseases
viruses persist for long periods before clinical signs
are manifest (but where and in what form?) and cell to
cell contact is often required for efficient spread; and
immunosuppression precedes CNS disease in certain
instances. In SSPE intranuclear inclusions were noted
in brain tissues, and isolation attempts have been
partially successful with recovery of a measles-like
virus and morphologic evidence for a second agent in the
cytoplasm resembling papova virions (histochemically DNA
but noninfectious) in original brain cells, in ferrets
following inoculation with SSPE material, and in target
cells after fusion with SSPE cells. Interestingly the
spread of the second agent to new cells seems to be
possible only through cell fusion by the measles virus,
representing a convenient mode of parasitism without
complete virus replication and without an extracellular
phase. An excellent example of modified viral expression
occurs in cell cultures of SSPE patients' brain tissues
(Table 1). Do these various levels of expression result
from infection of an unusual tissue by measles virus or

303

Table 1. Viral Expression in Cultures of Brain Cells from Patients
          with Subacute Sclerosing Panencephalitis

| | | | | Infectious virus | | |
| | | Nucleo- | | after fusion with | | without |
Patient	Syncytia	capsids	Antigen*	CV-1	self	fusion
ROB	+	+	−	−	−	−
JAC	+	+	+	+	−	−
LEC	+	+	+	+	+	−

* Antigen detected by patients' sera or measles immune serum.

by the presence of a second "agent"?

     Comments relating to viruses of the CNS are not only
appropriate for this panel discussion but have added
significance for this symposium since the agents are
capable of cell transformation. Examples include visna
virus, PML (human SV40), SSPE, where original brain cultures showed morphologic and chromosomal abnormalities
characteristic of transformation, and in JC disease, where
in addition to "spontaneous transformation" of brain cell
cultures, the effect was transmissible by cell fusion to
normal target cells in the absence of a detectable agent.
These approaches to latent infection of the CNS, where
virus cannot always be visualized or isolated, should
have wider application in cancer research; the brain
tissue of cancer patients may provide an environment for
localization of suspected oncogenic viruses.

                    * * * * *

                    DISCUSSION

CASALS: In so far as the modification of viruses a
number of examples come to mind of arboviruses which,
when passed in tissue culture or in a species other than
that in which they maintain themselves in nature, undergo
very profound changes that affect virulence rather than
antigenicity. That is really the basis for the development of a number of vaccines. For example, the 17D
yellow fever strain, which started as a very highly
virulent strain with high affinity for liver of monkeys
and man, finally developed into a strain which had no
affinity for liver but can, when injected intracerebrally
produce encephalitis, but when injected subcutaneously
will not reach the brain or the liver. It therefore has
been modified by passage in tissue culture. Venezuelan
equine encephalitis virus, a highly pathogenic group A

arbovirus, has also been observed to lose its pathogenicity during passage in tissue culture. Yet when
injected peripherally it will still protect against
virulent wild-type virus. These are examples of viruses
which have been modified by passage in tissue culture.

Now on the other hand there are many viruses that
are not pathogenic for their natural host but which
produce serious disease when introduced into another
host, such as man. For example, Bolivian hemorrhagic
fever virus infects certain rodents (Calomys callosus)
without producing disease, but if this virus gets from
this rodent to man it produces a very serious illness.

ALLISON: Dr. Gajdusek, as I understand it in your
cultures from primate brains you constantly recover
viruses of all sorts. How do you think that relates to
what we've just heard about the isolation of certain
viruses fairly consistently in certain conditions from
the human?

GAJDUSEK: Long-maintained tissue cultures, both primary
explants and trypsinized cell layers, of chimpanzee
brain and lymph node tissue usually yield masked viruses
which emerge sometimes only after many months of in
vitro cultivation. Two chimpanzee foamy viruses, Pan 1
and Pan 2, are usually isolated and there is always
specific antibody to the viruses isolated in the
chimpanzee from whose tissue the viruses were obtained.
Six new chimpanzee adenoviruses are the next most
frequent isolates. At times a single explanted brain
culture has produced both chimp foamy viruses, one
earlier, the second later.

The fact that latent or masked viruses are now
known to be present in most normal-appearing chimpanzee
cell cultures, even brain, in the absence of any electron
microscopically recognizable virions makes the significance of activating virion production and demonstrating
noncellular nucleic acid by annealing procedures
difficult to assess. It has been impossible to link any
of these agents with kuru or C-J disease etiology. From
man only rare papovavirus particles, reovirus or herpes
simplex virus isolations have thus far confused the
picture.

BERG: How subtle an immunosuppression do you believe
would significantly increase susceptibility to viral
induced tumor formation? Could, for example, the mild
immunosuppression resulting from infection or widespread

use of cold remedies increase susceptibility to viral oncogenesis?

ALLISON: I think it depends upon the situation. In the kind of circumstances I described, you're dealing with a strong immune response against a relatively strong antigen and so strong measures are needed. But it is a fact that with murine sarcoma virus, for example, cortisone treatment in what you'd call therapeutic doses can change the thing from a regressing tumor to one which will progress and kill the animal. So under certain circumstances fairly weak immunosuppression can do this.

HELLMAN: Monroe finds that he can induce tumors in primates when he treates them with progesteron. Immuno-suppression has also been reported when taking oral contraceptives. Earlier, women in specific phases of their menstrual cycle were reported to be more susceptible to polio. Do you have any comment to this?

ALLISON: I don't know how well authenticated the last thing you mentioned is. It is a fact that female sex hormones do have a marked effect on immunity under certain circumstances. Not only progesteron treatment but also cortisone did increase virus-induced tumors in primates.

HUANG: I think that it is important to be aware of the potential for pseudotype formation with unrelated viruses. This has now been demonstrated by the formation of pseudotypes of both murine and avian tumor viruses. Such pseudotypes are likely to have a markedly altered host range.

LEWIS: In view of the observations by Field that tem-perature-sensitive mutants of reovirus can produce chronic degenerative CNS disease in rats, I wonder if we should consider the implications and potential risks inherent in the escalating development of temperature-sensitive mutants of SV40, VSV, influenza, and human adenoviruses.

Session V

**What Can Be Done
To Control Biohazards
in Cancer Research?**

EPIDEMIOLOGIC STUDIES AND SURVEILLANCE
OF HUMAN CANCERS AMONG PERSONNEL
OF VIRUS LABORATORIES

Philip Cole
Department of Epidemiology
Harvard School of Public Health
Boston, Massachusetts

During the sessions of this conference, the dilemma
with which you are faced has emerged with reasonable
clarity. On one hand, most of you believe that the
viruses with which you work can--or at least may--cause
human cancer or other chronic disease, and consequently,
that you should protect yourselves and your contacts
from them. On the other hand, most of you also share
the opinion that it is at least possible that these
viruses pose no threat. That being so, it would be
foolish to waste money, and especially time, protecting
yourselves against a threat which would never materialize.
One of the first items of business, therefore, is to
resolve this dilemma. In my opinion resolution will
come, if at all, only from formal, careful epidemiologic
studies. In no sense do I wish to denigrate the
suggestion that complete reporting and full assessment
be made of interesting cases. By all means, this should
be done. But as you do it, be aware of its limitations.
A major limitation is that case reporting permits only a
positive conclusion. The absence of a problem cannot be
documented by case reports. Of even greater consequence
is the fact that, because of the long latent period
characteristic of cancer, the problem, if it exists, may
readily be missed by collecting only case reports.

It is clear from prior discussions that we agree
that a convenient way to approach the question at hand
is to divide it into its retrospective and prospective

components. With this frame of reference, the epidemiologic studies which I shall describe are addressed to
the following two questions:

1.   Have laboratory-related virus exposures caused
     cancer in man?
2.   If such exposures have not yet caused cancer
     but will in the future, how can we recognize
     this at the earliest time?

In addition, if laboratory exposures have caused
or will cause human cancer, we want information on the
identity of the responsible agents, host characteristics
associated with susceptibility, and the temporal aspects
of the disease. Such related issues may also be
answerable by epidemiologic studies. However, I will
address here only the two major questions, which, for
brevity I will refer to as the "What has happened?"
and "What is happening?" questions.

THE PAST EXPERIENCE

To answer the question, "What has happened?" we
would, ideally, like to know the incidence rate of
cancers of each type for some definable population of
exposed laboratory workers. Alternatively, if that
population is large enough, it may suffice to know the
mortality rate due to each type of cancer. However,
whether incidence or mortality is measured, the critical
element is the provision of some basis for a valid
comparison. Observed cancer rates will be meaningful
only if we also know the rates to be expected in a
population comparable to ours, but not exposed.

The type of study which would provide the necessary
information is a "retrospective cohort study" (1).
Basically, the plan is to characterize and enumerate, as
of one or more points in time in the past, a group of
persons who had been (and may or may not have continued
to be) "exposed." Exposure may be specified in a number
of ways, but the more objective and refined the specifi-
cation, the better. This group or population would then
be followed up to the present time. This follow-up
would permit the group's cause-specific mortality rates
or, if possible, its site-specific cancer incidence rates
to be determined. These rates would then be compared
with those expected based on sources such as national
mortality data or incidence data from the Connecticut
Cancer Registry. Internal comparisons of heavily vs.
lightly exposed persons might also be possible.

A retrospective cohort study of this type is probably

a feasible approach to our problem. Feasibility, however, is a function of the size and characteristics of
the cohort, the point(s) in time at which it could be
assembled and, of course, its disease experience. For
example, if a cohort of 10,000 laboratory workers with
an average age of 45 years and a sex ratio of 1:1 could
be identified as of, say, January, 1955 and followed
through 1972, there would be 180,000 person-years of
observation available (10,000 persons x 18 years).
In such a population, about 20 new cases (or deaths) of
leukemia would be expected (2). If 30 are observed, we
could be reasonably sure that a problem existed (lower
95% confidence limit of 30 is 22) even though a
relatively small one that might be related to some
characteristic of the cohort other than exposure. This
type of retrospective cohort study warrants serious
consideration because: (1) it is not extremely expensive
--a study such as I outlined might be done for $50,000
assuming the full cooperation of laboratory directors
throughout the United States, and (2) it can be accomplished relatively rapidly, probably within two years.
There is no way to obtain a faster answer to your
question.

    A second alternative to the "What has happened?"
issue is a case-control study. Such a study uses
information from a number of persons with a specific type
of cancer and a series of comparable control persons.
Typically both the case and control series are more-or-
less representative of a general population. For each
subject, information regarding exposure is sought; in
this case frequency and degree of exposure to specified
oncogenic viruses by history and, if possible, one or
more objective measures of virus infection such as
titers of specific antibodies. A typical case-control
study is not feasible in this case because of the low
frequency of exposed laboratory workers in the general
population and because a whole series of such studies
would be required to cover the cancers of interest.
Nevertheless, we could consider modifying the case-
control technique to adapt it to our needs. We could
identify a series of laboratory workers who have
developed cancer and select other laboratory workers as
controls. Comparisons would then be made between the
two groups with respect to the specific types of
exposures sustained. It might be found, for example,
that technicians who develop breast cancer have had a
higher frequency of exposure to B-type RNA viruses
than have their normal female controls. Such a study
would offset the problem presented by the rarity of
exposure, but a series of such studies would still be
required to cover all the cancer sites of interest.
This study could probably be done rapidly and economically.

However, it presents a serious pitfall which may be
difficult to avoid, namely, the possibility that cases
might be included in the study if they were known to
have exposures of special interest.

PRESENT AND FUTURE EXPERIENCE

The questions of what is happening and what will
happen are both more relevant and more difficult to
answer than the question of what has happened.   Three
possible approaches can be considered: a prospective
cohort study, a series of case-control studies done in
a more-or-less continuous sequence and a system of
surveillance.

A prospective cohort study would be done by
assembling a group of healthy laboratory workers,
ascertaining their exposures and then following them
up, into the future, to ascertain their disease and
mortality experience.   During the follow-up period it
would be valuable to collect further information on how
exposure has changed, if at all.   At least part of the
exposure information might be obtained at the study's
beginning but not actually interpreted until some time
in the future.   What I am referring to is the collection
of blood specimens which are then kept frozen for antibody determinations in the future.   When a member of the
study group develops cancer his specimen and that of one
or more unaffected persons is retrieved and assayed.
This approach, ideally, can permit the best features of
follow-up and case-control studies to be merged.   However
several significant issues must be addressed.   These
are (1) how large a group should be studied, (2) what
exposures should be measured, (3) how long should the
group be followed, and (4) what diseases or causes of
death should be measured?   These questions can not be
answered without careful planning, but let me try to
give an idea of the answers.   I would estimate that a
cohort of 20,000 persons would be necessary.   This is
based on the idea that the cohort members have an
average age of 45 years and a sex ratio of 1:1 at the
beginning of the study and that we would want to know
within five years if the exposed members of the cohort
are experiencing a leukemia risk which is two or more
times greater than that experienced by unexposed members
or by the general population.   If diseases less common
than leukemia are of great interest, if we want an
answer sooner than five years or if we insist upon
detecting a risk increase less than twofold, the size of
the cohort must be increased.   By the same token, if
we are willing to wait longer, if our emphasis is on
more common cancers and, especially, if we consider it

sufficient to detect only large-risk increases, the cohort can be smaller. Each of these factors has to be assessed individually, but the risk increase to be detected is critical.

The case-control studies that should be considered in the context of the current problem are similar to the retrospective case-control studies described earlier. Their application differs in that, to provide data as current as possible, a sequence of such studies must be done. Another way of thinking of this is simply that one continuous case-control study would be initiated and the data from it analyzed at specified time intervals.

Finally, in trying to assess what is happening, we could consider setting up a surveillance system. This would mean establishing a central bureau or office to which would be reported all cancers or cancer deaths occurring within some population. Unfortunately, the population usually is either unspecified or specified only in general terms. Further, the criteria for filing a report may not be universally agreed-upon or applied. Both of these problems could be overcome with care. On the positive side, we acknowledge that a surveillance system probably could be set up almost immediately and could be operated on a minimum budget. Also on the positive side the concept of surveillance implies action. That is, after some period of time adequate to establish a baseline as to how many reports are to be expected in a given time, it would be decided that action would be taken if that baseline were exceeded by a specified amount. I am uncertain as to what that action would be. On the negative side, we have to realize how little we may gain from a surveillance system. Perhaps, if a very striking increase in risk, say tenfold or greater, is present in laboratory workers, such a system may permit its recognition. This in turn would necessitate a formal study of a type which I have described. This is one type of action which such a system could stimulate. Another type of action might be the universal adoption of the most effective methods available to protect workers from laboratory exposures. In addition to being insensitive to real increases in risk, a surveillance system could suggest an apparent increase in risk when in fact there is none. This could result because we would not have a reliable estimate of the number of cancers to be expected in our population in the absence of an increase in risk, or because of a lack of constancy of the reporting mechanism. At worst, such surveillance might be no more valuable than case report collecting. At best, it could merge imperceptibly into a prospective cohort study. Surveillance mechanisms have an important place in the control of acute diseases, but play very

little role in the control of the chronic diseases whose
frequency changes only slowly.

RELATED POINTS

This presentation would be incomplete if several
somewhat unrelated points were not made, even though
these do not fit directly into the major scheme presented.

The first point has been mentioned repeatedly. It
is that the reference population for our epidemiologic
studies must not be limited to laboratory workers. It
should include their close contacts, especially those
who, for one reason or another, may be susceptible to
virus oncogenesis even if the laboratory workers themselves are not. The most obvious example of such a
group is the unborn children carried by pregnant
laboratory workers.

A second point has hardly been mentioned. It is
that the epidemiologic studies which I have described
could, especially if positive, have implications reach-
ing far beyond our own health. Clearly, they would
comprise one of the strongest arguments available that
viruses can cause human cancer. If the studies are
negative, I believe they would be less clear in their
implication. In any event, I can think of no more
valuable epidemiologic studies which could be done now
to evaluate the possible association between human
cancer and horizontally transmitted oncogenic viruses.

RECOMMENDATION

I had hope that out of this Conference there would
come the formation of a body of responsible persons
committed to evaluating the many suggestions that have
been made. This group might then formalize and
disseminate those suggestions which it can recommend.
Irrespective of what epidemiologic studies, if any, are
recommended, I would urge that we begin promptly to do
the one thing which would enhance any future investigations, epidemiologic or virologic. That is, to establish
a population register to include all persons whose work
brings them into contact with known or suspected
oncogenic viruses (and chemicals) and, possibly, their
contacts as well. The information to be registered can
be decided after discussion but must include some
measures of exposure, health status and identifying
information adequate to allow the individuals to be
followed-up. Such a population register could serve as
the base from which a prospective cohort study could be

initiated. Later, if deemed desirable, the register could be extended back in time as a definitive, or as the initial effort in a definitive, retrospective cohort study. In any case, a feasibility study should be undertaken now of the retrospective cohort approach.

I wish to conclude by emphasizing that our uncertainties today stem from two epidemiologic aspects of cancer. These are first, that cancer of any given site is an uncommon occurrence, and second, that the latent period of these diseases, at least among adults, is probably measured in decades. The fact that you are all exposing yourselves to these viruses may eliminate the problem of the rarity of cancer--I hope not. But the only way to come to grips with the long latency is to get busy.

References

1. MacMahon, B. and T. F. Pugh. Epidemiology: principles and methods. Little, Brown & Company, Boston.
2. Doll, R., C. Muir and J. Waterhouse. 1970. Cancer incidence in five continents, vol. 2. Springer-Verlag, New York.

* * * * *

DISCUSSION

DIMMICK: As a result of this meeting, a great many laboratory people may change their methodology. If so, won't a future survey fail to show increased risk from the past?

COLE: That difficulty should not deter application of safety methods. And, if we learned of an increased risk in the past, it would be a strong argument that these agents can cause human cancer.

FRIEDMAN: Do you think that cancer should be a reportable disease?

COLE: It is reportable in Connecticut. It would be very expensive to do this on a national level and it would be fifteen years before the data would be worthwhile. Several more state or regional registries are needed.

ANIMAL FACILITIES

R. A. Griesemer and J. S. Manning
California Primate Research Center and
Departments of Veterinary Pathology and Veterinary Microbiology
University of California
Davis, California

Increased animal experimentation involving human and nonhuman primate cancer tissues and viruses has increased the need for animal facilities in which the risks to personnel and experimental animals can be reduced. The principles of environmental control and containment in the animal-holding facility are the same as those in the virology laboratory. The requirements of animal experiments, however, present additional problems for containment. Animals must be handled and examined frequently with blood and other biopsies collected as part of most experiments. The animals must be fed and exercised and their liquid and solid wastes disposed of safely. They must be cared for humanely with attention to their quality of life. At the same time, attempts must be made to standardize the experiments by controlling the external environment, the diet, and the microflora. While it is possible to confine animals in sterile air-tight chambers as one would tissue cultures, this is usually impractical. The goal, therefore, is to assess the risks in each animal experiment and to apply appropriate minimal containment methods.

The objectives of this report are (1) to promote awareness of the risks in the animal laboratory, (2) to assist in assessing risks, and (3) to suggest methods to reduce or eliminate risks.

HAZARDS ASSOCIATED WITH USE OF EXPERIMENTAL ANIMALS

LABORATORY ACCIDENTS.   Accidents are the major personnel
hazards encountered in animal laboratories.   Falls on
wet, slippery floors and cuts from cages that are
improperly designed or in poor repair are frequent
occurrences.   With the increasing use of steam to clean
animal cages within infectious disease laboratories
rather than in centralized facilities, and with the
tendency to use concentrated chemical disinfectants on
animal room equipment, thermal and chemical burns are
increasingly frequent.   Electric shocks are not uncommon
in animal laboratories because heat and moisture often
lead to rapid deterioration of the insulation on electric
cords.   Disposable paper gowns may pose a fire hazard
when open flames are employed in the animal laboratory.
Finally, the quality of life of individuals who work in
animal laboratories (as well as that of the animals) is
adversely affected by the noise levels which are often
excessive.

Most hazardous is accidental self-inoculation with
needle and syringe caused by carelessness, lack of
experience, or improper restraint of animals during
animal inoculations.   Wedum et al. (1) have called the
needle and syringe the most dangerous instrument in
research.

Animal bites and scratches are a daily occurrence.
When properly cared for, bites and scratches rarely result
in significant clinical illness.   Exceptions are the
penetrating bite wounds of cats and monkeys and the
tearing bite wounds of dogs.   Cat bites are important
causes of wound infections due to the nature of the
normal microbial flora of the cat's mouth.   When working
with nonhuman primates, it is recommended that hydrogen
peroxide be kept in the animal room for immediate cleans-
ing of animal bites until medical attention can be
obtained.   However, the effectiveness of hydrogen
peroxide in preventing wound infection has not yet been
established.

Hypersensitivity to animals and their products are
not uncommon.   When recognized, one can limit the time
of exposure, wear a face mask to filter out allergens,
take antihistamines, or undergo a series of desensitiz-
ation injections.   It is frequently necessary to change
the job of the affected individual so that he no longer
comes in contact with animal allergens.

INFECTIOUS DISEASES. Most animal colonies have
experienced devasting outbreaks of spontaneous disease
at one time or another. More often affected are
laboratories where rabbits, hamsters and nonhuman
primates are used rather than specific pathogen-free
rodents. Such disease outbreaks tend to be spectacular,
are easily recognized and can be dealt with readily.
Of greater importance are the latent infections of
animals which become activated by stresses or by such
procedures as experimental immunosuppression. For
example, mice infected with mouse hepatitis virus
ordinarily present no clinical signs, but when they are
stressed they may develop fatal hepatitis. Similarly,
hemobartonellosis in rats is ordinarily a clinically
insignificant, chronic infection, but when infected rats
are splenectomized or immunosuppressed, hemobartonella
organisms multiply in the erythrocytes producing a
precipitous, fatal anemia. Protozoa of the genus
Encephalitozoon persist for months in the brains and
muscle tissues of rodents and most rabbits. Activation
by a wide variety of stresses results in disseminated
fatal disease. Spontaneous encephalitozoonosis is a
major impediment to aging studies and may also be
expected to be a problem in long-term cancer studies.
The dangers to the animal colony from spread of
indigenous nononcogenic infective agents is far greater
than that from horizontal spread of oncogenic viruses.
Wedum and Kruse (2,3) have demonstrated cross-
contamination by 77 microorganisms.

Aside from the spontaneous spread of polyoma virus
in rodent colonies, Marek's disease virus and leukosis
viruses in chicken flocks, and feline leukemia virus in
cat colonies, there is little evidence for natural
horizontal transmission of oncogenic viruses. Rauscher
murine leukemia (4) and Yaba pox viruses (5) have been
transmitted experimentally by the aerosol route, and
Rauscher murine leukemia virus has been transmitted to
cage mates. Cats can be infected intranasally with
feline leukemia virus (6). Although horizontal spread
of leukemia viruses in animal colonies is not likely to
be significant in most experiments, the possibility
exists where open caging is used.

ZOONOSES. Spontaneous infectious diseases of experimental
animals may also pose a hazard to the investigator (7,8).
In Table 1 are listed some of the more important of the
nearly 200 zoonotic diseases. It should be noted that a
number of infectious diseases, including tumors induced
by Yaba virus (9,10), can be acquired from nonhuman
primates (11,12). In a recent survey of personnel in
primate centers, the conversion rate for tuberculosis in

Table 1. Important Zoonoses

Species	Disease	Frequency in U.S. Animal Colonies	Severity in Man
Chickens,	Ornithosis	++	+
Turkeys,	Newcastle disease	++	+
Japanese quail	Equine encephalomyelitis	+	+
	Salmonellosis	+	+
Mice, rats,	Lymphocytic choriomeningitis	+	+++
Hamsters,	Encephalomyocarditis	+	++
Guinea pigs	Leptospirosis	+	++
Rabbits	Tularemia	+	++
Opposum, skunk,	Rabies	++	+++
Fox	Leptospirosis	+	++
Cats	Toxoplasmosis	++	++
	Cat scratch disease	+	+.
	Ringworm	++	+
Dogs	Rabies	+	+++
	Leptospirosis	+	++
	Visceral larva migrans (Toxocara canis)	+++	+
Cattle, sheep,	Louping ill	+	++
Goats, pigs	Q-fever	+	+
	Anthrax	+	+
	Tuberculosis	+	++
	Brucellosis	+	++
	Listeriosis	+	+
	Contageous ecthyma	+	+
	Ringworm	++	+
	Vesicular stomatitis	+	+
	Cow pox	+	+
	Erysipelas	++	++
	Leptospirosis	+	+
Nonhuman	Tuberculosis	++	++
Primates	Hepatitis	+	++
	Marburg viral disease	±	+++
	Salmonellosis	++	+
	Shigellosis	+++	++
	Herpes simiae infection (B virus)	++	+++
	Malaria	+	++
	Yaba and Tanapox	+	+
	Measles	++	+
	Amebiasis	+	++
	SV40	++	±
	Rabies	±	+++

1971 was 320 per 10,000 compared with 3 per 10,000 for
the population in general (13). A tuberculosis infection
rate in macaques was recently reported as 17/1000 within
90 days after capture (14).

ASSESSING RISK OF INFECTION

The hazard to man from handling diseased animals
has never been accurately ascertained. The presumptive
evidence, however, is impressive. In one study (15)
21% of trained scientific personnel working with
infectious agents became infected in one year, while in
the same institution 12% of animal caretakers also
became infected. In a review Sulkin et al. (16) reported
that of 2262 infections acquired in the laboratory, 221
occurred in animal caretakers. Of 1303 bacterial
infections, 126 resulted from contact with animals and
animal ectoparasites, 28 from the bite of an animal or
ectoparasite, and 95 while performing autopsies. Thus
19% of the bacterial infections were acquired from
animal studies. Although the causes of most biological
accidents are unknown, it seems that animals or animal
tissues are involved in 30-40% of the infections acquired
in the laboratory.

To date there has been no demonstrated increase in
the risk of contracting cancer from experimental animals.
There is no firm evidence that veterinarians or pet
owners (17) have increased risk of cancer. It is known,
however, that tumor tissues and other tissues of many
diseased animals contain oncogenic viruses and that
oncogenic viruses are sometimes found in urine, feces,
milk, saliva and semen. It is also known that several
oncogenic viruses are capable of crossing species
barriers under experimental conditions, especially when
large doses of virus are used or the recipient is
immunosuppressed. Thus even though there is as yet no
evidence of transmission to man, the possibility of a
long latent period prompts us to be cautious. At present,
there appears to be a greater danger to man from nononcogenic indigenous agents in animal tissues and fluids
than from recognized oncogenic viruses.

PROCEDURES ASSOCIATED WITH INCREASED RISK. Close contact
with newly arrived animals, especially primates, during
the quarantine period should be avoided. About 2% of
newly arrived rhesus monkeys (Macaca mulatta) have oral
lesions and are shedding Herpesvirus simiae (18), a
virus associated with high mortality in man (7).
Performing an autopsy generates aerosols of large
droplets as the tissues are manipulated and sliced. The

autopsy procedure also introduces the possibility of penetrating wounds from sharp instruments, especially when the pathologist is relatively inexperienced or fatigued. Improperly labeled or unidentified infective animals, bedding, carcasses, or tissues are another risk. Animal wastes must also be identified and properly handled so that exposure of custodial or maintenance personnel is prevented.

Disposal of animal wastes and bedding presents a special problem. When bedding is being changed in the conventional animal room, the bacterial count in the room air often goes from a resting level of less than 5 colony forming units (cfu) per cubic foot to 50 or even 200 cfu/ft^3. The level stays high for 15-20 minutes after changing of bedding has been completed, and smaller particles remain in the air for hours. Where bedding is known to be contaminated with a hazardous agent, one should consider (a) using no bedding, if this does not interfere with the health of the animals, (b) removing bedding by means of a vacuum system to prevent aerosols, (c) autoclaving cages and bedding before cleaning, or (d) using disposable cages that can be sterilized without cleaning. Fecal material may contain more than one billion organisms per gram, and when mixed with bedding is difficult to sterilize by autoclaving. Thus it is essential that autoclaving procedures be carefully monitored. For moderate-risk agents, the use of flammable litter, which is collected by vacuum and sterilized by incineration, is recommended. All animal waste should either be incinerated or sterilized by autoclaving before disposal.

Disposal of contaminated tissues is also potentially hazardous. Autoclaving carcasses for 30 minutes does not insure sterility. The only safe way to dispose of infected animals is by incineration in a properly operating incinerator. Contaminated liquid wastes should be collected through a separate sewage system and treated in a steam sterilizing tank before release into the community sewage system.

PRECAUTIONS TO MINIMIZE RISKS

Wherever possible one should purchase experimental animals that are known to be free of contagious diseases. They should be quarantined and conditioned before use. After quarantine the animals should be housed in isolation facilities where they are less likely to become diseased. In Table 2 are listed some of the features of well-designed animal laboratories. A program should be instituted for continuous surveillance for contagious

Table 2.   Features of Well-Designed Animal Laboratory
           Facilities

---

1. Primary barrier at the cage level.
2. Secondary barriers in room layout and air flow.
3. Graded air pressures that are more negative as the
    risk increases.
4. Provision of clothes-change rooms and showers.
5. Ultraviolet or laminar-flow air locks.
6. Provision for steam or chemical sterilization of
    effluents.
7. An intercommunication system.
8. Service corridors.
9. Adequate storage space.
10. Separate traffic flow for animals, people, supplies.
11. Space and equipment for washing and sanitizing.
12. Quarantine facilities.
13. Diagnostic laboratory support.
14. Accessibility to users.
15. Refrigerated feed storage.
16. Incinerator.

---

animal diseases. This will require the combined efforts
of laboratory animal veterinarians, microbiologists, and
pathologists. It is important to realize that animals
infected with agents hazardous to man usually serve as
sources of low level, discontinuous contamination rather
than as the source of a burst of high level contamination
(such as may be produced by accidents occurring in the
virology laboratory). In both the animal laboratory and
the virology laboratory, however, the major emphasis in
containment should be at the source of contamination in
the form of a primary barrier. The better the primary
barrier, the less important the need for secondary
barriers in the design and operation of the animal
building. Before starting an experiment in a new
facility, one may wish to apply a biologic test for
containment in which susceptible animals are used as
sentinels of disease spread. A rigorous biologic test
is the use of Newcastle disease virus and chicks (3).

    Animals and people can be separated when necessary
by the application of containment equipment such as
isolation cages, safety cabinets and personnel suits.
Visitors should be excluded from the animal laboratory
as they are from the virology laboratory. Finally and
most important, a continuous educational program must be
instituted to remind personnel of the principles of
safety in the animal laboratory and the proper use of
containment equipment and methods.

CAGE DESIGN FOR BIOLOGIC SAFETY. When working with
highly hazardous infectious agents, the agents and the
animals can be confined in gastight safety cabinets.
When properly functioning such units provide complete
containment in which the entire experiment can be
conducted. The unit can then be sterilized at the end
of the experiment. However, modular safety cabinets are
expensive and time consuming to use. It is difficult
to restrain animals for examination or biopsy when
working through fixed glove ports and difficult to work
with large animals that require manual restraint.

Animals can also be confined in germ-free isolators.
The rigid plastic isolator can be operated under negative
pressure and provides good visibility. Units of this
type, however, are made in two pieces with a gasketed
assembly that is hard to seal and eventually leaks.
They are moderately expensive (in the $1000 to $1500
range) and as in the gastight safety cabinets one has
only a limited reach through fixed glove ports. Flexible
film germ-free isolators are relatively inexpensive, easy
to use and provide good containment. However, unless
provided with a support frame, they must be operated
under positive pressure, with the attendent risk of
disseminating the infectious agent if there are any leaks.

Ventilated cages that are solid-sided with a manifold, filtered air supply are available commercially.
They can be operated under positive or negative pressure
and the entire unit can be autoclaved between experiments.
The major problem with the use of ventilated cages is
that the barrier cannot be maintained when the door is
opened to examine the animal or to remove the bedding.
Ventilated cages work well for small animals where the
entire cage unit can be moved to an ultraviolet or
laminar flow hood before opening the cage and conducting
the work. However, not all commercially available
ventilated cages are gastight, especially around the
doors; thus when working with high-risk agents, one must
test the cage and vacuum manifold to eliminate leaks.

Cages equipped with filter tops (Fig. 1) can be used
in the same way as ventilated cages. They have the
advantages that they are inexpensive, commercially
available and the entire unit can be autoclaved. As
with ventilated cages, the barrier cannot be maintained
when the cage is open and it is necessary to use a hood
to protect the operator and other animals. In filter-top
cages pathogen-free mice housed in the same room with
infected mice have remained free of viral disease (1).

Another means of personnel protection is to enclose
personnel in airtight suits and house the animals in open

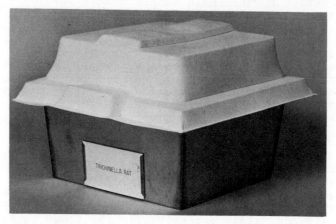

FIGURE 1. One of several commercially available filter-
topped cages that are widely used where
minimal containment is required.

cages. Airtight rubber or plastic suits are operated
under positive pressure and give good personnel protection. The individual is supplied with filtered air from
a back-pack or a manifold of conditioned air which
provides a stream of downward air over the face.
Personnel suits are uncomfortable when worn for long
periods even with conditioned air. The suit may tear
and destroy the barrier. Provision must be made for a
disinfectant shower of peracetic acid, formaldehyde or
detergent to decontaminate the outside of the suit when
entering and leaving the animal area. A modification of
the principle of isolating personnel is the use of a
plastic head hood with a sterile air supply to avoid
exposure to aerosols.

Another principle of containment that awaits
biological and engineering evaluation is the use of
laminar air flow applied to the cage openings. Horizontal
laminar-flow benches with air flow directed outward are
suitable for housing animals to maintain them free of
disease (19) but are hazardous to personnel. Experiments
at Fort Detrick (1) indicate that a curtain of ultraviolet
light or inward flowing air at the entrance to the cage
may provide a suitable barrier to prevent spread of
disease. One would presume that a vertical laminar airflow room would provide suitable protection for both
animals and personnel when a single row of cages is used.
We can anticipate the rapid development of new techniques
for isolating animals by controlled, directed air flow
in the next several years. Prototype caging and rooms

should not be used until tested with animals and infectious agents.

It should be apparent that no single animal cage and room design can be recommended for all virus research. Experiments with small laboratory animals can usually be conducted in filter-top cages. Nonhuman primates can be housed in ventilated cages. If desired, personnel can be further protected with inexpensive ventilated head hoods. While these relatively simple precautions may be suitable for oncogenic virus experiments, it should be emphasized that they may not be suitable to protect animals or personnel from indigenous agents in the animal colony. It is essential, therefore, that each investigative unit establish a health care program, for the animals as well as for personnel, with the goal of providing continuous surveillance for spontaneous diseases.

GUIDELINES FOR HUMANE TREATMENT OF ANIMALS

Users of experimental animals must comply with the legal requirements in the Animal Welfare Act (Public Law 89-544 and Public Law 91-579). The criteria for proper care and humane treatment of animals in DHEW publication 73-23 (Guide for the Care and Use of Laboratory Animals) should be followed. Those investigators supported by grants and contracts from the National Institutes of Health must also comply with the policy statement in NIH Guide for Grants and Contracts No. 7, Care and Treatment of Laboratory Animals, June 14, 1971. Biomedical research institutions are required to provide assurance of accreditation or establish a committee to assure NIH that they will evaluate their animal facilities with regard to the care, use, and treatment of experimental animals. Adherence to the guidelines will not only provide appropriate care and humane treatment of animals but contribute to the obtaining of high quality, disease-free, standardized, experimental animals.

References

1. Wedum, A. G. et al. 1972. Handling of infectious agents. J. Amer. Vet. Med. Ass. 161:1557.
2. Wedum, A. G. and R. H. Kruse. 1969. Assessment of risk of human infection in the microbiological laboratory. Misc. Publ. 30, Dept. of the Army, Ft. Detrick, Maryland.
3. Kruse, R. H. and A. G. Wedum. 1970. Cross infection with eighteen pathogens among caged laboratory animals. Lab. Animal Care 20:541.

4. McKissick, G. E. et al. 1970. Aerosol transmission of Rauscher murine leukemia virus. J. Nat. Cancer Inst. 45:625.

5. Wolfe, L. G. et al. 1968. Experimental aerosol transmission of Yaba virus in monkeys. J. Nat. Cancer Inst. 41:1175.

6. Hoover, E. A. et al. 1972. Intranasal transmission of feline leukemia. J. Nat. Cancer Inst. 48:973.

7. Hull, T. G. 1963. Diseases transmitted from animals to man, 5th ed. Charles C. Thomas, Springfield, Illinois.

8. Graham-Jones, O. 1968. Some diseases of animals communicable to man in Britain. Pergamon Press, London.

9. Ambrus, J. L. et al. 1963. A virus-induced tumor in primates. Nat. Cancer Inst. Monogr. 10:447.

10. Grace, J. T. and E. A. Mirand. 1963. Human susceptibility to a simian tumor virus. Ann. N.Y. Acad. Sci. 108:1123.

11. Fiennes, R. 1967. Zoonoses of primates. Cornell Univ. Press, Ithaca, N. Y.

12. Perkins, F. T. and P. N. O'Donoghue, ed. 1969. Hazards of handling simians. Laboratory Animals, Ltd., London.

13. Primate Zoonoses Surveillance Report, No. 8. 1971. Center for Disease Control, Atlanta, Georgia.

14. Stunkard, J. A. et al. 1971. A review and evaluation of tuberculin testing procedures used for Macaca species. Amer. J. Vet. Res. 32:1873.

15. Wedum, A. G. 1964. Laboratory safety in research with infectious aerosols. Public Health Rep. 79:619.

16. Sulkin, S. E. et al. 1963. Laboratory infections and accidents. Diagnostic procedures and reagents, 4th ed., Chap. 3, pp. 89-104. Amer. Public Health Ass., N. Y.

17. Schneider, R. 1972. Human cancer in households containing cats with malignant lymphoma. Int. J. Cancer 10:338.

18. Keeble, S. A. et al. 1958. Natural virus B infection in rhesus monkeys. J. Path. Bact. 76:189.

19. Beall, J. R. et al. 1971. A laminar flow system for animal maintenance. Lab. Animal Sci. 21:206.

FACILITIES AND EQUIPMENT AVAILABLE
FOR VIRUS CONTAINMENT

W. Emmett Barkley
Office of Biohazards and Environmental Control
Viral Oncology, Division of Cancer Cause and Prevention
National Cancer Institute
Bethesda, Maryland

Biological safety and environmental control programs in tumor virus research laboratories are being implemented to reduce worker exposure to tumor viruses and to protect the integrity of research experiments from contamination. The risk of accidentally acquiring cancer from exposure to tumor viruses is not known; however, the potential for exposure to tumor viruses is an ever-present possibility. Many laboratory procedures produce respirable aerosols. The potential for accidental injection or ingestion cannot be totally avoided. Accidental contamination or cross-contamination of research materials, on the other hand, is a recognized problem in tumor virus research and can adversely affect the validity of experimental results. At the present time, the elimination of this hazard to experiments is the principal objective of biological safety and environmental control programs (1). Fortunately, success in accomplishing this objective will also contribute to reducing the tumor virus exposure potential for the laboratory worker.

In descending order of importance, the success of a biological safety and environmental control program is dependent on (1) the safety awareness and techniques of the investigator, (2) the availability and correct use of safety equipment, and (3) the design and operation of the tumor virus research facility. The scope of this article will be limited to the role of facilities and equipment in controlling laboratory exposures and contamination of research experiments.

327

Before this subject matter is treated in detail, it is important to emphasize that facilities and safety equipment alone cannot create a research environment that is safe and free from the potential for cross-contamination. Safety will always be dependent on the ability of the investigator to understand and use effectively facilities and safety equipment which have been well designed. The single most important factor is a genuine concern for developing and implementing good laboratory habits. No facilities and equipment, however well designed, take the place of the investigator's responsibilities for safety.

THE RESEARCH FACILITY

The primary function of a research facility is to provide a physical environment in which scientific investigation can be undertaken efficiently and safely. Tumor virus research facilities can be classified as either conventional virus facilities or containment virus facilities. The conventional virus facility can accommodate all research activities with tumor viruses not known to cause cancer in man or that do not create a serious potential hazard to the laboratory worker. The containment virus facility is capable of preventing the escape of hazardous biological materials from the laboratory environment. Tumor viruses which have a known potential for causing cancer in man should be confined to a containment facility. Activities which may create a serious potential hazard with nonhuman tumor viruses, such as large-scale virus production and aerosol transmission studies, require the use of containment facilities. Both types of facilities should minimize the potential for cross-contamination.

FEATURES OF CONVENTIONAL VIRUS FACILITIES. Few investigators have the opportunity to occupy laboratories designed specifically to support their research programs. More often, existing space must be modified or used without alteration. The most important considerations in a conventional virus laboratory include (1) the location and arrangement of space for all functions required to support the research objectives, (2) the physical characteristics of the individual laboratories, and (3) the mechanical systems which maintain the quality of the environment and support the research activities.

Adequate space for storage, office activities, glassware and media preparation, janitorial and mechanical equipment support is necessary in order to achieve optimum safety and environmental control. Consolidating

these functions into an individual laboratory space can
increase the potential for contamination and can create
overcrowding, which may then increase the incidence of
accidents. Common laboratory space also contributes to
the potential for cross-contamination.

In a recent survey of 16 tumor virus laboratories
the lack of sufficient space was found to be a common
problem. Therefore it is important that the available
space be considered in planning research programs. The
use of corridors as research space should be prohibited.
This practice can cause widespread dissemination of
hazardous materials.

The location and arrangement of different research
functions within the available space of a facility is
important. The arrangement of space should promote
efficient operation of the research program and reflect
a carefully thought-out routing of personnel, experimental
animals, and clean and contaminated materials. The
research functions and direct research support activities,
such as glassware and media preparation and waste
disposal, should not be easily entered from public access
or corridor areas. Controlled access to the research
areas can reduce the potential for introducing extraneous
contaminants into the research environment and can
eliminate the exposure of non-laboratory personnel to
laboratory accidents. Animal research should be
physically separated from other laboratory facilities.

Laboratories should be designed to promote cleanli-
ness and a pleasant atmosphere. All interior surfaces
should be monolithic and nonporous to facilitate
cleaning and decontamination. Exposed pipes and ducts
and other horizontal surfaces which are difficult to
clean should be kept to a minimum. Care should be
taken to seal the points at which utilities and ducts
penetrate the walls, floors and ceilings, thereby
controlling entry of vermin and migration of airborne
contamination.

Mechanical systems are an integral part of the
laboratory. They provide utilities and ventilation.
The ventilation requirements are of utmost importance.
In addition to providing temperature and humidity
control, the air-conditioning and ventilation system
must prevent the dissemination of accidentally
aerosolized contaminants. This is accomplished by
balancing the air supply and exhaust so that the flow
is in the direction of increasing potential laboratory
hazard. Air supplied to the corridors should, therefore,
always flow into the individual laboratories and this
is accomplished by a slight excess of exhaust over supply

for each laboratory (2). In this way aerosols will be
contained within each laboratory until removed through
laboratory exhaust air. Improper ventilation in
laboratories has resulted in widespread contamination
from localized accidents (3-6).

The removal of accidentally aerosolized contaminants
from the laboratory space is dependent on the relative
air volume changes within the laboratory space. Adequate
dilution can be achieved with 10-15 air volume changes
per hour. This rate of air exchange will also provide
the capability for exhausting air from ventilated
biological safety cabinets.

The ventilation system should also provide for
filtration of air entering the laboratory to reduce
incoming contaminants. Supply air should pass through
conventional 85-95% efficient filters. Filtration of
exhaust air from conventional virus facilities is not
required as long as the final facility exhaust location
is at least 100 feet away from the nearest supply air
inlet.

Adequate utilities are required to support contamination control equipment. Special design
considerations are required for autoclaves, dry-air
sterilizers, glassware washers and animal cage washers.

All central vacuum services should include pipeline
absolute filters placed between service cocks and pump
receiver. The filters should be installed with piping
arrangements which allow for decontaminating used
filters and checking new filters. Filters should also
be installed between different functional areas serviced
by the same vacuum pump.

CONTAINMENT VIRUS FACILITIES. The technology has been
developed to design and construct laboratories to
prevent the escape of hazardous microbiological materials.
Contamination control is provided by barriers which
contain, remove or inactivate biologicals that may
accidentally contaminate the research environment. These
barriers include (1) physical barriers, such as floors,
walls and ceilings, (2) ultraviolet air locks and
personnel clothes change and shower areas, (3) directional
airflow so that air flows from areas of lowest hazard
toward areas of highest hazard, (4) exhaust air
filtration, (5) sterilization systems for the treatment
of contaminated liquid wastes and (6) double-door pass-
through autoclaves for the sterilization of laboratory
refuse. All of these barriers isolate the laboratory
from the non-research areas. This physical isolation is

the main distinction between containment facilities and
conventional facilities (7-10).

There is a serious misconception that containment
facilities provide absolute protection for the laboratory
worker. The purpose of these barriers is to prevent
hazardous biological materials from escaping the
containment facility. Reported laboratory-acquired
infections among workers who occupy containment facilities attest to the fact that these facilities provide
only limited protection for the investigators (11).

Containment facilities have limited, but important,
application in biomedical research. The Center for
Disease Control requires these facilities for the containment of microbiological agents classified as
"extremely hazardous to laboratory personnel or may cause
serious epidemic disease" (12). Among viral agents
included in this classification are Herpes simiae,
Lassa virus, Marburg virus and tick-borne encephalitis
viruses.

While containment facilities are recommended for
large-volume production of tumor viruses, most other
tumor virus research can be performed within conventional
facilities. The National Cancer Institute recognizes
that in the event that a virus of known potential for
inducing cancer in man is isolated, containment facilities
would be required.

Therefore the National Cancer Institute has
constructed a virus containment facility at NIH in
Bethesda, Maryland (Building 41, Fig. 1). There are six
major functional zones within this building: (1) the
contained laboratory and animal zones, (2) the laboratory
support zone, (3) the administrative zone, (4) the utility
distribution and mechanical penthouse, (5) a liquid
waste sterilization plant and (6) a biohazard research
and development area.

Four identical virus laboratory suites are surrounded
by a network of clean corridors which connect the suites
with the laboratory support zone. This special arrangement increases operation efficiency.

The usable laboratory space of the NCI virus containment facility is approximately 20% of the total area.
The utility distribution, mechanical penthouse and
laboratory support zones represent more than three times
the laboratory area. For conventional virus facilities,
the usable laboratory space may approach 50% of the total
area. Containment is achieved at a great sacrifice in
usable laboratory space.

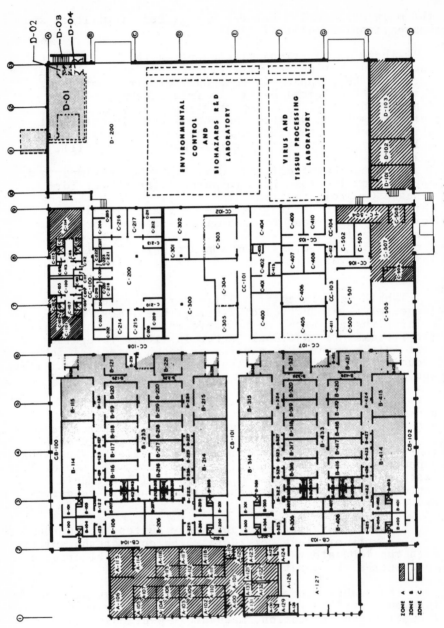

FIGURE 1. NCI virus containment facility.

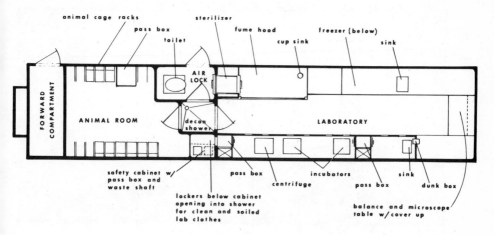

FIGURE 2.   Mobile virus containment laboratory.

In the event that containment is required, a practical solution is to isolate only hazardous activities and not conventional or less hazardous activities.   NCI has developed three approaches for providing such containment space.   A mobile laboratory was constructed to provide containment to any investigator anywhere. This facility is totally self-sufficient and provides the equipment and biological barriers for complete isolation of research materials (Fig. 2).   This mobile laboratory was used to isolate the Marburg virus at the National Center for Disease Control (13).

The second approach is the construction of pre-fabricated containment laboratories.   They are constructed rapidly at a construction contractor's plant to reduce engineering and construction time and to obtain more reliable labor capability.   The prefabricated facilities (Fig. 3) are easily transported and relocated.

The third approach, which is currently under development, may offer the most practical method for creating containment space in existing conventional laboratories.   An "off-the-shelf" modular approach allows assembly within a variety of institutional settings, while off-site component fabrication allows close quality control.   Completed facilities may therefore be either expanded or reduced in area without extensive cost.   Figure 4 illustrates this approach.

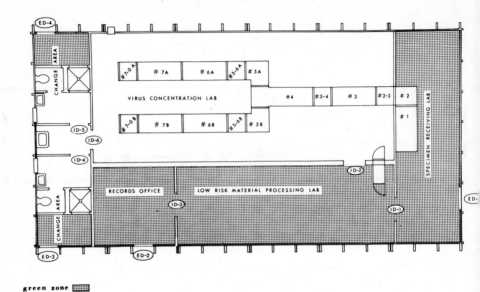

green zone ▨
blue zone ☐

FIGURE 3. Prefabricated virus concentration laboratory.

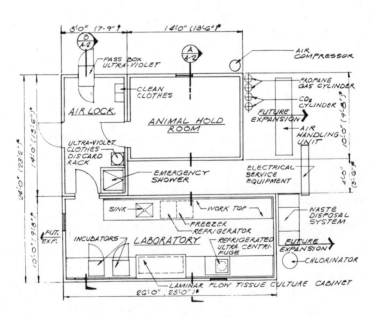

FIGURE 4. Modular virus containment laboratory.

SAFETY EQUIPMENT

Primary containment devices protect laboratory
workers from exposure to aerosols produced by routine
procedures. These devices control contamination at its
source. The importance of primary containment is
emphasized by the fact that the majority of reported
laboratory-acquired infections are those which occur
among personnel who directly handle infectious materials
(14,11).

The microbiological safety cabinet is the most
useful primary containment device (15,2). It can be
designed to provide absolute or partial containment.
Absolute containment cabinets are capable of preventing
both aerosol and contact exposures. Partial containment
cabinets are designed to reduce exposures to accidentally
created aerosols. The selection of cabinet type is
dependent on the relative hazard of the microbiological
agent and the potential of the activity to create
infectious aerosols.

Absolute containment cabinets (Class III). The Class III
cabinet is used to contain highly infectious microbiological agents. Most current cancer research would not
appear to warrent this degree of containment. This
enclosure is gas-tight and maintained under negative
air pressure. Work is performed through rubber gloves.
The cabinet air supply is exhausted through HEPA filters
and may be incinerated. Separate exhaust duct systems
are required. Such cabinets have been designed to
enclose incubators, refrigerators, freezers and centrifuges. Figure 5 illustrates a Class III cabinet system.

Partial containment cabinets (Class I). The most commonly
used safety cabinet in microbiological research is a
local exhaust ventilation hood. Personnel safety is
provided by the flow of room air into the cabinet. This
airflow can prevent the escape of aerosols. The Class I
cabinet has a limited front opening. The air velocity
through the front opening should be between 75-100 feet
per minute (fpm). The exhaust air should be HEPA
filtered or incinerated and then exhausted from the
laboratory.

Class I safety cabinets should be limited to
routine procedures employing agents of low infectivity
or risk (2). A drawback of Class I cabinets is that
interruption in ventilation can lead to direct exposure
of the user.

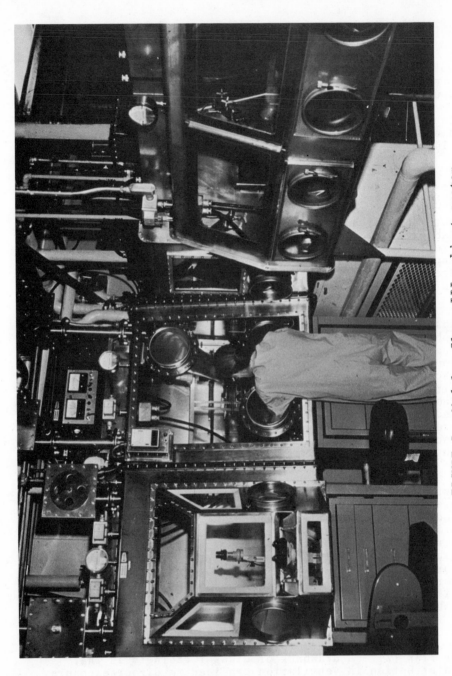

FIGURE 5. Modular Class III cabinet system.

Because of the inward flow of room air, Class I safety cabinets are not suitable for experimental systems which are highly susceptible to airborne contamination (15).

Laminar flow biological safety cabinets. This is a partial containment cabinet (Fig. 6) developed to protect the worker and the experiment from airborne contamination (16,17). Personnel protection is again dependent on the inward flow of room air and collection efficiency of the exhaust air filter. Unlike Class I cabinets, the inward airflow does not cross the work surface; this air is drawn into a front lateral suction grille which is contiguous with the forward leading edge of the work surface. To protect the experiment from airborne contamination, the work surface is bathed by a downward flow of filtered air. The appropriate air volume is recirculated through supply HEPA filters.

An optimum air balance between the inward airflow and the filtered cabinet supply air is necessary to achieve both personnel safety and experiment protection. The average velocity of the filtered cabinet supply air

FIGURE 6.   Laminar flow biological safety cabinet.

should be 80 fpm. The inward airflow average velocity
should not be less than 75 fpm. A fixed front access
opening height of 8-12 inches is required to maintain
the air balance. An increase in the access opening
will cause a reduction in the average inward airflow
velocity and thereby compromise personnel safety.
Therefore, currently designed vertical laminar flow
cabinets with sliding windows are not recommended for
containment of microbiological procedures. However a
prototype unit which is now under construction gives
great promise of providing personnel protection with the
convenience of an adjustable access opening.

Personnel protection is also dependent on obtaining
the rated HEPA filter efficiency. HEPA filters are
capable of collecting 99.97% of airborne particles in
the respirable particle size range; their efficiency is
essentially 100% for airborne particles above and
below this size range (18).

For maximum personnel safety the exhaust air from
the laminar flow biological safety cabinets should be
ducted out of the facility.

Present laminar flow biological safety cabinets
have limitations which must be recognized by the
cabinet user. The recirculation of approximately 75%
of the cabinet supply air precludes the use of the
cabinet for containment of flammable solvents. In
those installations where the cabinet exhaust air is
returned to the room environment, toxic, radioactive and
highly odorous materials cannot be safely contained.
These deficiencies limit the use of the cabinet to the
containment of particulate contaminants.

The difficulty of monitoring the performance of
the laminar flow biological safety cabinet is a serious
deficiency of present units. Experience has shown that
the majority of cabinets that are field tested fail to
meet the air balance and filter efficiency requirements
and must be corrected to provide adequate safety (Table
1). Even after the cabinets are certified, the user
has no convenient way of knowing whether or not adequate
performance is being maintained.

Conventional laminar flow hoods. Many laminar flow
cabinets now in use provide only product protection.
Examples are conventional horizontal and vertical clean
benches (19). These devices are often used by microbiological investigators to protect tissue cultures from
airborne contamination. The user must recognize that
this type of laiminar flow cabinet affords no personnel

Table 1.   LAMINAR FLOW BIOLOGICAL SAFETY CABINETS
FIELD TESTING AND CERTIFICATION DATA

Manufacturer	Number Tested	Units with Serious Filter Leaks	Units with Unacceptable Velocity Profile	Calculated Inward Airflow Velocity fpm
A	29	18	11	20-40
B	22	10	18	7-40
C	12	8	11	17
D	10	7	7	20-40
E	5	1	0	37
F	8	0	8	-
G	6	3	6	-
H	2	2	0	30
TOTALS	94	49	61	

protection.   The attractiveness of a contamination-free
environment for tissue culture and the minimal costs of
such units mask the potential hazard of these cabinets.

EQUIPMENT USE.   Containment equipment can only benefit
the laboratory worker if it is (1) properly designed
for the intended function, (2) properly installed, (3)
certified to be operating correctly and (4) used in an
efficient and safe manner.   Installation and certification should be performed by qualified personnel.
Operating protocols should be prepared by the equipment
user and diligently observed.   Most important, the user
must know and understand the capabilities and limitations
of the containment equipment he is using.

In conclusion, it is necessary to recognize that the
safe handling of potentially hazardous viruses depends
on the awareness of the laboratory worker and his
meticulous attention to safe laboratory practices.
Facilities and containment equipment can do much to
protect the experiment and the laboratory worker;
nevertheless the attitude of the laboratory worker is
the most important element in our precautionary program.

References

1. Wedum, A. G., W. E. Barkley and A. Hellman. 1972. Handling of infectious agents. J. Amer. Vet. Med. Ass. 161:1557.
2. Chatigny, M. A. 1961. Protection against infection in the microbiological laboratory: Devices and procedures. Adv. Appl. Microbiol., vol. 3, p. 131. Academic Press, New York.
3. Huddleson, I. F. and M. Munger. 1940. A study of an epidemic of brucellosis due to Brucella melitensis. Amer. J. Public Health 30:944.
4. Huebner, R. J. 1947. Report of an outbreak of Q fever at the National Institutes of Health. Amer. J. Public Health 37:431.
5. Phillips, G. B. 1965. Causal factors in micro- biological laboratory accidents and infections. Misc. Publ. 2, p. 144, U. S. Army Biological Laboratories, Fort Detrick.
6. Sullivan, J. F. and J. R. Songer. 1966. Role of differential air pressure zones on the control of aerosols in a large animal isolation facility. Appl. Microbiol. 14:674.
7. Wedum, A. G. and G. B. Phillips. 1964. Criteria for design of a microbiological research laboratory. J. Amer. Soc. Heat, Refrig., Air Cond. 6:46.
8. Phillips, G. B. and R. S. Runkle. 1967. Laboratory design for microbiological safety. Appl. Microbiol. 15:378.
9. Darlow, M. 1969. Safety in the microbiological laboratory. Methods Microbiol., vol. 1, p. 169. Academic Press, London.
10. Runkle, R. S. and G. B. Phillips. 1969. Micro- biological contamination control facilities. Van Nostrand-Reinhold, New York.
11. Wedum, A. G. 1964. Laboratory safety in research with infectious aerosols. Public Health Dept. 79:619.
12. Classification of etiological agents on the basis of hazard. National Center for Disease Control, Atlanta, Georgia (1972).
13. The green monkey disease. PHS World 3:11 (1968).
14. Sulkin, S. E. and R. M. Pike. 1951. Survey of laboratory-acquired infections. Amer. J. Public Health 41:769.
15. Wedum, A. G. 1953. Bacteriological safety. Amer. J. Public Health 43:1428.
16. McDade, J. J., F. L. Sabel, R. L. Akers and R. J. Walker. 1968. Microbiological studies on the performance of a laminar airflow biological cabinet. Appl. Microbiol. 16:1086.
17. Coriell, L. L. and G. J. McGarrity. 1968. Biohazard hood to prevent infection during microbiological procedures. Appl. Microbiol. 16:1895.

18. Harstad, J. B. and M. E. Filler. 1969. Evaluation
    of air filters with submicron viral aerosols and
    bacterial aerosols. Amer. Ind. Hyg. Ass. J. 30:280.
19. Brewer, J. H. and G. B. Phillips. 1971. Environ-
    mental control in the pharmaceutical and biological
    industries. Critical Rev. Environ. Control 1:467.

* * * * *

DISCUSSION

OXMAN: How much difference does active use make in the
outfall from these vertical laminar flow biohazard hoods?

BARKLEY: The increase in outfall of contaminants with
activity has not been accurately quantitated. Performance studies of fume hoods and ventilated cabinets have
shown an increase in outfall with activity from 2- to
100-fold. Safety cabinets must be designed to compensate
for activity conditions. In our studies we assumed a
10-fold increase in outfall with activity. We found that
the face velocities of laminar flow biological safety
cabinets must be 75 feet per minute in order to contain
this outfall potential.

Convection currents caused by a laboratory burner
within the cabinet can disrupt the unidirectional airflow
of the cabinet. This airflow disruption will not reduce
containment as long as the burner is located at the back
of the cabinet work space.

MILLER: I wonder about the concept of informed consent
among people employed to work in virus laboratories. Are
they advised that the risk may be different from work
elsewhere? Also, do employers advise women workers to
notify them when they are pregnant? Dr. Hellman, you
mentioned that you keep a log of new agents as they are
introduced into the laboratory--but what about agents
that are introduced into the workers as drugs? I
realize that it would be difficult to acquire and record
this information. The questions suggest the possibility
that if a formal registry of workers is developed, it
might contain identifying information on the worker,
his progeny, and past medical history, looking for
diseases for which drugs might be given on a continuing
basis. Perhaps new employees could be asked about
medication which they take frequently, because the
responses might draw attention to diseases, such as
severe asthma, which might be grounds for excluding
someone from a position in a virus laboratory.

APOSHIAN: There is one simple recommendation I would like to see this group make and that is that each institution establish a biohazard committee.

YATVIN: Perhaps a unique set of circumstances is required to provide the impetus for a university to establish a biohazard committee. Last year we had an outbreak of Q fever in people working with sheep in the medical school complex. In addition a number of research groups on our campus were studying mink encephalopathy by adapting it to grow in various species. Concern was expressed that growing the material in primates might result in natural species barriers being overcome and thus present a biohazard. As a result of these two events and a number of less dramatic ones over the years, our University responded to the problem of biohazards and established a biological safety committee.

ROIZMAN: Local biohazard committees can be a blessing to some institutions and a disaster for others. As an example of this, some 13 years ago a local biohazard committee decided that polio was much too dangerous and made an edict which stopped me from working with polio from then on. So some institutions may have a good, sensible committee, another may not. Unless there is a national guideline it's not going to be very useful. Another comment I want to make is that there is a clear period between the time of initial exposure to a virus and the time the tumor will develop. For example in the studies which have been done on cervical cancers it is clear that cervical cancer appears some 10 years or so after first infection with virus--if indeed the virus has anything to do with it--and 10 years is of course a very long interval. From that point of view retrospective studies of the kind that have been suggested may not be very meaningful because you will really have to go back a very long time and it is not clear that the conditions that were being worked with 10 or 15 years ago are the same as today. Anyone working in viral biochemistry now has to produce tremendous amounts of materials for doing anything of meaningful nature and of course the prospect of getting an infection that eventually leads to cancer may be dose-related. Doses that we are being exposed to today are very much higher. Therefore any decision based on retrospective studies alone may not be terribly meaningful 10 or 15 years from now.

BARKLEY: I think we ought to stress one thing: Safety is not dependent alone on equipment and facilities but is primarily dependent on the attitude of the scientist. It is our responsibility to reduce laboratory exposures, to train our staff in relevant safety methods, and to assess the risk of potentially hazardous research materials.

PANEL V

COMMON SENSE IN THE LABORATORY: RECOMMENDATIONS AND PRIORITIES

Alfred Hellman
National Cancer Institute

Michael N. Oxman
Harvard Medical School

Robert Pollack
Cold Spring Harbor Laboratory

COMMENTS BY

Alfred Hellman

The Special Virus Cancer program of the National Cancer Institute has established certain minimal standards that can be strengthened by individual laboratories, if necessary. We consider them to be appropriate and not restrictive for the normal operation of an efficient biological laboratory. (These standards are included in the Appendix.)

Along with these standards our institute offers consultation in regards to biological and engineering problems that arise in our contract laboratories. When time permits we are pleased to extend these services to any requesting member of the scientific community.

* * * * *

COMMENTS BY

Robert Pollack

I hope this meeting has accumulated enough facts to get a few people concerned and to get a few concerned people to work more safely. I would like to address my comments to the agencies that support our research.

First, almost every recommendation we have heard will cost money to implement. Any agency, private or governmental, that supports our research has the responsibility of providing funds to make that research safe. Without their cooperation either the work will continue with unnecessary risks, or it will not go on at all. Since both of these possibilities are cheaper than safe research, we must accept that obtaining safer facilities will require economic, political, and scientific decisions, and we must be prepared to present our opinions in all three areas.

We should be aware that safety can be an excellent excuse for cutting off a laboratory's research funds. In particular, coercive policies making certain laboratory designs an absolute prerequisite to work with viruses, cells or animals are dangerous.

Second, I think it is clear that prospective statistical studies on the incidence of disease--obviously not just cancers--in laboratory workers and paired

345

controls must be initiated immediately. This is
probably the cheapest way to find out whether any of the
laboratory agents under study are harmful to humans.

Finally, I was struck by the unevenness of central
certification for the materials we use. I can get
primary monkey cells, SV40, or adeno-SV40 hybrid virus
just by a phone call to any of a number of people, and
I have no idea what other adventitious similar viruses
accompany them. At the same time, I cannot purchase
polio vaccine, which has been tested for biological
purity, because I am not licensed to write a prescription.

I think the time is ripe for a central clearing
house to test and certify primary cell lines, sera and
virus seed stocks for purity with the best available
techniques.

* * * * *

COMMENTS BY

Michael Oxman

Almost any form of biological research involves some
potential biohazard. Though this has always been the
case, it has only recently become of concern to many
people outside of those few laboratories that are directly
involved with agents of known pathogenicity for man and
other animals. This sudden expansion of concern in the
absence of adequate information has resulted in a good
deal of fear and confusion. Consequently we have in this
conference sought to assemble some of the relevant
information that is available and to initiate the formulation of a rational approach to this difficult problem. I
shall here outline a few principles which I believe
should be followed in dealing with three broad aspects of
the problem of biohazards in biological research. These
aspects are (1) the degree of risk that is justifiable to
take and who should take it, (2) procedures whereby the
risk can be reduced or eliminated, and (3) surveillance
and the acquisition of further knowledge.

Once we recognize that an experiment involves a
potential biohazard we must try to define the risk. This
will require that we answer a number of questions. For
example: How infectious is the agent for man? How
susceptible are those who may be exposed? What are the
likely consequences of infection if it should occur
(e.g., a common cold, fatal encephalitis, or malignancy
after a twenty year latent period)? Are we dealing with

an agent already present in the community, or with some exotic import (e.g., Marburg virus) or laboratory creation (e.g., adenovirus-SV40 hybrid viruses)? What methods are available for the detection of the agent and how sensitive are they? How adequate is our assessment of the risk and what additional data do we need? Once these questions have been answered we are in a position to decide whether the potential benefits of an experiment justify the risk involved. But in making such an assessment, it is necessary to separately consider (1) the investigator himself, (2) other workers in the laboratory who are interested and knowledgeable (including technicians and graduate students), and (3) the uninvolved and uninformed public at large (as well as glassware washers, housekeeping personnel, secretaries and workers in adjacent laboratories).

Curiousity, faith in the biological significance of the expected result, humanitarian impulses or even personal ambition may convince an investigator to take certain risks. I believe he should be free to do so, as long as only he is at risk. Yet it may be difficult to limit the risk to the investigator himself, for if he becomes infected with an agent from the laboratory he may transmit it to others. Thus, for example, several of the victims of Marburg virus disease were hospital personnel who acquired the disease from infected laboratory workers for whom they were caring. Whereas an investigator may himself decide to assume certain risks, he does not have the right to make that decision for anyone else. In fact, it seems to me that the decision to assume a risk can only legitimately be made by the individual who will be in jeopardy. This principle of "informed consent" can readily be applied to other investigators, graduate students and technicians, as long as they are informed of the nature and magnitude of the risk and are then free to decide whether or not they are willing to take it. However the principle of "informed consent" cannot realistically be applied to glassware washers, secretaries, housekeeping personnel, workers in adjacent laboratories or the public at large. Consequently we must insure that under no circumstances will these people be exposed to risks as a result of our research.

These considerations demand that we remain constantly alert to the possibility of biohazards, that we insure that nothing hazardous leaves the laboratory, and that we carefully limit access to hazardous areas. In addition we must take steps to minimize the risk to these workers who, by their own election, will be exposed. The most obvious solution is to work with innocuous agents which are already prevalent in the community and to which all or most individuals are insusceptible, either because of natural resistance or as a consequence of immunity induced by prior infection. This approach has been widely used by workers studying the biology and biochemistry of virus replication. However it requires a good deal of prior knowledge and is clearly not applicable to many current investigations, particularly those concerned with the etiology and pathogenesis of such diseases as cancer. Thus most of us will have to erect some barriers to protect ourselves, and others, from our experiments.

There are two types of barriers: physical and biological. We have already devoted a good bit of time to discussions of physical barriers, and I would only emphasize the need to consider all materials which have been in contact with living cells as potentially infectious. Thus spent medium, used pipettes, etc., should be autoclaved or otherwise decontaminated before being discarded. The biological barriers, on the other hand, have received too little attention. This is regrettable because, whereas physical barriers may be expected to reduce the possibility of infection by 3 to 6 logs, biological barriers may offer much greater degrees of protection. For example, because chicken cells lack receptors for poliovirus, it is impossible to initiate infection in this species with poliovirus stocks containing an excess of 10^{10} plaque-forming units per ml. On the other hand, the same cells are fully susceptible to infection by poliovirus RNA. Such natural resistance provides one of the strongest barriers against virus infection, and we should make use of it whenever possible. Moreover we must recognize the risks involved when such barriers are breached, either by infection via unnatural routes or as a consequence of physical or genetic alterations of the virus. In addition in the case of most viruses, immunity induced by prior infection provides solid protection, even against massive reexposures. Thus prior exposure or immunization can provide much more protection than biohazard hoods or UV lights. Most investigators experienced in handling agents pathogenic for man have already come to recognize that biological barriers are far more reliable than physical ones. Thus individuals are routinely immunized against rabies and yellow fever virus before being permitted to work in laboratories in which these agents are handled, no matter how elaborate the containment facilities may be.

Although the intelligent application of available knowledge should often permit an investigator to select the least hazardous biological system for study, or at least to maximally protect his personnel, this will only be possible if he is aware that his experiments may

involve some biohazard. Unfortunately workers trained
in areas other than microbiology often do not think in
these terms. Thus a laboratory engaged in serological
studies of cancer patients may not be concerned with
biohazards until someone develops serum hepatitis.
Furthermore I'll bet there is at least one person in
this room who has been working with human lymphocytes or
doing fusion experiments with mouse cells in a horizontal
laminal flow hood. Consequently I would like to
reiterate the recommendation, already made by Drs.
Hsiung, Lennette and Rowe, that all individuals planning
research which may involve biohazards receive training
in microbiology--not simply to learn techniques for
handling potentially infectious materials, but also to
learn to think in microbiological terms. I suspect, for
example, that such training, together with plain old
common sense, would cause many investigators to think
twice before initiating experiments that might broaden
the host range of a virus.

If viruses play a significant role in the etiology
of human neoplastic diseases, laboratory workers engaged
in cancer research are likely to have more than average
exposure. Thus in spite of the low incidence and long
latent period to be expected, it is important that we
act now to establish a formal registry of laboratory
workers and a mechanism for periodic surveillance of this
population for neoplastic and slowly progressive
degenerative diseases. Serum should be obtained and
stored from all of these individuals at 6 to 12 month
intervals, and all suspicious illnesses should be
promptly reported to some independent agency or committee,
which could then initiate the appropriate investigations
and ask the necessary hard-nosed questions. The establishment of such an independent agency is essential
because it is extremely difficult for an investigator to
deal open-mindedly and unemotionally with illness in his
own laboratory. A system of surveillance such as this
will also provide the basis for both the prospective and
retrospective studies which Dr. Cole has outlined.

DISCUSSION

STANSLY: A policy is actively being developed at the NIH for the guidance of staff and grant applicants regarding biohazards. Applicants will be instructed to describe the safety measures that have or will be taken if a potentially hazardous situation is involved in the research. Further, where a potential hazard has been identified by reviewers but appropriate safety measures have not been mentioned or are inadequate, assurance from the applicant institution will be required, prior to issuance of an award, that suitable safety measures will be taken.

Finally, I should like to assure this audience that there is no prohibition to requesting funds in research grant applications for necessary equipment or facilities to reduce biohazards. I know of no grant application that has been turned down merely because such a request was made. It is advisable to request this in the initial application or in a competitive renewal. In this regard, two aspects have particularly to be considered: It is the investigator's responsibility to determine what is appropriate to request, and it is the reviewer's responsibility to be prepared to evaluate such a request. It is hoped that one function of this conference will be to develop standards so that these two considerations can be adequately met.

BLACK: We are finally discussing the costs of the safety precautions and have come to the realization that they will inevitably reduce the number of grants available and increase the time required to reach our ultimate goal. If we do believe in our mission of trying to control cancer, it behooves us to accept some risk. Even if, as has been suggested, five or ten people were to lose their lives, this might be a small price for the number of lives that would be saved. We have been self-centered in this session in emphasizing absolute safety. We heard Dr. Huebner say that, in spite of extensive searches, there is no evidence that the avian or murine viruses have been responsible for human disease. Where we have such negative data, let's stop worrying and move ahead with the main project.

MILLER: I am alarmed at the thought that some people are willing to accept some deaths in order to accomplish their work. I do not see how you could have predicted in 1950 how many deaths would occur 15-25 years after stilbestral was given to pregnant women. It was thought then that there was no hazard, and now there are almost

a hundred young women who have developed cancer as a
result. Also, it was suggested that the carcinogenicity
of viruses is the only measure of health hazard or
mortality, but we have been told here that viruses are
related to cerebral degenerative diseases, which are
just as fearsome to me as cancer is.

WATSON: I'm afraid I can't accept the five to ten
deaths as easily as my colleague across the aisle. They
could easily involve people in no sense connected with
the experimental work, and most certainly not with the
recognition and fame which would go to the person or
group that shows a given virus to be the cause of a human
cancer. When one works at a Detrick-type installation or
in building 41, the assumption of shared, unclear risk
can be made obvious to anyone entering the labs, but
this is not true for tumor virus work conducted in a
building not specifically devoted to this task.

    Yet, if we wish to broaden the base of tumor virus
research by bringing it into academic environments, we
will generate a situation where, unless "safe labs" are
constructed, many outsiders will be exposed to the many
viruses that we want to study at the molecular level.
But as we all know "safe labs" will require considerable
expense and cannot be funded from research money given
to do experiments. They will only be built if specific
funds exist for biohazard prevention, say on the order
of $125,000 for a moderate-sized lab and much more if
we are dealing with a major lab. Now I'm afraid that
the NCI avoids facing up to its moral, if not legal,
responsibility by declaring almost all the viruses we
work with as unlikely to be of sufficient long-term
danger so as to require specially designed rooms with
their own air supply, Baker-type hoods, and exit filters
to remove any viruses that might be dispersed as aerosols
into the laboratory air. Of course everyone working with
a given virus hopes very much that his particular virus
is indeed safe. But I think we must now help create the
situation where the real reason for a decision is the
awfulness of the alternative possibility--which I suspect
is how the AEC calculates the low probability of a
catastrophic accident to a nuclear power plant.

DARNELL: It seems to me that an important issue has
gotten too little attention from the people assembled
here. The viruses with which we are now working that
may be dangerous are the SV40-polyoma group and the RNA
tumor viruses. It seems to me that we ought to return
as calmly as possible to the issue of whether or not
there is a biohazards problem with these viruses. I

would like to try to phrase a question and a possible
approach to this. What are we going to accept as
evidence that these viruses have caused a human tumor?
It seems to me that positive antigenic proof in the
tumor would be good enough. If there were only some
quasi-statistical change in total tumor occurrence in
people who have been exposed to SV40-RNA tumor virus,
for example, but no indication that the tumors they
developed had virus antigens, I for one wouldn't be
safe in assuming that viruses have caused the tumor.
It seems reasonable to me that it would be possible to
mount a campaign, which would be relatively less
expensive than anything being talked about here, to find
100 or 500 or some reasonable number of laboratory
workers who have or have had tumors in the past five
years, cells from which are still available, and test
these cells with the best gs antisera and with the best
SV40 T antiserum available. We should then find out
sometime in the next few months or year what the
reactions to these sera are. If the reactions are
uniformly negative, and I would bet a case of whiskey
that they will be, then I think we ought to stop
worrying about these viruses and call them safe for at
least a period of 5 years. If no cases of SV40 or RNA
virus transformation can be proven in such human tumors
from lab workers, then the number of laboratory workers
who may become infected in the course of the next 5
years is likely to be small. It's terrible to say that
even a small number of people are expendable; but I think
if no present laboratory worker's tumor can be shown to
have SV40 or RNA virus antigens, then it is unlikely
that even a small number of infections will occur.
Without some attempt of this sort, we may have restric-
tions imposed which will really bring research to a
halt.

ROWE: I think a very important concept is that much of
microbiological safety consists simply of having good
habits. For example, it would never cross the mind of
a trained microbiologist to touch anything that a drop
of virus had fallen on, even if it was the most harmless
virus around. Lab workers should have the operating
room mentality, that there are clean and dirty areas
with clearly defined but constantly changing boundaries.
There is no reason why this mental approach should be
restricted to microbiology, but it should be part of
the training and lifelong habits of every lab worker;
it is just as easy to work using good sterile technique
as it is to use bad technique.

LENNETTE: I should like to reinforce Dr. Rowe's emphasis on the need for good laboratory habits and to remind you of other considerations.   It is one thing for an individual to say, "I am working with a known agent whose nature, properties and characteristics I know quite well and so I know how to handle it and to take the necessary precautions for its containment."   But he is confronted with something quite different when he deals with human or animal tissues which come directly from man or animals suffering from infection with an unknown agent, i.e., now there are no specific guidelines and absolutely no allowance for careless or sloppy or indifferent techniques or shortcuts in safety precautions.

The physical aspects of agent containment and personnel protection are mandatory and air locks, laminar flow hoods, ultraviolet lamps, etc., are fine, but human carelessness can override the inherent safety factor of such physical safeguards.   The human element is the most important in the equation.   Speaking from my own experience, we find that individuals graduating from college microbiology departments, or who have finished graduate work in such departments, may have an excellent foundation in biophysics, biochemistry and the genetics of microorganisms, but little or no perception of the hazards involved in handling pathogenic agents or how to protect themselves and innocent bystanders against such pathogens.   Such people coming into our laboratories have to be trained or retrained to the point where aseptic techniques and safety precautions become virtually reflex and are carried out without conscious attention being paid to them.   I am appalled at some of the techniques and procedures I see used by people who work with known or potential pathogens, but with little or no exposure to medical microbiology, and I think the time has come, if it is not long overdue, to remedy such situations.

* * * * *

CLOSING REMARKS

Paul Berg

As this meeting draws to a close, I'd like to offer a few observations and comments on behalf of the people who provided the funds and who planned the program. From the beginning we felt the most likely outcome of the papers and discussion would be educational.   It was our feeling that having the information and wisdom of acknowledged experts in the respective fields aired,

discussed and eventually incorporated into a written
record would provide a good starting point for dealing
rationally with the question of biohazards in oncogenic
virus research. I think the Conference has accomplished
its purpose and I'd like to express our collective thanks
to the agencies that provided financial support, the
speakers and panelists who contributed so willingly, and
especially to the Organizing Committee of Al Hellman,
Michael Oxman and Robert Pollack for the superb program
they put together.

I've been asked several times since coming here:
"Where do we go from here? What is the next step after
Asilomar?" There are several here who would argue that
there need be no next step--there is no problem!
Admittedly there is no clear-cut (scientifically provable)
indication that the oncogenic viruses or cell culture
systems currently being studied in laboratories round
the world constitute a serious threat to the scientists
working with them or to the population at large.
That's reassuring but, nevertheless, I'm persuaded by
what I've heard that prudence demands caution and some
serious effort to define the limits of whatever potential
hazards exist. To do less, it seems to me, is to play
russian roulette, not only with our own health, but also
with the welfare of those who are less sophisticated in
these matters and who depend on our judgment for their
safety.

What steps can we take, then? I was persuaded that
epidemiologic studies may well be the best way we have
of detecting a low-order risk; moreover, Dr. Miller
pointed out that the best chance of detecting such a
correlation would be to go to the individuals where the
risk is greatest, the laboratories where people are in
contact with these agents most frequently. Dr. Cole
even laid out a "bargain-rate" prospective study of such
individuals which could conceivably assess the magnitude
of the hazard, if there is, in fact any hazard at all.
I would, therefore, like to call for such a study, most
logically sponsored by or even carried out within the
NIH, to determine if there is increased risk of cancer
(or other diseases) stemming from current and projected
laboratory researches with biologic oncogenic agents.
I'm quite confident that such a study, but particularly
the data collected in the course of the study (histories,
medical examinations, specimen collection of selected
scientific personnel, their families and suitable
controls), would be useful and interesting.

I should also like to suggest that the Biohazards
Branch of NCI consider publishing a periodic newsletter
or other publication that would disseminate new exper-

imental and technical data related to containment
procedures; it could serve as the communication link for
the scientific community concerned with biohazards.

Finally, I should like to urge the Federal and
private funding agencies that are encouraging and
funding research in both basic and applied aspects of
animal cell biology, and particularly with known
oncogenic agents, to consider sympathetically and realistically the funding requirements for establishing
even minimal containment safeguards. New laboratories
working with tumor viruses cannot be set up as inexpen-
sively as those which investigated T4 and λ phages!

APPENDIX I

MINIMUM STANDARDS OF BIOLOGICAL SAFETY

AND ENVIRONMENTAL CONTROL FOR

CONTRACTORS OF THE SVCP

Prepared by

Office of Biohazard and Environmental Control
National Cancer Institute
National Institutes of Health
Bethesda, Maryland

July, 1972

Occupational Health Program

Pre-employment serum samples from no less than 10 ml of blood
shall be collected from all new employees to establish a base
line reference. Blood shall be collected annually from each
employee. The blood will be processed and the serum will be
packaged and shipped to the SVCP central repository.

Persons currently employed or transferred to the contractual
effort shall be urged to participate in this precautionary
medical measure. This part of the Occupational Health
Program applies to all persons handling potentially oncogenic
biological or chemical materials and to those employees
cleaning laboratory glassware, handling or caring for
experimental animals or their tissues, or doing janitorial
duties.

Pregnant women employed in oncogenic virus laboratories shall
be counseled at the earliest possible time as to the potential
risk involved in their continued presence in the laboratory.
The option of remaining in the laboratory, transferring to a
non-virological unit, or terminating their employment should
be decided by the principal investigator with the consent of
the employee concerned.

Laboratory Safety and Environmental Control

The laboratory management shall establish and implement an
environmental control and personnel safety program that will
include biological safety and the complementary aspects of
radiological, chemical, and industrial safety related to the
laboratory activities. The management's program shall include
the formulation of a safety policy and publication of a safety
manual that shall be made available to all laboratory personnel.
The general practices and procedures that directly affect laboratory safety and contamination control shall be described.
Both mandatory and advisory provisions shall be included. A
description of the program and a copy of the manual shall be
submitted to the NCI, VO, OB & EC for review and comment prior
to implementation.

A Safety and Environmental Control coordinator or officer, who
will report directly to laboratory management, shall be appointed.
The coordinator shall discuss the objectives of the safety program with all employees at the time of their initial employment
and once a year thereafter.

Laboratory Access

(a)   Children and pregnant women visitors shall not enter
      oncogenic virus laboratories.

(b)   Appropriate signs shall be located at points of access to
      laboratory areas directing all visitors to a receptionist
      or receiving office for access procedures.

(c)   The universal Biohazard symbol shall be displayed at
      specific laboratories in which moderate and high risk
      virus manipulations are being performed (Table I).   Only
      authorized visitors shall enter laboratories displaying
      the Biohazard symbol.   Doors displaying the Biohazard
      symbol shall not be propped open, but shall remain closed
      except when in use.

Clothing

(a)   All employees and visitors in laboratories where moderate
      or high risk materials are being used shall wear laboratory
      clothing and full-bodied laboratory shoes or shoe covers.
      A full length solid front gown or pullover surgical smock
      and pants are preferable.   A completely fastened (buttoned)
      full length laboratory coat is acceptable.

(b)   Disposable gloves shall be worn whenever radiological,
      chemical, carcinogenic materials, or virus preparations
      of moderate or high risk are handled.

(c)   Laboratory clothing including laboratory shoes or shoe
      covers shall not be worn outside the work area.

(d)   Animal handlers shall be provided a complete clothing
      change including pants and shirt or jumpsuit, shoes or
      boots, and gloves.   Head covers and filter masks shall be
      worn when handling animals, changing bedding, or working
      in animal rooms which house animals inoculated with moderate
      or high risk viruses.   This clothing shall not be worn
      outside the animal area.

(e)   Laboratory clothing from laboratories in which moderate and
      high risk virus manipulations are being performed shall be
      autoclaved before it is sent to a laundry.

Eating, Drinking, and Smoking

There shall be no eating, drinking, or smoking in any virus
work areas.   However, the use of water fountains is permitted
(fountains with foot pedals are recommended over those with hand
levers).   Food shall not be stored in a laboratory containing
high, moderate, or low risk materials.

Pipetting

There shall be no mouth pipetting in _any_ of the laboratories.

Use of Containment Equipment

(a) Adequate precautions, with considerations for both equipment and operating practices, will be taken to reduce
potential hazards of aerosol generating processes such
as centrifugation, lyophilization, sonication, grinding,
etc. for both low and moderate risk materials.

(b) Moderate risk viruses (Table I) and cell culture manipulations
involving these viruses, shall be confined in primary
barriers, i.e. Class I ventilated cabinets or laminar flow
biological safety cabinets of approved design.

(c) The necropsy and inoculation of animals inoculated with
moderate risk materials shall be confined in primary barrier
systems, i.e. Class I ventilated cabinets, where appropriate.

(d) High risk virus manipulations will be confined to absolute
primary barrier systems, i.e. Class III cabinets (gas
tight glove boxes). Definition of such agents will be
determined by the contractor and OB & EC at times of
need.

(e) Low risk viruses (Table II) and cell culture manipulations
in specific laboratory areas where two or more viruses or
cell culture lines are being maintained shall be confined
in laminar flow biological safety cabinets of approved
design to minimize cross-contaminaion.

Transportation of Virus Materials Within the Laboratory Area

Care should be taken when moving breakable containers of
biological agents from work sites to cold boxes, incubators,
centrifuges, etc., and in moving contaminated glassware, etc.
to a sterilizer prior to cleaning or discard. Pans with solid
bottom and side walls and/or instrument carts should be
utilized for the transfer. Low, four-wheel, hand trucks
should be used to transfer larger vessels (e.g. carboys).

Use of Liquid or Gaseous Disinfectants

(a) All contaminated laboratory ware and similar items to be
removed from a laboratory to a sterilization site will be
placed in containers and immersed in appropriate chemical
disinfectant. The containers will be covered (e.g. with
foil) during transport and autoclave staging.

(b)   Contaminated items too large for autoclaving or hot air
sterilization will be hand wiped with disinfectant or
sterilized with gas.

Housekeeping

(a)   Dry contaminated wastes from laboratories will be collected
in impermeable bags which will be sealed at the collection
site before removal to the autoclave or incinerator. Metal
cans with tight sealing covers may be used in lieu of bags.
The seals of both containers should be loosened appreciably,
at the autoclave staging area, to insure sterilization of the
contents.

(b)   No dry mopping of work areas shall be permitted. Vacuum
cleaners may be used, provided they are equipped with a
HEPA exhaust filter.

(c)   Laboratory floors shall be wet mopped with a disinfectant/
detergent solution.

Sterilization of Used Materials

(a)   All contaminated materials shall be decontaminated or
sterilized before disposal or recycling. Preferably, this
should be done before these materials leave the laboratory
area, unless they can be moved in sealed or covered containers.

(b)   Tissue culture or other virus-containing liquid wastes
shall be decontaminated, either chemically or by heat,
before being discharged to the community sanitary sewer
system.   (Water from toilets, hand wash basins and personnel
showers in a change room requires no special treatment.)

(c)   All animal wastes from rooms in which animals have been
inoculated with moderate or high risk potentially oncogenic
biological preparations will be incinerated or autoclaved
prior to other disposal. Precautions will be taken to
contain the wastes to minimize aerosols during collection,
transfer, and autoclave (or incinerator) staging. Impermeable sealed bags are suitable.

Laboratory Vacuum

A liquid trap of concentrated disinfectant and an absolute
biological filter will be installed in the vacuum line between
the point of application and the vacuum source.

Testing of Barrier Systems

All primary barrier systems will be tested annually to certify correct containment and operation. Consult with the NCI Office of Biohazard and Environmental Control for assistance in testing primary barrier systems.

Animal Holding and Care

Compliance with the regulations of the Animal Welfare Act and the "Guide for Laboratory Animal Facilities and Care", National Institutes of Health, D.H.E.W.

RESOURCES AVAILABLE TO THE SCIENTIFIC COMMUNITY

A.   WHO Regional Reference Center
                        for
                  Simian Viruses

                Southwest Foundation
            for Research and Education
                San Antonio, Texas

                S. S. Kalter, Ph.D.
        Director, Simian Virus Reference Center

The Simian Virus Reference Center is sponsored and funded
by the National Institutes of Health and the World Health
Organization.

                  Purposes of the
            Simian Virus Reference Center

1. To develop and maintain a working repository for
    simian viruses.
2. To act as a source of certified reference seed
    virus strains and specific antisera.
3. To provide consultation services.
4. To provide diagnostic services, including
    identification and characterization of viruses
    for users of nonhuman primates unable to perform
    this task. (This also includes screening for
    human viruses.)
5. To provide information and exchange of organisms
    among primate centers and other health organizations.
6. To train interested students in virological
    laboratory procedures associated with primate
    investigations.

B.   Testing of murine viruses

Dr. J. Parker
Microbiological Association
Bethesda, Maryland

C.   PPLO (mycoplasma) testing (limited)

Dr. L. Hayflick
Stanford Medical School
La Jolla, California

D.
1.   Biohazard consultation and certification of
     laminar flow hoods.
2.   Training of laboratory technicians in proper
     techniques for microbiological research.
3.   Laboratory facility design for biohazards
     control.

Dr. A. Hellman
National Cancer Institute
Bethesda, Maryland

CONFERENCE PARTICIPANTS

Allen, David W., Thorndike Memorial Laboratory, Boston
City Hospital, Boston, Mass.

Allison, Anthony C., Clinical Research Center, Harrow,
England.

Aposhian, H. V., Dept. of Pharmacology, University of
Maryland School of Medicine, Baltimore, Md.

Baltimore, David, Dept. of Biology, Massachusetts
Institute of Technology, Cambridge, Mass.

Barkley, W. Emmett, National Cancer Institute, NIH,
Bethesda, Maryland.

Berg, Paul, Dept. of Biochemistry, Stanford University
School of Medicine, Stanford, Calif.

Bigner, D. D., Dept. of Microbiology, Duke University
School of Medicine, Durham, N. C.

Bishop, J. M., University of California School of
Medicine, San Francisco, Calif.

Black, Francis, Yale University School of Medicine,
New Haven, Conn.

Bond, Edward, Electro-Nucleonics Laboratories, Inc.,
Bethesda, Md.

Capecchi, M., Dept. of Bacteriology, Harvard Medical
School, Boston, Mass.

Casals, Jordi, Dept. of Epidemiology, Yale University
School of Medicine, New Haven, Conn.

Clarke, G., Imperial Cancer Research Fund, London, England.

Cole, Philip T., Dept. of Epidemiology, Harvard School
of Public Health, Boston, Mass.

Coriell, L. L., Institute for Medical Research, Camden,
N. J.

Darnell, J. E., Dept. of Biological Sciences, Columbia
University, New York, N. Y.

Davidson, Richard, Children's Hospital Medical Center,
Harvard Medical School, Boston, Mass.

Davis, A., The American Cancer Society, New York, N. Y.

Dieckmann, M., Stanford University School of Medicine,
Stanford, Calif.

Dimmick, Robert L., The Naval Biomedical Research
Laboratory, Oakland, Calif.

Dixon, F. J., Dept. of Experimental Pathology, Scripps
Memorial Hospital, San Diego, Calif.

Edelman, Gerald M., Dept. of Biochemistry, The Rockefeller
University, New York, N. Y.

Fialkow, P., University of Washington, School of Medicine,
Seattle, Wash.

Friedman, T., Dept. of Pediatrics, University of
California at San Diego, La Jolla, Calif.

Gajdusek, D. Carleton, National Institute of Neurological
Diseases and Stroke, NIH, Bethesda, Md.

Gardner, Murray B., Dept. of Pathology, University of
Southern California School of Medicine, Los Angeles,

Calif.
Gartler, S., University of Washington School of Medicine,
   Seattle, Wash.
Gerin, J., Union Carbide, New York, N. Y.
Gerone, P., Delta Primate Center, Tulane University, New
   Orleans, La.
Gibbs, C. J., Jr., National Institute of Neurological
   Diseases and Stroke, NIH, Bethesda, Md.
Girardi, Anthony J., Wistar Institute, Philadelphia, Pa.
Granoff, A., St. Jude Children's Research Hospital,
   Memphis, Tenn.
Griesemer, Richard A., California Primate Research
   Center, University of California, Davis, Calif.
Gordon, Judy, Cold Spring Harbor Laboratory, Cold Spring
   Harbor, N. Y.
Hackett, A., The Naval Biomedical Research Laboratory,
   Oakland, Calif.
Hardy, W. D., Sloan-Kettering Institute for Cancer
   Research, New York, N. Y.
Heath, Clarke, Center for Disease Control, Atlanta, Ga.
Hellman, Alfred, National Cancer Institute, NIH,
   Bethesda, Md.
Hilleman, Maurice R., Merck Institute for Therapeutic
   Research, West Point, Pa.
Holland, J., Dept. of Biology, University of California
   at San Diego, La Jolla, Calif.
Hopkins, N., Dept. of Biology, Massachusetts Institute of
   Technology, Cambridge, Mass.
Hsiung, G. D., Virology Laboratory, Yale University School
   of Medicine, New Haven, Conn.
Huang, A., Dept. of Microbiology, Harvard Medical School,
   Boston, Mass.
Huebner, Robert J., National Cancer Institute, NIH,
   Bethesda, Md.
Hull, Robert N., The Lilly Research Laboratories,
   Indianapolis, Ind.
Jackson, D., Dept. of Microbiology, University of
   Michigan, Ann Arbor, Mich.
Jacobs, Leon, National Institutes of Health, Bethesda, Md.
Jordon, J., Dept. of Chemistry, University of California,
   Los Angeles, Calif.
Kaizer, H., Dept. of Microbiology, Johns Hopkins
   University, Baltimore, Md.
Kalter, S. S., Dept. of Microbiology, Southwest Research
   Institute, San Antonio, Texas.
Kaplan, H., Stanford University School of Medicine,
   Stanford, Calif.
Keefer, G., National Cancer Institute, NIH, Bethesda, Md.
Kissling, R., Center for Disease Control, Atlanta, Ga.
Lennette, Edwin H., California State Department of Public
   Health, Berkeley, Calif.
Lerner, R., Scripps Memorial Hospital, San Diego, Calif.
Lewis, Andrew M., Jr., National Institute of Allergy and

Infectious Diseases, NIH, Bethesda, Md.
Lewis, H., Genetic Biology Program, National Science
   Foundation, Washington, D. C.
Lorenz, D., Division of Biological Standards, NIH,
   Bethesda, Md.
Lucas, Z., Stanford University School of Medicine,
   Stanford, Calif.
Manaker, R., National Cancer Institute, NIH, Bethesda, Md.
Maniatis, G. M., Dept. of Human Genetics, College of
   Physicians and Surgeons, Columbia University, New York,
   N. Y.
Manning, J. J., Dept. of Veterinary Microbiology,
   University of California, Davis, Calif.
Marcus, P., University of Connecticut, Storrs, Conn.
Martin, G., Dept. of Pathology, University of Washington,
   Seattle, Wash.
Mason, R., The American Cancer Society, New York, N. Y.
Miller, R., National Cancer Institute, NIH, Bethesda, Md.
Morris, J., National Institutes of Health, Bethesda, Md.
Newman, P., National Cancer Institute, NIH, Bethesda, Md.
Oshoroff, B., National Institutes of Health, Bethesda, Md.
Oxman, M., Children's Hospital Medical Center, Harvard
   Medical School, Boston, Mass.
Ozer, H. L., Worcester Experimental Foundation,
   Shrewsbury, Mass.
Parker, John C., Microbiological Associates, Bethesda, Md.
Pollack, Robert, Cold Spring Harbor Laboratory, Cold
   Spring Harbor, N. Y.
Rickard, C., Dept. of Pathology, NYS Veterinary College,
   Cornell University, Ithaca, N. Y.
Robb, J. A., Dept. of Pathology, University of California
   at San Diego, La Jolla, Calif.
Robbins, P., Massachusetts Institute of Technology,
   Cambridge, Mass.
Robinson, W. S., Stanford University School of Medicine,
   Stanford, Calif.
Roizman, B., University of Chicago School of Medicine,
   Chicago, Ill.
Rowe, W., National Institute of Allergy and Infectious
   Diseases, NIH, Bethesda, Md.
Salzman, N., National Institute of Allergy and Infectious
   Diseases, NIH, Bethesda, Md.
Schlesinger, M., Dept. of Microbiology, Washington
   University Medical School, St. Louis, Mo.
Schlesinger, S., Dept. of Microbiology, Washington
   University Medical School, St. Louis, Mo.
Schneider, Jerry A., Dept. of Pediatrics, University of
   California at San Diego, La Jolla, Calif.
Sedwick, D., Stanford University School of Medicine,
   Stanford, Calif.
Stansly, P. G., National Cancer Institute, NIH, Bethesda,
   Md.
Steeves, R., Dept. of Developmental Biology and Cancer,

Albert Einstein College of Medicine, Bronx, N. Y.
Strauss, J., California Institute of Technology,
  Pasadena, Calif.
Takemoto, K. K., National Institute of Allergy and
  Infectious Diseases, NIH, Bethesda, Md.
Talbot, B., National Institutes of Health, Bethesda, Md.
Tennant, R. W., Atomic Energy Commission, Oak Ridge, Tenn.
Thormar, H., State Institute of Basic Research, New York,
  N. Y.
Ting, R., Bionetic Research Laboratories, Bethesda, Md.
Todaro, George J., National Cancer Institute, NIH,
  Bethesda, Md.
Udry, William, Cold Spring Harbor Laboratory, Cold Spring
  Harbor, N. Y.
Watson, James, Cold Spring Harbor Laboratory, Cold Spring
  Harbor, N. Y.
Wedum, A., Dept. of the Army, Ft. Detrick, Frederick, Md.
Weismann, I., Stanford University School of Medicine,
  Stanford, Calif.
Weissbach, H., Roche Institute of Molecular Biology,
  Nutley, N. J.
Wolfe, L. G., Dept. of Microbiology, Rush Hospital,
  Chicago, Ill.
Yatvin, M., Dept. of Radiology, University of Wisconsin,
  Madison, Wis.
Zinder, N. D., The Rockefeller University, New York, N. Y.